Distributed Event-Based Systems

T0216721

Gero Mühl · Ludger Fiege
Peter Pietzuch

Distributed Event-Based Systems

With 158 Figures and 17 Tables

 Springer

Authors

Gero Mühl

Fakultät IV Elekrotechnik und Informatik
Technische Universität Berlin
Einsteinufer 17
10587 Berlin, Germany
g_muehl@acm.org

Ludger Fiege

Siemens AG
CT SE2
Otto-Hahn-Ring 6
81730 München, Germany
ludger.fiege@siemens.com

Peter Pietzuch

Div. of Engineering and Applied Sciences
Harvard University
33 Oxford Street
Cambridge, MA 02138, USA
prp@eecs.harvard.edu

ACM Computing Classification (1998): C.2.4, C.3, D.2.11, D.2.12

ISBN:13 978-3-642-06912-3   e-ISBN:13 978-3-540-32653-3

Springer is a part of Springer Science+Business Media

springer.com

© Springer-Verlag Berlin Heidelberg 2010
Printed in Germany

Cover design: KünkelLopka Werbeagentur, Heidelberg

For Regina.
— Gero

For Biggi,
Franzi, Flori, and Johanna.
— Ludger

For Bohdan Bańkowski.
— Peter

Preface

The field of event-based systems is surprisingly broad. In many scientific communities, technical talks, commercial products, and industrial projects people think about asynchronous computations and messaging, scalability and maintainability, stepwise evolution and loose coupling. Most likely, these people are discussing event-based systems, even if they use other terms.

When we began investigating event-based systems some years ago, we were surprised to see that *eventing* was scattered among many disciplines of computer science. There were no workshops or conferences dedicated to this topic, for example, although many aspects of event-based systems cannot be assessed from a database, network, or software engineering perspective alone. In the same sense, commercially available products that could help solving problems of event-based architectures are often bundled and marketed in solutions of a specific domain.

In order to channel some of the attention, the Distributed Event-Based Systems (DEBS) workshop series was created. It attracts people from distributed computing, database, and software engineering audiences, and it demonstrates the wide variety of facets event-based systems have. After having heard about and being engaged in interesting discussions about allegedly "academic" and "real-world" problems, in investigating many findings, and after creating many solutions in both academic and industrial environments, we decided to write this book to present both the current state-of-the-art and its base concepts.

The book takes a distributed system's point of view. This is, of course, partly due to our own background, but more importantly we believe a solid understanding of distributed event-based systems is a good starting point for building modern computing systems. It lets you integrate sophisticated filter and data processing capabilities as well as new network topologies and routing algorithms.

Acknowlegements

We want to thank our colleagues, coauthors, and friends who discussed and developed most of the ideas presented in this book with us. Without being able to name all, we want to thank Jean M. Bacon, Alejandro P. Buchmann, Frank Buschmann, Mariano Cilia, Felix C. Freiling, Rachid Gerraoui, Michael A. Jaeger, Arno Jacobsen, Mira Mezini, Ken Moody, Joe Sventek, Andreas Ulbrich, Andreas Zeidler, and many others we worked and talked with in universities and companies, at conferences, and via email. The good thing about writing a book is that you gain so many new insights into already known topics.

We also want to thank Ralf Gerstner from Springer Verlag for his patience and continuous support, and the reviewers and proofreaders who helped us improve the book.

Last but not least, we are grateful to our families and friends for their patience and understanding for yet another evening being occupied with this "nonsense". Thanks.

Berlin, Germany   *Gero Mühl*
Munich, Germany   *Ludger Fiege*
Cambridge, MA, USA   *Peter Pietzuch*

May 2006

Contents

1 **Introduction** 1
  1.1 Networked Computing 1
  1.2 Middleware 2
  1.3 Event-Based Systems 3
  1.4 Application Scenarios 4
      1.4.1 Information Dissemination 4
      1.4.2 Network Monitoring 4
      1.4.3 Enterprise Application Integration 5
      1.4.4 Mobile Systems 6
      1.4.5 Ubiquitous systems 6
  1.5 Putting Event-Based Systems Into Context 7
  1.6 From Centralized to Internet-Scale Event Systems 8
  1.7 Structure of the Book 8

2 **Basics** 11
  2.1 Terminology 11
      2.1.1 Events and Notifications 11
      2.1.2 Producers and Consumers 12
      2.1.3 Subscriptions and Filters 13
      2.1.4 Event Notification Service 13
  2.2 Models of Interaction 14
      2.2.1 Request/Reply 15
      2.2.2 Anonymous Request/Reply 15
      2.2.3 Callback 16
      2.2.4 Event-Based 16
      2.2.5 Comparison 17
      2.2.6 Interaction vs. Implementation 17
  2.3 Notification Filtering Mechanisms 19
      2.3.1 Channels 19
      2.3.2 Subject-Based Filtering 19
      2.3.3 Type-Based Filtering 19

2.3.4   Content-Based Filtering 20
2.4   A Model Distributed Notification Service 20
2.4.1   System Model 20
2.4.2   Architecture 21
2.4.3   Distributed Notification Routing 22
2.5   Specification of Event Systems 23
2.5.1   Formal Background 24
2.5.2   A Simple Event System 26
2.5.3   A Simple Event System With Ordering Requirements .. 30
2.5.4   Simple Event System With Advertisements 31
2.6   Further Reading 33

3   Content-Based Models and Matching 35
3.1   Content-Based Data and Filter Models 35
3.1.1   Tuples 35
3.1.2   Structured Records 36
3.1.3   Semistructured Records 52
3.1.4   Objects 56
3.2   Matching Algorithms 57
3.2.1   Brute Force 59
3.2.2   Counting Algorithm 59
3.2.3   Decision Trees 60
3.2.4   Binary Decision Diagrams 61
3.2.5   Efficient XML Matching 63
3.3   Further Reading 64

4   Distributed Notification Routing 67
4.1   System Model 67
4.2   Routing Algorithm Framework 69
4.2.1   Atomic Steps of the Implementation 69
4.2.2   Notification Forwarding and Delivery 72
4.2.3   Avoidance of Duplicate and Spurious Notifications 73
4.2.4   Routing Table Updates 73
4.3   Valid and Monotone Valid Routing Algorithms 74
4.3.1   Valid Routing Algorithms 74
4.3.2   Monotone Valid Routing Algorithms 76
4.4   Valid Framework Instantiations 77
4.5   Content-Based Routing Algorithms 80
4.5.1   Flooding 81
4.5.2   Simple Routing 82
4.5.3   Identity-Based Routing 85
4.5.4   Covering-Based Routing 91
4.5.5   Merging-Based Routing 98
4.5.6   Discussion 104
4.6   Extensions of the Basic Routing Framework 107

        4.6.1   Routing With Advertisements........................107
        4.6.2   Hierarchical Routing Algorithms 112
        4.6.3   Rendezvous-Based Routing 115
        4.6.4   Topology Changes 117
        4.6.5   Joining and Leaving Clients 119
        4.6.6   Routing in Cyclic Topologies.......................120
        4.6.7   Exploiting IP Multicast 122
        4.6.8   Topology Maintenance 123
    4.7 Further Reading...125

5   Engineering of Event-Based Systems.......................129
    5.1 Engineering Requirements 129
        5.1.1   Application Examples..............................130
        5.1.2   Requirements 132
        5.1.3   Existing Support 136
    5.2 Accessing Publish/Subscribe Functionality..................137
        5.2.1   Generic APIs 137
        5.2.2   Domain-Specific APIs..............................139
    5.3 Using the API 140
        5.3.1   Patterns and Idioms 141
        5.3.2   Emitting Notifications 143
    5.4 Further Reading..147

6   Scoping..149
    6.1 Controlling Cooperation 150
        6.1.1   Implicit Coordination and Visibility 150
        6.1.2   Explicit Control of Visibility 151
        6.1.3   The Role of Administrators.........................151
    6.2 Event-Based Systems With Scopes.........................152
        6.2.1   Visibility and Scopes.............................152
        6.2.2   Specification 153
        6.2.3   Notification Dissemination.........................156
        6.2.4   Duplicate Notifications 158
        6.2.5   Dynamic Scopes...................................159
        6.2.6   Attributes and Abstract Scopes 161
        6.2.7   A Correct Implementation..........................161
    6.3 Event-Based Components 164
        6.3.1   Component Interfaces..............................164
        6.3.2   Scope Interfaces 164
        6.3.3   Event-Based Components 167
        6.3.4   Example 167
    6.4 Notification Mappings...................................169
        6.4.1   Specification 169
        6.4.2   A Correct Implementation..........................173
        6.4.3   Example 176

6.5   Transmission Policies 176
       6.5.1   Publishing Policy 177
       6.5.2   Delivery Policy 179
       6.5.3   Traverse Policy 180
       6.5.4   Influencing Notification Dissemination 181
6.6   Engineering With Scopes 182
       6.6.1   Development Process 182
       6.6.2   Scope Graph Handling 183
       6.6.3   Scope Graph Language 187
6.7   Implementation Strategies for Scoping 196
       6.7.1   Scope Architectures 197
       6.7.2   Comparing Architectures 209
       6.7.3   Implement Scopes as Event Brokers 210
       6.7.4   Integrate Scoping and Routing 213
6.8   Combining Different Implementations 225
       6.8.1   Architectures and Scope Graphs 226
       6.8.2   Bridging Architectures 227
       6.8.3   Integration With Other Notification Services 228
6.9   Further Reading 228

7   Composite Events 231
7.1   Application Scenarios 231
7.2   Requirements 234
7.3   Composite Events 234
7.4   Composite Event Detection 236
       7.4.1   Composite Event Detectors 236
       7.4.2   Composite Event Language 238
7.5   Detection Architectures 242
       7.5.1   Centralized Detection 243
       7.5.2   Distributed Detection 244
7.6   Further Reading 250

8   Advanced Topics 253
8.1   Security 253
       8.1.1   Application Scenarios 254
       8.1.2   Requirements 255
       8.1.3   Access Control Techniques 256
       8.1.4   Secure Publish/Subscribe Model 258
       8.1.5   Further Reading 264
8.2   Fault Tolerance 264
       8.2.1   Fault Masking 265
       8.2.2   Self-Stabilizing Publish/Subscribe Systems 265
       8.2.3   Self-Stabilizing Content-Based Routing 266
       8.2.4   Generic Self-Stabilization Through Periodic Rebuild ... 273
       8.2.5   Further Reading 276

8.3 Congestion Control 276
     8.3.1 The Congestion Problem 277
     8.3.2 Requirements 277
     8.3.3 Congestion Control Algorithms....................279
     8.3.4 Further Reading 285
8.4 Mobility..287
     8.4.1 Mobility Issues in Publish/Subscribe Middleware 289
     8.4.2 Physical Mobility................................290
     8.4.3 Logical Mobility.................................295
     8.4.4 Further Reading 302

9 Existing Notification Services 305
9.1 Standards 305
     9.1.1 CORBA Event and Notification Service 305
     9.1.2 Jini..310
     9.1.3 Java Message Service (JMS) 311
     9.1.4 Data Distribution for Real-Time Systems (DDS) 313
     9.1.5 WS Eventing and WS Notification..................317
     9.1.6 The High-Level Architecture (HLA) 317
9.2 Commercial Systems 318
     9.2.1 IBM WebSphere MQ 318
     9.2.2 TIBCO Rendezvous 320
     9.2.3 Oracle Streams Advanced Queuing 322
9.3 Research Prototypes 324
     9.3.1 Gryphon 324
     9.3.2 SIENA...326
     9.3.3 JEDI 329
     9.3.4 REBECA 331
     9.3.5 Hermes 334
     9.3.6 Cambridge Event Architecture (CEA)................337
     9.3.7 Elvin 340
     9.3.8 READY..340
     9.3.9 Narada Brokering 340

10 Outlook 343

References 349

Index 379

List of Figures

1.1 A news story dissemination system 5
1.2 The Active Office ubiquitous environment 6
1.3 The structure of the book 9

2.1 Event-based systemss: interaction versus implementation 12
2.2 Taxonomy of cooperation models 15
2.3 The router network of REBECA 21
2.4 A simple event system 26

3.1 Identity of filters consisting of attribute filters............... 44
3.2 $F_1 \sqsupseteq F_2$ although neither $F_1^1 \sqsupseteq F_2^1$ nor $F_1^1 \sqsupseteq F_2^2$ (two examples) 45
3.3 Covering of filters consisting of attribute filters.............. 46
3.4 Disjoint filters consisting of attribute filters 47
3.5 Overlapping filters consisting of attribute filters 47
3.6 Matching algorithm based on counting satisfied attribute filters 50
3.7 Covering algorithm that determines all *covering* filters 51
3.8 Covering algorithm that determines all *covered* filters 51
3.9 Merging algorithm based on counting identical attribute filters . 52
3.10 A simple notification..................................... 54
3.11 Implementation of a `ClassFilter` in Java 58
3.12 Implementation of a `QuoteFilter` in Java 58
3.13 Using a multilevel index structure for the counting algorithm .. 60
3.14 An exemplary decision tree 60
3.15 An exemplary binary decision diagram 61
3.16 Evaluating a filter using a binary decision diagram 62
3.17 Evaluating an ordered binary decision diagram.............. 62
3.18 XPath Queries and their corresponding finite state automaton . 63
3.19 Combined nondeterministic finite state automaton 64

4.1 Content-based routing framework, part I 70
4.2 Content-based routing framework, part II 71

4.3   Diagram explaining notification forwarding 73
4.4   Flooding 81
4.5   Simple routing 83
4.6   Diagram explaining simple routing (new subscription)........ 84
4.7   Relation among α and β for simple routing 84
4.8   Identity-based routing 87
4.9   Identity-based routing: Processing a new subscription from a
      neighbor 87
4.10 Identity-based routing: Processing a new subscription from a
      client 88
4.11 Relation among α and β for identity-based routing 88
4.12 Covering-based routing.................................. 90
4.13 Covering-based routing: Processing of a new subscription from
      a client.. 93
4.14 Covering-based routing: Processing of a new subscription from
      a neighbor.. 94
4.15 Covering-based routing: Processing of an unsubscription from
      a neighbor.. 94
4.16 Covering-based routing: Processing of an unsubscription from
      a client.. 95
4.17 Covering-based routing: Processing of an unsubscription from
      a client.. 95
4.18 Covering-based routing: Processing of an unsubscription from
      a neighbor, example 2.................................. 96
4.19 Relation among α and β for covering-based routing 97
4.20 Merging-based routing 99
4.21 Merging: deletion of covering filters 99
4.22 Merging: searching for a covering merger 100
4.23 Merging: handling of subscriptions........................ 101
4.24 Merging: handling of unsubscriptions 103
4.25 Circular evolution of CBR algorithms 106
4.26 Routing using advertisements, part I...................... 108
4.27 Routing using advertisements, part II 109
4.28 **prune** for simple routing 110
4.29 **prune** for identity-based routing........................ 111
4.30 Hierarchical covering-based routing 112
4.31 Hybrid routing 113
4.32 Rendezvous-based routing 116
4.33 Managing connects and disconnects....................... 119
4.34 Simple routing in cyclic topologies: algorithm 121
4.35 Example of simple routing in cyclic topologies............. 122
4.36 Routing a message in a Pastry network.................... 125

5.1   Data flow graphs of applications: bipartite single (a) and mult
      source (b), and a general group (c) 130

5.2   An example stock trading application 133
5.3   Generic publish/subscribe interface 138
5.4   The structure of the observer pattern 141
5.5   Event and notification in a UML class diagram 144

6.1   A metamodel of scopes 153
6.2   An exemplary scope graph 154
6.3   Outgoing and incoming notifications 157
6.4   Two ways of generating duplicates 158
6.5   A possible implementation of a scoped event system 162
6.6   Different scope interfaces 165
6.7   The graph of the stock application 168
6.8   Interfaces of the components in the example application 168
6.9   Recursive definition of the relation $(n_1, X) \rightsquigarrow (n_2, Y)$ 170
6.10 Transformation of mappings into components 174
6.11 Architecture of scoped event system with mappings 174
6.12 Three important transmission policies in scope graphs 177
6.13 Scope definition accuracy 196
6.14 Design dimensions of scope architectures 197
6.15 Implicit implementation shifts visibility control into
      application components 201
6.16 A comparison of scope architectures 203
6.17 Steps of scoped notification delivery 207
6.18 Types of architectures, their characteristics, and examples 208
6.19 Comparison of scope architectures 210
6.20 An exemplary scope graph 214
6.21 Scopes as overlays within the broker topology 214
6.22 A flat routing table for broker B_1 215
6.23 Enhanced routing tables of B_1 incorporating scopes 216
6.24 Scope lookup tables 217
6.25 Overall routing algorithm 220
6.26 The naïve matching algorithm with mappings 221
6.27 Interscope forwarding 222
6.28 Duplicate scopes to separate QoS requirements 226

7.1   The Active Office with different sensors 232
7.2   A system for monitoring faults in a network 233
7.3   The components of the composite event detection service 236
7.4   The states in a composite event detection automaton 237
7.5   The transitions in a composite event detection automaton 237
7.6   A composite event detection automaton 238
7.7   The architecture for the composite event detection service 243
7.8   Illustration of centralized composite event detection 243
7.9   Illustration of distributed composite event detection 244

7.10 Two cooperating composite event detectors for distributed detection 245
7.11 The life cycle of a mobile composite event detector 245
7.12 The design space for distribution policies 247

8.1   An event type hierarchy for the Active City 255
8.2   Illustration of the secure publish/subscribe model 258
8.3   An event type hierarchy with attribute encryption 262
8.4   Subscription coverage with attribute encryption 263
8.5   Deriving the minimum leasing time 269
8.6   Notification bandwidth saved by doing filtering instead of flooding 272
8.7   Choosing π such that "old" and "new" update messages do not interleave 274
8.8   Derivation of the maximum stabilization time 276
8.9   Flow of DCQ and UCA messages 280
8.10 Processing of DCQ and UCA messages at IBs 283
8.11 Consolidation of UCA messages at IBs 283
8.12 Missing notifications in a flooding scenario 291
8.13 Moving client scenarios with one and multiple producers 293
8.14 Blackout period after subscribing with simple routing 297
8.15 Blackout period with flooding and client-side filtering 297
8.16 Defining the quality of service for logical mobility 298
8.17 Network setting for the example 299
8.18 Movement graph defining movement restrictions of a consumer . 299
8.19 Total number of messages generated for flooding and two scenarios of the new algorithm 301

9.1   Internal structure of an object request broker (ORB) 305
9.2   Push mode vs. pull mode (typed event communication) 308
9.3   Typed event communication using an event channel 308
9.4   The structure of a structured event (from [287]) 309
9.5   Conceptual overview of data-centric publish/subscribe (DCPS) . 314
9.6   A Gryphon network with virtual event brokers 325
9.7   A hierarchical topology in SIENA 327
9.8   An acyclic peer-to-peer topology in SIENA 327
9.9   A generic peer-to-peer topology in SIENA 328
9.10 Hierarchical event routing in JEDI 329
9.11 Substituting one link with another link 330
9.12 An exemplary router network of REBECA 331
9.13 The filtering framework of REBECA 332
9.14 Layered networks in HERMES 335
9.15 Overview of the HERMES architecture 336
9.16 The publish–register–notify paradigm in the CEA 338
9.17 An ODL definition of event types in ODL-COBEA 339

List of Tables

2.1   Some exemplary temporal formulas and their informal meaning   26
2.2   Interface operations of a simple event system 27
2.3   Changes of the state variables caused by interface operations... 27
2.4   Additional interface operations for advertisements 32
2.5   Changes of the state variables caused by the additional
      interface operations for advertisements 32

3.1   Covering among notification types 40
3.2   Covering among (in)equality constraints on simple values 40
3.3   Covering among comparison constraint on simple values 41
3.4   Covering among interval constraints on simple values 41
3.5   Covering among constraints on strings 41
3.6   Covering among set constraints on simple values 42
3.7   Covering among set constraints on multi values 42
3.8   Perfect merging rules for attribute filters 49

4.1   Portfolio of content-based routing algorithms 104

7.1   Example of five distribution policies 248

8.1   Values of $ploc(x,t)$ for the example setting.................. 300
8.2   Values of filters in example setting........................ 301

1
Introduction

1.1 Networked Computing

The speed at which business is conducted continues to increase. Customer service is important, and mergers as well as joint ventures require flexibility and adaptability of business infrastructures. With the reduction of coordination and communication costs, organizational structures are changed more easily and more frequently. So even after the end of the hype about a New Economy, the trend toward more volatile business structures has neither ceased nor lost its importance [247]. To foster processes and applications that cross traditional modules of enterprise systems, SAP, the major enterprise resource planning (ERP) company, has recently identified an "adaptive business" strategy to be the key to competitive advantage [38]. Services and data are integrated in ever new constellations so that application architectures are getting more volatile. The transition to loosely integrated distributed systems requires IT infrastructures that facilitate both *scalability* and *system evolution*.

Consequently, the development of today's computer systems is mainly influenced by the effects of networking. Increasing connectivity and the size of networked systems give rise to a number of issues. A basic requirement is the availability of scalable communication mechanisms, which are crucial for building and maintaining these systems. The mechanisms not only have to support large numbers of components, but also face complex application environments that are dynamic and subject to unexpected and recurrent change.

A second important aspect of today's systems is the automation of data processing. While systems were traditionally designed to respond to interactive user requests, the aim today is to provide increasingly autonomous data processing to improve functionality and utility. Instead of having human operators mediate between applications, e.g., to replenish an inventory by manually reordering goods, directly connected applications are able to initiate replenishment automatically. In this example, low supplies initiate activity. In general, for a computation to be automated, it must be provided with the data necessary to check for such conditions. Applications are driven by information available in the system, they are *data-* or *information-driven*.

1.2 Middleware

The concept of a *middleware* was introduced to facilitate communication between entities in a heterogeneous distributed computing environment. Middleware is an additional layer between operating systems of individual nodes and a distributed application. It deals with communication issues and attempts to provide a homogeneous view of the world to the application. As such, it is widely used and has proved to be a successful abstraction that helps with the design and implementation of complex distributed systems.

The variability of dynamic networked environments and the automation of data exchange shifts the focus when dealing with the delivery of data and services, moving from a stationary world to one that is in a state of flux. Traditionally, middleware has viewed data and services as being stationary in a collection of objects or databases, with inquiries directed at them in a request/reply mode of interaction. This concept has led to client/server middleware architectures that emphasize explicit delegation of functionality, where system components access remote functionality to accomplish their own goal. *Remote procedure calls (RPC)* and derivative techniques are classic examples [44, 269, 371]; even the incipient Web services mainly rely on sending requests with the Simple Object Access Protocol (SOAP) [347]. These techniques deliberately draw from a successful history of engineering experience, their principles are well understood, and they have been an appropriate choice for many well-defined problems.

In the context of dynamic networked systems, however, request/reply has serious restrictions. The direct and often synchronous communication between clients and servers enforces a tight coupling of the communicating parties and impairs scalability [158]. Clients poll remote data sources, and they have to trade resource usage for data accuracy, especially in chains of dependent servers. Unnecessary requests due to short polling intervals waste resources, whereas long intervals increase update latency. In addition, request/reply restricts system evolution. The control flow is encoded in application components, which makes it accessible to engineers but also mixes the actual configuration of the system with the application logic of individual components. Consequently, the capability to orchestrate the whole system is limited by the means available to adapt application components at runtime. Finally, delegating functionality inevitably implies a functional dependency on the called service, and on its presence.

The need for asynchronous and decoupled operation led to various extensions of existing middleware. For instance, CORBA and Java 2 Enterprise Edition (J2EE) were extended with asynchronous invocation methods and notification services [279, 336, 338, 364], and similar features are available in Microsoft's COM+ and in the language model of the new .Net platform [235, 315], too. Database research, software engineering, and coordination theory corroborate the advantages of loosely coupled interaction as well [80, 171, 295, 356].

1.3 Event-Based Systems

Instead of stepwise amending the conventional request/reply mode of interaction, *event-based computing* takes a contrasting approach and inherently decouples system components. In an event-based mode of interaction *components* communicate by generating and receiving *event notifications*, where an *event* is any occurrence of a happening of interest, i.e., a state change in some component. The affected component issues a *notification* describing the observed event. An *event notification service* or *publish/subscribe middleware* mediates between the components of an *event-based system* (EBS) and conveys notifications from *producers* (or *publishers*) to *consumers* (or *subscribers*) that have registered their interest with a previously issued *subscription*.

The power of an *event-based architectural style* [68] is that neither the published notifications nor the subscriptions are directed toward specific components. The notification service decouples the components so that producers are unaware of any consumers and consumers rely only on the information published, but not on where or by whom it is published. Event-based components are not designed to work with specific other components, which facilitates the separation of communication from computation. The event-based style carries the potential for easy integration of autonomous, heterogeneous components into complex systems that are easy to evolve and scale [32, 355].

In view of the above arguments, the use of events is superior to request/reply in many information-driven scenarios [157]. In fact, many improvements of tightly coupled communication converge to an asynchronous approach. For instance, caching data in network nodes [322], callback handling according to the observer pattern [161], asynchronous remote invocations [338] introduce some form of indirection, decoupling interaction from computation. The loose coupling makes applications easier to adapt and integrate, and it allows a specialized mediator, the notification service, to achieve scalability.

As a consequence, the potential of the event-based style has been recognized both in academia and in industry. The event-based architectural style is becoming an essential part of large-scale distributed systems' design, and many applications and their underlying infrastructures have incorporated event-based communication mechanisms. *Information buses* are the basis of many systems [27, 289], and a number of event notification services were developed (e.g., [71, 92, 172, 353, 364, 381]) as well as integrated into modern component platforms such as CORBA Component Model (CCM) [278] and Enterprise JavaBeans (EJB) [362].

The aim of this book is to provide the reader with an overview of the rich area of event-based systems. We cover a broad spectrum of topics, ranging from a formal treatment of local and distributed event matching algorithms, through a more practical discussion of software engineering issues raised by the event-based style, to a presentation of state-of-the-art research topics in event-based systems, such as composite event detection and security. Our hope is that our presentation shows the power of event-based systems in modern systems design and encourages both researchers and practitioners to exploit the event-based style in next-generation large-scale distributed applications.

1.4 Application Scenarios

The range of application scenarios for event-based systems is broad. Often, applications use event-based communication to improve scalability or to achieve adaptability. In order to understand the power of event-based system, we consider application scenarios, in which the use of traditional request/reply communication would be prohibitively expensive in terms of efficiency or usability. Next, we describe several application scenarios to motivate the use of an event-based style for systems design.

1.4.1 Information Dissemination

Information dissemination, in general, is the apparent application domain of notification services, which includes news story dissemination, real-time control systems, and stock market monitoring applications. The timely and efficient dissemination of information to many consumers is a prerequisite in these systems. In addition to simple unidirectional data distribution scenarios, in which the focus is on the direct communication of few producers with many subscribers, more sophisticated applications require the set-up of complex information flows, in which an event-based style is used to drive advanced processing workflows of real-time data.

In its most basic form, Internet-wide distributed systems involve the exchange of information among a large number of nodes. A system for *news story dissemination* is depicted in Fig. 1.1. News reports that are generated by local news agencies are distributed worldwide among many news corporations. News corporations desire to receive relevant information only, and news agencies prefer to avoid the complexity of having knowledge about all news corporations. The loose coupling of producers and consumers in an event-based system for this application leads to a flexible and robust system design.

1.4.2 Network Monitoring

In general, any form of system monitoring is very compatible with an event-based style. Information about the current status of the system's components

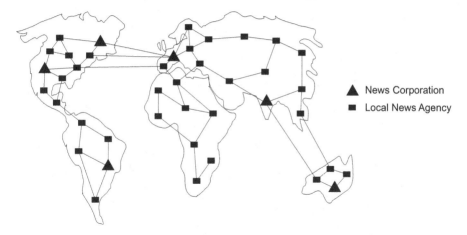

Fig. 1.1. A news story dissemination system

is logged with only a minimal influence on the control flow of the running system as it is published as state changes. Network management has a strong need for sophisticated monitoring capabilities of runtime statistics, alerts, and configuration changes [216, 339]. Especially, distributed network intrusion detection [162, 163] has gained widespread attention in recently with high-profile distributed denial of service attacks and the compromise of personal data.

In particular, network monitoring applications often require a high-level view of the system [330, 348]. A network failure or intrusion attempt may lead to a multitude of low-level events being triggered. The challenge for the network administrator is to track down the root cause of these events as quickly as possible. Real-time processing of events to detect patterns in the form of composite events are a powerful technique for this; they will be described in Chap. 7.

1.4.3 Enterprise Application Integration

Many business environments are characterized by their variability and need to facilitate change. Enterprise application integration (EAI) is about connecting custom-built, third-party, or legacy systems to share data and join business processes. However, the integrated applications are often independently developed, deployed, and maintained. To avoid tightly coupled dependencies between any pair of these applications, the resulting system architecture usually relies on a mediator to achieve the loose coupling that is necessary to achieve scalability and flexibility [191].

Information buses, messaging, and the source-driven distribution of data is an inherent characteristic of EAI [215, 289]. A mediator approach decouples interfaces, allows for independent evolution, and extracts communication and coordination tasks into an extra component. The event-based paradigm

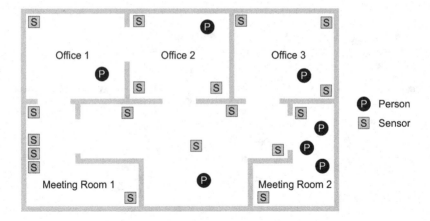

Fig. 1.2. The Active Office ubiquitous environment

directly addresses the EAI objectives and publish/subscribe middleware is therefore an important candidate for implementing such a mediator [27, 215].

1.4.4 Mobile Systems

Mobile systems are distributed systems in which a subset of the nodes (e.g., the clients in a client/server system) or even all nodes of the system are mobile leading to a more or less dynamic network topology. These systems are obviously dynamic environments where central servers, synchronous communication, and static bindings are inappropriate and not available [60]. An infrastructure for mobile systems always has to cope with reconfigurations, thus making the event-based style pertinent here, too [90, 142]. Similar arguments hold for wireless sensor networks [9, 205, 254].

1.4.5 Ubiquitous systems

A different type of large-scale distributed system is a ubiquitous sensor-rich environment concentrated on a small area, such as the *Active Office* building shown in Fig. 1.2. In such a building, sensors that are installed in offices provide information about current occupancy, environmental conditions, and equipment health to interested devices, applications, and users. The Active Office is aware of its inhabitants' behavior and enables them to interact with it in a natural way.

Since this ubiquitous environment is highly dynamic with components constantly entering and leaving, the loose coupling of an event-based system vastly simplifies system design. In addition, the large number of sensors potentially produces a large amount of data. As a result, information consumers prefer a high-level view of the primitive sensor data, making efficient data aggregation a necessity.

1.5 Putting Event-Based Systems Into Context

In this section, we point out areas related to event-based systems. By doing so, we want to put event-based systems into the context of better known areas in computer science. This also shows how event-based techniques were independently developed in different areas to address similar challenges introduced by scale, system evolution, and real-time requirements.

The database community has quickly recognized the need for databases to react to data changes. **Active databases** [304, 394], such as HiPAC [103] and Postgres [352], follow an event-based style using *database triggers* [101] that are expressed in the form of *event–condition–action (ECA)* rules. The action of a trigger is executed when an event occurs, such as the modification of a database table, and a predicate conditions holds. This enables the database to react to user actions, tying computation to external events.

In **software engineering** the event-based mode of interaction is also known as *implicit invocation* [115, 164]. It is defined as an architectural style that determines how components of a software architecture communicate in principle [165]. Events can be found in enterprise architectures [151, 191] as well as in software patterns like the observer pattern [161]. In fact, Garlan et al. [166] early identified the prominent importance of using events for the construction of flexible software architectures. The book by Luckham [242] thoroughly elaborates the important relation between enterprise applications and complex event processing. Consequently, eventing is part of most modern component and container frameworks. The concept is also employed in graphical user interfaces (GUIs), where the model view controller paradigm [178] and observer patterns are applied. More recently, aspect-oriented programming uses events to identify points in the execution of programs at which aspect code is activated [82, 147].

Stream processing systems [63] such as *Aurora* [1], *Borealis* [23], *Hi-Fi* [156], and *TelegraphCQ* [75] combine event-based, push-based data dissemination with transformation capabilities implemented by *stream operators* [14]. Data sources continuously produce events, which are transformed according to existing *continuous queries* [17], thus delivering a result stream to consumers. The event-based style ensures that producers are decoupled from consumers, which is important in common application scenarios such as supply chain management and financial market analysis.

Distributed monitoring and debugging [249] systems, for example GEM [251]), are used to gather changes in the distributed state of a system. Composite event detection [339] is then used to derive more abstract events from the primitive ones. Monitoring can be either done online or offline. With *offline* monitoring, primitive events are collected locally until, for example, an error occurred. Then, the collected events are transferred to a central facility that does the analysis, e.g., to find the source of the error. Contrarily, in the case of *online* monitoring, primitive events are collected and composite events are detected in realtime. This approach is much more complicated but it is also more powerful since it allows to build adaptive distributed systems.

In **sensor networks**, algorithms such as *directed diffusion* [204, 205] are used that are very similar to content-based routing algorithms. In sensor networks event sources publish events that are then routed hop-wise through the network to reach event sinks. While events are routed, they may be aggregated on intermediate nodes if possible. This is done to reduce the network traffic. In the case of direction diffusion, gradients are derived (e.g., hop-based on the distance from the sink) by flooding the subscriptions into the network. Path reinforcement is used to enforce single path delivery.

Until recently, **workflow systems** [176] were often built around centralized data stores, but distributed execution environments led to generalizations relying on ECA rules and events [177, 242]. Event notification services are used as building blocks for distributed activity services [80, 190, 217]. However, such high-level application domains have requirements that typical publish/subscribe middleware can hardly fulfill today [118], and which led to the discussion about engineering issues in Chap. 5.

1.6 From Centralized to Internet-Scale Event Systems

The first generation of event-based systems were centralized systems with integrated active functionality. Examples of these systems are active databases and toolkits for graphical user interfaces. With the introduction of distributed system middleware, such as CORBA, the idea of incorporating asynchronous event-based communication into this middleware arose. This led to notification services such as the CORBA Event Service. However, most implementations of these services were still centralized and had only limited filtering capabilities.

The next step was to integrate more expressive filtering leading to, e.g., the CORBA Notification Service. Still, most distributed systems were rather closed, and event forwarding based on IP multicast was sufficient. Then, the need to integrate several distributed systems to form a larger system arose. This raises the need for security mechanisms, content-based forwarding, mechanisms to cope with data heterogeneity, and scoping. The next step will be an Internet-scale notification service that enables information to flow from one node in the Internet to another. To achieve this, optimized content-based routing, scoping with input and output interface, and refined security mechanisms (e.g., to enable trust) are needed. This book discusses the techniques that lie on the road from centralized to Internet-scale event systems and which are needed to make this vision a reality.

1.7 Structure of the Book

The structure of the book is as follows (see Fig. 1.3):

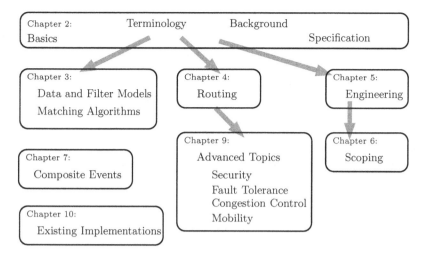

Fig. 1.3. The structure of the book

Chap. 2: Basics. We introduce the basic terminology and from introduced interaction models we derive our notion of event-based interaction. We describe notification filtering mechanisms including channels, subjects, types, and content-based filtering. A specification of event system is given that forms the basis of our further investigations.

Chap. 3: Content-Based Models and Matching. We describe data and filter models that allow notification to be described and filtered as well as matching algorithms that match published data to registered subscriptions. This chapter can be skipped if matching is considered as black box functionality.

Chap. 4: Distributed Notification Routing. This chapter presents details about the distributed implementation of publish/subscribe systems. The system model is described and framework for content-based routing is introduced. Based on the framework we discuss a number of routing algorithms and we present several extensions of the basic framework.

Chap. 5: Engineering of Event-Based Systems. We describe the engineering issues related to event-based systems and how one can build event-based applications.

Chap. 6: Scoping. This chapter introduces a scoping concept for event-based systems. We discuss how to restrict the visibility of events by scopes, event-components based on scopes with input and output interfaces, notification mapping, and transmission policies.

Chap. 7: Composite Events. We detail composite event detection by discussing composite event detectors based on automata, a composite event language, and detection architectures.

Chap. 8: Advanced Topics. This chapter collects more advanced topics (such as security, fault tolerance, congestion control, and mobility) to which only an introduction is given because a full coverage is out of the scope of this book.

Chap. 9: Existing Notification Services. Here, we discuss existing standards (e.g., JMS, CORBA Notification Service), commercial products (e.g., Oracle Streams Advanced Queuing, TIBCO Rendezvous), and research prototypes (e.g., SIENA, JEDI, REBECA).

Chap. 10: Conclusions. This chapter summarizes the main insights of this book and give an outlook to potential future research directions.

2

Basics

2.1 Terminology—Constituents of Event-Based Systems

An *event-based system* consists of the following constituents (see Fig. 2.1): events and notifications as means of communications, producers and consumers as interacting components, subscriptions signifying a consumer's interest in certain notifications, and the event notification service responsible for conveying notifications between producers and consumers.

2.1.1 Events and Notifications

Any happening of interest that can be observed from within a computer is considered an *event*. This may be a physical event such as the appearance of a person detected by sensors, a timer event that indicates progression of real time, or generally an arbitrary detectable state change in a computer system. We consider only the third kind of events here, because event detection is out of the scope of this book.

A *notification* is a datum that reifies an event, i.e., it contains data describing the event. A notification is created by the observer of the event and may just indicate the plain occurrence, but often may carry additional information describing the circumstances of the event. For instance, in the active badge system of Bacon et al. [18] events are raised when persons wearing a badge approach a sensor, and the published notification carries the detected ID of the badge and the time of observation. In general, different notifications can be created that describe the same underlying event, but from multiple viewpoints. This may be done due to application or security reasons, or simply because notifications are encoded in different data models. The most common data models are name/value pairs [71], objects [18, 104, 127], and semistructured data [12, 264], e.g., XML.

On the lowest level considered here, notifications are conveyed via *messages*, which are data containers on the network level transmitting data be-

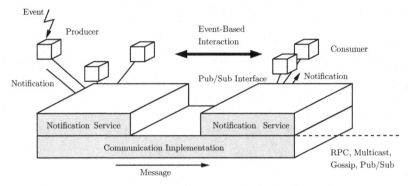

Fig. 2.1. Event-based systemss: interaction versus implementation

tween the endpoints of the underlying communication mechanism. The distinction between events, notifications, and messages is used to clearly separate the underlying communication technique from the mode of interaction, cf. Sect. 2.1.4.

2.1.2 Producers and Consumers

The software components of an event-based system act as producers and consumers of notifications. *Producers* are components that publish notifications. A producer's implementation is "self-focused" in the sense that it observes only its own state. The decision to publish a state change is made by the component's internal computation and is a core part of its function. What changes are published and how this decision is configured/programmed into the producer are issues of past and ongoing research in areas like debugging [33] and monitoring [231], reflection [223], and aspect-oriented programming (AOP) [82]. Published notifications are not addressed to any specific (set of) receivers; they are rather forwarded to the event notification service for further distribution. Producers are unaware of any other components and they do not anticipate any reaction on the receiver side; this is detailed in Sect. 2.2.

Consumers react to notifications delivered to them by the notification service. They, too, are unaware of their specific communication peers. Not knowing the actual producers of notifications, consumers issue subscriptions to describe the kinds of notifications they are interested in; different classes of subscriptions are depicted in the next section. If a component is both consumer and producer, it reacts to both incoming notifications and observed internal state changes, and the resulting computation may lead to newly published notifications.

2.1.3 Subscriptions and Filters

A *subscription* describes a set of notifications a consumer is interested in. Consumers register their interest in receiving certain kinds of notifications by submitting subscriptions to the notification service. The service evaluates the subscriptions on behalf of the consumer and delivers those notifications that match one of the consumer's subscriptions. Subscriptions are *filters*, which are basically Boolean-valued functions that test a single notification and return either *true* or *false*. Indeed, filters are a common way to implement subscriptions, although, in general, subscriptions may comprise more than only a filter function. They can additionally include (meta)data to govern notification selection beyond a per-notification level; for example, security credentials for accessing certain classes of notifications [34] or timing information to get past notifications [81]. Subscriptions can be seen as input interfaces of consumers, describing the data they are prepared to process.

Advertisements are issued by producers to declare the notifications they are willing to send. They also describe sets of notifications and may be of the same form as subscriptions. From a network level point of view, advertisements help to improve routing decisions, because the notification service knows which notifications can be expected from where. From a software engineering viewpoint, advertisements comprise a component's output interface.

The expressiveness of subscriptions in terms of filtering capabilities depends on the filter model and the data model employed. The combination of a filter model and a data model is called a *notification classification scheme*. In distributed notification services, essentially four *filter models* are distinguished: channels, subjects, types, and content-based, which are described in Sect. 2.3. Tuples, sets of name/value pairs, and semistructured documents are the most prominent data models for distributed notification services. They are described in Chap. 3.

2.1.4 Event Notification Service

The *event notification service*, or notification service for short, is the mediator in event-based systems that decouples producers from consumers. It alone is responsible for conveying notifications, and it must deliver every published notification to all consumers having registered matching subscriptions. It implements a publish/subscribe interface, providing *adv*, *pub*, *sub*, *unsub*, and *notify* operations; the last being an output operation called on a registered consumer to deliver a notification. The notification service gets notifications from producers via the *pub* operation, and they must match the advertisements issued with the *adv* operation. The service tests notifications it got from producers against subscriptions it got from consumers via the *sub* operation, and delivers the notifications to those consumers that have a matching subscription with *notify*. In essence, it separates communication responsibility from components in the sense that the mediating service is responsible for subscription evaluation on behalf of the consumers and for delivering notifications on behalf of the producers. Note that we just described a basic notification service; a more advanced service may exhibit more operations.

From the perspective of application components, the notification service is a black box; its function does not depend on it being distributed. However, its nonfunctional attributes, such as efficiency, scalability, and availability, are influenced by the architecture and the communication techniques used to distribute notifications.

In addition to the notification service, event-based systems often contain further event handling capabilities, such as event and notification type repositories, descriptions of available data and filter models, and other "metadata" as well as programming language bindings beyond service invocations. To reflect the broader functionality the collection of notification service plus any additional event handling is termed *event system*.

2.2 Models of Interaction

From a technical point of view, an event notification service just provides publish/subscribe functionality, which *may* be used for transporting notifications, but also for sending requests to groups of servers. The essence of event-based systems is not found in the Application Programming Interface (API) or the techniques used for transmitting notifications. Event-based interaction is mainly a characteristic of the components, and not of the underlying communication technique [54, 265].

In order to provide a fundamental and simple characterization, four *interaction models* are distinguished by the way interdependencies between components are established. The four models are differentiated by two attributes (Fig. 2.2). The first attribute, *initiator*, describes whether consumer or provider initiates the interaction, where the former depends on data or functionality provided by the latter. The second attribute, *addressing*, distinguishes whether the addressee of the interaction is known or unknown, i.e., whether the peer component is directly or indirectly addressed.

The resulting four interaction models are independent of any underlying implementation technique. Any interaction between a set of components can be classified according to these models. Even though interaction may show more nuances in practice, the models are complete in the sense that they essentially cover all major paradigms.

Furthermore, the interaction models characterize the inner structure of components, because the models determine how dependencies between the components are established. From an engineering point of view, this helps to identify constraints and requirements posed by a given component on its usage scenarios and on the underlying infrastructure. Architectural mismatches are disclosed early; they would otherwise have to be tackled by an integrating

		Initiator	
		Consumer	Provider
Addressee	Direct	*Request/Reply*	*Callback*
	Indirect	*Anonymous Request/Reply*	*Event-Based*

Fig. 2.2. Taxonomy of cooperation models

implementation, which impedes system evolution and scalability sooner or later [167].

2.2.1 Request/Reply

The most widely used interaction model is request/reply. Any kind of remote procedure call or client/server interaction belongs to this class. The initiator is the consumer (i.e., client) that requests data and/or functionality from the provider (i.e., server), and it expects data to be delivered back or relies on a specific task to be done. The provider is directly addressed, its identity is known, and the caller is able to incorporate information about the callee into his own state and processing, resulting in a tight coupling of the cooperating entities. Replies are (in most cases) mandatory in this model.

2.2.2 Anonymous Request/Reply

The anonymous request/reply model also uses request/reply as basic action, but without specifying the provider that should process the request. Instead, requests are delivered to an arbitrary, possibly dynamically determined set of providers. The consumer does not know the identity of the recipient(s) a priori, yet it expects at least one reply—one request may result in an unknown number of replies.

This model is eligible when redundant providers are available or when the appropriate provider may be different for each request. For instance, load balancing selects a provider either arbitrarily or based on the content of the request; cf. the IP Anycast mechanism [301] tries to route a packet to the nearest member of a group of destinations without resolving the IP address in advance. Similarly, component models and containers decouple component instances and allow for runtime binding of references, cf. JavaBeans [359] and the *Dependency Injection Pattern* [153, 252]. However, this often only means providers are resolved just before the call, making the identity known to the caller and potentially leading to tight coupling as in classic request/reply.

This cooperation model is besides the event-based model the second model that is directly implemented by publish/subscribe services, which often confuse these two models. Anonymity of providers adds more flexibility to the request/reply model, but the dependency on externally provided data or functionality persists.

2.2.3 Callback

In the callback model, which is employed in the well-known *observer* design pattern [161], consumers register at a specific, known provider their interest to be notified whenever some condition becomes true. The provider repeatedly evaluates the condition and if necessary calls the registered component back. The provider is responsible for administering its callback list of registered consumers. If multiple callback providers are of interest, a consumer must register separately for all of them. The identity of the components is known and must be managed on both sides, leading to a tight coupling with no coordination medium in between.

On the other hand, knowing the identities of consumers, callback processing can be customized so that only subsets of consumers are notified in an application-dependent way. However, it would be each component's responsibility to apply callback handlers that implement current application needs, which is an issue of integration rather than of component implementation. In any case, a sophisticated implementation of callback handlers leads to the event-based approach, described next.

2.2.4 Event-Based

The event-based interaction model has characteristics inverse to the request/reply model. The initiator of communication is the provider of data, that is, the producer of notifications. Notifications are not addressed to any specific set of recipients, as was described earlier. A consumer can receive notifications from many providers, because subscriptions are, in general, neither directed nor limited to a particular producer. If a notification matches a subscription, it is delivered to the registered consumer. Providers are not aware of the consumers. In contrast to the callback model, providers are relieved from the task of interpreting and administering registrations, i.e., subscriptions.

The essential characteristic of this model is that producers do not know any consumers. They send information about their own state only, precluding any assumptions on consumer functionality. A component "knows" how to react to incoming notifications and it publishes changes to its own state, but it must not publish a notification with the intention of triggering other activity. A component's implementation is "self-focused" in that the knowledge encoded in the program, and used by the programmer, is limited to the component's own task. This approach completely separates the internals of different parts of an application.

Of course, the overall functionality of the system still depends on the proper interaction of all the components, but this is no longer a matter of individual components. It is rather the composition of components and their interaction that determine the functionality. But event-based interaction with-draws the control of interaction from the participating components, and the necessary coordination has to be handled externally. So, in addition to the role of specifying and implementing individual components, the orchestration of an event-based system demands extra support. Currently, no such support is available.

2.2.5 Comparison

The complexity of a decomposed system is characterized by the degree of de-pendence between its components. Software reliability analysis formally cor-roborates a result that is informally apparent: If a component relies on other components to accomplish its own goal, its correctness is degraded by fail-ures of others [2, 253]. Conversely, the correctness of individual components is not affected if they process available data only, which is exactly the case in event-based systems. The event-based style clearly separates computation from communication and offers the potential of easily evolvable systems. On the other hand, engineering complexity is considerably affected by the quality of the abstractions and tools available for coordinating the components.

The dichotomy of request/reply and event-based interaction is marked by the simplicity of the former and the flexibility of the latter. Request/ reply is easy to handle, implement, and understand, and consequently is well established. It corresponds to the imperative nature of common programming languages and component models. Some of its shortcomings are alleviated by a long list of supplementary techniques such as caching, asynchronous request/ reply, container-controlled operation, dependency injection, etc., that are used to enhance scalability and system evolution.

However, if interaction becomes less coupled, it gets more indirect. And this raises the question whether the use of events would be a more appro-priate solution. In fact, without being formally corroborated, it appears that request/reply and event-based interaction form a duality in the sense that for most problems there exist solutions based on either model. Classic request/ reply examples can be rebuilt using events. Event-based interaction typically relies on a reversed software architecture, reversing activity and data flows, but the same function can be implemented in both paradigms. The involved tradeoff is between scalability and flexibility, on the one hand, and simplic-ity on the other. System engineers have to decide whether they opt for a simple implementation or for an extensible one. One goal of this book is to make choosing the extensible solution less costly, and thus eligible for more scenarios.

2.2.6 Interaction vs. Implementation

The mode of interaction influences the design of components and is difficult to change. It is a prerequisite of good design to choose an interaction model that matches the function a component has to accomplish. Otherwise, architectural mismatches would inevitably impede system composition and evolution [167]. For this reason, this basic but principal distinction of interaction models helps system designers to identify the core structure of components, and it avoids mixing interaction and implementation issues [137].

Unfortunately, the mode of interaction is often confused with the choice of implementation techniques currently available. In particular, event-based interaction is often equated with using general publish/subscribe services. While being obvious candidates for implementing notification dissemination, they are not the only ones; other techniques may as well be employed, like point-to-point messaging, IP multicast, Linda tuple space engines, or even classical remote procedure calls. For instance, if a system engineer *knows* that a set of event-based components interacts only within a small group, nothing speaks against using RPC. In fact, if the communication happens to be sensitive to eavesdropping, RPC even becomes the most appropriate choice. Note that producers still publish notifications as before, only the underlying implementation is considered here. Conversely, a publish/subscribe service can also be used to implement anonymous request/reply interaction.

Generally, there is no best implementation technique for a certain interaction model. The technique must be chosen in view of the deployment environment, the demanded quality of service, and the overall need for flexibility and scalability. Event-based interaction facilitates the distinction of interaction and implementation due to its separation of computation from communication. And while traditional publish/subscribe services focus on unidirectional delivery (Sect. 5.1), many different techniques can be exploited in building event-based systems.

The preceding description of event-based interaction basically refines the one given in literature, e.g., [68, 165, 295]. The discussion makes it now possible to unambiguously define the involved terminology. The system outline given in Fig. 2.1 spans several levels of abstraction. On the lowest level, *messages* are *sent* and *received*. Arbitrary asynchronous messaging techniques can be used, be it connectionless point-to-point network protocols, IP multicast mechanisms, or publish/subscribe implementations.

On the next level, the publish/subscribe interface is implemented. It is used to *publish* data that is *delivered* to *subscribers*. As part of its implementation, messages containing the data are sent and received. From a technical point of view, the publish/subscribe interface implements both anonymous request/ reply and event-based interaction.[1]

On the highest level, where event-based interaction finally takes place, *producers* publish notifications that are delivered to *consumers*. Only this level is of concern when assessing the characteristics of event-based interaction and its effect on system engineering.

---

[1] Although all arguments made here explicitly target event-based systems, they are equally applicable to any general publish/subscribe scenario.

2.3 Notification Filtering Mechanisms

2.3.1 Channels

Channels are the simplest form of identifying sets of notifications. In this model, producers select a named channel into which a notification is published. Consumers, on the other hand, select a channel and they will get all notifications published therein. An example of this approach is the CORBA Event Service [280]; the CORBA Notification Service [287] also relies on channels but additionally offers filters on notification content.

2.3.2 Subject-Based Filtering

Subject-based filtering uses string matching for notification selection [289]. Publishers annotate each notification with a subject string that denotes a rooted path in a tree of subjects. For example, a stock exchange application publishes new quotations of *FooBar Ltd.* under the subject `/Exchange/Europe/London/Technology/FooBar`, classifying it to be traded in London and to belong to the technology sector of the stock market. Consumers subscribe for `/Exchange/Europe/London/Technology/*` to get all technology quotations. It is implementation-dependent whether `/Exchange/Europe/London/*` already includes notifications of subsubjects or not. In principle, arbitrary pattern matching can be executed on subjects.

The simplicity of this approach has deficiencies that limits its applicability. The requirement to use a single path in a tree to classify a notification severely constrains the expressiveness of this model. The subject hierarchy is a tree—multiple super-subjects are not allowed—and it classifies only from a single point of view. Alternative classifications, e.g., `/Exchange/Europe/Technology/London`, are only possible if different subtrees permute the order of subjects. This leads to repeated publications and an exponential growth of tree size if several alternative viewpoints shall be reflected.[2]

2.3.3 Type-Based Filtering

Type-based filtering uses path expressions and subtype inclusion tests to select otherwise opaque notifications [32, 127]. With multiple inheritance, the subject tree is extended to type lattices that allow for different rooted paths to the same node. Often, type-checking is complemented with content-based filters to improve selectivity [311].

---

[2] Similarly, from a software engineering point of view such hierarchies have been criticized as restrictive and impeding integration and evolution [188].

2.3.4 Content-Based Filtering

Content-based filtering is the most general scheme of notification selection [69, 262]. Filters are evaluated on the whole content of notifications, where the data model of the notifications and the applied predicates determine the expressiveness of the filters. Available solutions range from template matching [92], simple comparisons [71], or extensible filter expressions [264] on name/value pairs, to XPath expressions on XML [12] and arbitrary programs and mobile code [117].

Concept-based Publish/Subscribe is orthogonal to the above approaches and is proposed by Cilia et al. [83]. It employs semantic mappings between data and filter models to transform subscriptions from one model to another.

2.4 A Model Distributed Notification Service

This section describes the system model and the basic characteristics of the REBECA notification service [136]. It implements the publish/subscribe interface described in Sect. 2.1 and conforms to the preceding definition of simple event systems. Its basic architecture is a representative example of a distributed notification service, which is comparable to that of other services like SIENA, JEDI, etc. REBECA is different from other services with regard to its support for different routing algorithms and data and filter models [263, 267], and the visibility control extensions presented in this book. REBECA serves two roles: first, its system model is the basis for investigating visibility issues, and second, the available implementation acts as testbed for publish/subscribe functionality.

2.4.1 System Model

The model assumed in REBECA and this book is a process model in which computational activity is represented by the concurrent execution of processes [230]. Processes interact by passing messages via links between them. A link connects a pair of processes and forwards messages asynchronously so that there is a delay between sending a message and receiving it. Links are assumed to exhibit no failures and to obey first-in-first-out (FIFO) ordering of messages. This means that no messages are lost or corrupted due to link failures and that messages are received in the same order they were sent. Although being impractical in general, it is a reasonable assumption in the present context, because it simplifies the discussion. In Sect. 8.2 we discuss fault tolerance issues. In fact, initial solutions for both problems exist elsewhere and may be used later to extend the model, e.g., [86, 263].

More concretely, the considered distributed system consists of a set of physical nodes interconnected by a communication network and each node runs one or more processes. Communication links are point-to-point connections in this network, and their failure model is easily matched by TCP/IP connections, for instance. This is the basic model that is broadly applicable, and which nevertheless is open for implementation-dependent options, like using multicast, to improve communication performance (cf. Sect. 6.7.4).

2.4.2 Architecture

The system constituents are illustrated in Fig. 2.3 and both the application components and thenotification service itself are implemented by the aforementioned processes. Each component is executed by a separate process, which is linked to a process of the notification service. The service is accessed as a black box that is conceptually centralized, but its implementation is distributed across several processes and nodes to split the load and exploit locality in notification delivery.

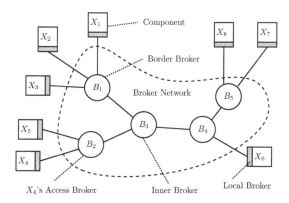

Fig. 2.3. The router network of REBECA

The notification service forms an overlay network in the underlying system. An *overlay network* is a virtual network of processes that communicate by means of a second underlying (physical) network, employing routing strategies different from the underlying ones. Here, the overlay consists of *event brokers* that run as processes on some of the physical nodes. The communication topology of the overlay is described by a graph. Currently, only acyclic graphs are supported. Edges are process links and as such are mapped to point-to-point connections in the underlying network, namely, TCP/IP connections. The acyclic graph used is comparable to the single spanning tree approach of multicast algorithms [106]. Obviously, the single tree is a bottleneck of the system, but, again, it is an adequate model in this context, and extensions exploiting redundancy are available to tackle problems of scalability and single points of failure [86, 311, 374].

Three types of brokers are distinguished: local, border, and inner brokers. *Local brokers* are access points to the middleware. They are typically part of the communication library loaded into application components; they are not represented in the graph, but are only used for implementation issues. A local broker is connected to one border broker. *Border brokers* form the boundary of the distributed communication middleware and maintain connections to local brokers, i.e., the components of the service. *Inner brokers* are connected to other inner or border brokers and do not maintain any connections to components.

Local brokers implement the publish/subscribe interface of the notification service and initially put the first message containing a newly published notification into the network. Border and inner brokers forward the messages to neighbor brokers according to filter-based routing tables and respective routing strategies. At the end the messages are sent to the local brokers of the consumers and from there the notifications are delivered to the application components. Routing notifications from producers to consumers through a broker network is also called *distributed notification routing*.

2.4.3 Distributed Notification Routing

The function of distributed notification routing is rather simple: just match all notifications with all subscriptions and deliver the notification to all clients and neighbor brokers with a matching subscription. In a centralized implementation the problem is reduced to efficient matching algorithms [266, 404]. A centralized implementation, however, not only concentrates all computational efforts but also becomes a bottleneck of communication bandwidth. Hence, REBECA distributes matching on multiple brokers.

Flooding is the simplest approach to implement routing: brokers forward notifications to all neighboring brokers and only those brokers to which components are connected test on matching subscriptions. Flooding guarantees that notifications will reach their destination, but many unnecessary messages (e.g., notifications that do not have consumers) are exchanged among brokers. The main advantage of flooding is its simplicity and that subscriptions become effective instantly since every notification is processed by every broker anyway.

Filter-based routing depends on routing tables (RT), which are maintained by the brokers and consist of routing entries. A routing entry is a filter/destination pair indicating to which local client or neighbor broker matching notifications have to be delivered or forwarded, respectively. The entries are updated by sending control messages corresponding to new or canceled subscriptions through the broker network. New subscriptions add (F, D) entries with D denoting the destination from which they were received, and unsubscriptions delete the respective entries. Every incoming notification is tested against the routing table entries to determine the set of destinations with matching filters, omitting the originating destination if it is a neighbor broker to prevent loops. If the incoming notifications of each destination are routed sequentially, end-to-end FIFO-producer ordering holds. In the case of a acyclic broker network also causal ordering holds.

Different flavors of filter-based routing exist, which differ in their strategy to update the routing tables. *Simple routing* assumes that each broker has global knowledge about all active subscriptions. It minimizes the amount of notification traffic, but the routing tables may grow excessively. Moreover, every (un)subscription has to be processed by every broker, resulting in a high filter forwarding overhead if subscriptions change frequently. In large-scale systems more advanced routing algorithms must be applied to exploit commonalities among subscriptions in order to reduce routing table sizes [267]. REBECA includes three of them [263]. *Identity-based routing* avoids forwarding of subscriptions that match identical sets of notifications. *Covering-based routing* [71] avoids forwarding of those subscriptions that only accept a subset of notifications matched by a previously forwarded subscription. Note that this implies that it might be necessary to forward some of the covered subscriptions along with an unsubscription if a subscription is canceled. *Merging-based routing* [266] can be implemented on top of covering and goes even further. In this case, each broker can merge existing routing entries into a broader subscription, i.e., the broker creates a new cover for the merged routing entries that replaces the old ones. Only the resulting merged filter has to be forwarded to neighbor brokers, where it covers and replaces existent base filters. Merging can be done either in a perfect or an imperfect way. Perfectly merged filters only accept notifications that are accepted by at least one of its base filters, whereas imperfectly merged filters accept notifications besides their base filters. Imperfect routing table entries increase network traffic but allow for lazy updates, hiding frequent reconfigurations in covered parts of the network.

Advertisements are an additional mechanism to optimize subscription forwarding. Subscriptions need only be forwarded into those subnets of the overlay network where a producer has issued an overlapping advertisement, i.e., where matching notifications can be produced at all. If a new advertisement is issued, overlapping subscriptions are forwarded appropriately. Similarly, if an advertisement is revoked, it is forwarded, and remote subscriptions that can no longer be serviced are dropped. Advertisements can be combined with all routing algorithms discussed above.

2.5 Specification of Event Systems

A considerable amount of work on event-based systems and notification services exists, and many concrete systems have been designed and implemented. Unfortunately, understanding and comparing these systems is very difficult because of different and informal semantics. Section 2.5.1 presents a formalism that helps to specify the semantics of an event-based system unambiguously.

In Sect. 2.5.2 this formalism is used to specify a simple event system that captures the requirements considered mandatory for the basic level of service. This specification is extended in Sect. 2.5.3 and Sect. 2.5.4 to include ordering requirements and advertisements. In later chapters of this book, the basic specification is further extended. In Sect. 6.2.2 the basic specification is extended to construct scoped event systems, and in Sect. 8.2.2 it is extended to derive self-stabilizing publish/subscribe systems.

2.5.1 Formal Background

In the literature there exist well-developed methods to specify and validate concurrent systems. The aim of the proposed formalisms is to precisely describe the behavior of a system as a "black box", i.e., without referring to its internal (implementation) issues. The aim of the formalisms is to precisely describe the intended behavior of an interactive system. Usually, the formalisms model a system as *state machine* which moves from one *state* to another by means of an *action*. Formally this corresponds to the definition of a *labeled transition system (LTS)*. The black box view entails defining the correct behavior of such a system at its *interface*. In the literature this is termed *observation semantics*, and there are many different possibilities of defining observation semantics for concurrent systems. Intuitively, system evolution can be written as a sequence [53]:

$$\bar{s}_0 \xrightarrow{\bar{a}_0} \bar{s}_1 \xrightarrow{\bar{a}_1} \bar{s}_2 \ldots$$

, which denotes that starting from the *initial state* $\bar{s}_0$ the system reaches state $\bar{s}_1$ by executing action $\bar{a}_0$. Similarly, the system reaches (for $i \geq 0$) the state $\bar{s}_{i+1}$ from state $\bar{s}_i$ by executing action $\bar{a}_i$. Hence, it must be specified for each action how it changes the current state of the system.

To be able to do reasoning about sequences of states using temporal logic, we eliminate the actions from the trace by extending the states of the system to include the next action to be executed. Thus, we define a state s_i of the system to the pair $(\bar{s}_i, \bar{a}_i)$. This allows us to define traces to be sequences of only states. When we talk about the system to execute an action in the following, we thus mean that the part of the current state of the system that corresponds to actions equals the respective action.

Note that trace semantics can not only be used to describe the behavior of a single process but also be used to describe the behavior of concurrent systems such as distributed systems. The global state space of a set of concurrent processes is defined by the cross product of the state space of the individual processes. The system's evolution can then be viewed as a sequence of global states that occur by fairly interleaving the individual process traces such that every process can execute infinitely often.

One might argue that defining a trace as a total order is unrealistic in a distributed system because it is not possible or desirable to enforce total ordering of states. Indeed, it is possible to give specifications that are not (efficiently) implementable because of the inherent characteristics of distributed system such as the lack of a global time. However, the specifications we give are implementable because they impose ordering relations only on states that intentionally should be causally related in any sensible implementation.

Definition 2.1 (Trace). *A trace σ is a sequence of states*

$$\sigma = s_0, s_1, s_2, \ldots.$$

Definition 2.2 (Subtrace). *Let $\sigma = s_0, s_1, \ldots$ be a trace. Then, for $i \geq 0$ the* subtrace *$\sigma_{|i}$ is the trace $s_i, s_{i+1}, \ldots$.*

Definition 2.3 (Specification). *A* specification *Σ is a set of traces. A system satisfies a specification Σ if it only exhibits traces which are in Σ.*

In order to implement a specification, the implementation of a system usually has to execute internal actions in addition to the interface actions. To model this, any finite number of internal actions in between two interface actions is allowed. This is sometimes called *weak equivalence* [35] or *stuttering equivalence* [2, 229]. Inference rules and other proof techniques can then be used to formally derive the satisfaction relation.

In most cases, a specification is given as a set of predicates on traces. We utilize temporal logic [317] to express such predicates. The formal language is built from atomic predicates; the quantifiers $\forall$, $\exists$; the logical operators $\vee$, $\wedge$, $\Rightarrow$, $\neg$; and the "temporal" operators $\Box$ ("always"), $\Diamond$ ("eventually"), and $\bigcirc$ ("next"). The atomic predicate P is true for every trace whose first state satisfies P. The formula $\neg p$ is true for every trace whose first state does not satisfy P. The other logical operators and quantifiers are defined in the obvious analogous way. Manna and Pnueli [248] discuss the semantics of many temporal operators. The semantics of the temporal operators that we need in this book are defined as follows:

Definition 2.4 (Temporal Operators).   *Let Ψ be an arbitrary temporal formula and $\sigma = s_0, s_1, \ldots$ be an arbitrary trace.*

- *$\Diamond\Psi$ is true for trace σ iff there exists an $i \geq 0$ such that Ψ is true for the trace $\sigma_{|i}$.*
- *$\Box\Psi$ is true for trace σ iff for all $i \geq 0$, Ψ is true for the trace $\sigma_{|i}$.*
- *$\bigcirc\Psi$ is true for trace σ iff Ψ is true for the trace $\sigma_{|1}$.*

Note that the temporal operators have higher precedence than the logical operators. Intuitively, $\Diamond\Psi$ means that Ψ will hold eventually, i.e., there exists a subtrace for which Ψ holds. For an atomic predicate P, $\Diamond P$ means that P holds for at least one place of the trace. $\Box\Psi$ means that Ψ always holds, i.e., for all subtraces. $\Box P$ means that P holds for all places of the trace. Finally, $\bigcirc\Psi$ means that Ψ holds for the subtrace starting at the second place of the trace. $\bigcirc P$ means that P holds for the second place of the trace.

The meaning of nested temporal formulas is often not easy to see. Table 2.1 depicts some some exemplary temporal formulas and their informal meaning.

Table 2.1. Some exemplary temporal formulas and their informal meaning

$\Box\Diamond P$	P is satisfied by infinitely many places
$\Diamond\Box P$	From some place on, P holds forever
$\Box[P \Rightarrow \Box P]$	Once P holds, it continues to hold forever
$\Box[P \Rightarrow \Diamond Q]$	Every P is followed by a Q
$\Box[P \Rightarrow \bigcirc\Box\neg P]$	P is true for at most one place
$\Diamond P$	P is true for at least one place
$\Box\neg P \;\lor\; \Box\neg Q$	No trace satisfies both $\Diamond P$ and $\Diamond Q$
$P \Rightarrow \Diamond\Box Q$	If initially P holds, then eventually Q holds forever

2.5.2 A Simple Event System

In the following, we give a specification of a simple event system. First, we introduce the interface operations (i.e., actions) and the state of a simple event system. Then, we present a specification of simple event systems using the formalism introduced in the previous section.

Interface Operations and State

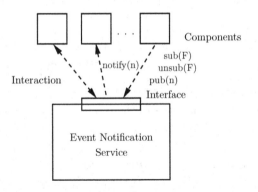

Fig. 2.4. A simple event system

A simple event system consists of a set of components (acting as producers and consumers) and of an event notification service (Fig. 2.4). For the purpose of specifying how a correct event system should behave, we view the event notification service as a black box. The components and the event notification service interact via an interface that offers several operations (Table 2.2). Note that from the viewpoint of the event notification service *sub*, *unsub*, and *pub* are *input operations*, while *notify* is the only *output operation*.

The operations take parameters from different domains: the set of all components $\mathcal{C}$, the set of all notifications $\mathcal{N}$, and the set of all filters $\mathcal{F}$. We make

Table 2.2. Interface operations of a simple event system

$sub(X, F)$	Component X subscribes to filter F
$unsub(X, F)$	Component X unsubscribes to filter F
$pub(X, n)$	Component X publishes n
$notify(X, n)$	Component X is notified about n

the following assumptions: First, notifications are unique, i.e., each notification $n \in \mathcal{N}$ can be published at most once. Second, every filter is associated with a unique identifier in order to enable the event system to distinguish subscriptions.

The state of the event system comprises three *specification variables* for every component $X \in \mathcal{C}$:

1. a set S_X of *active subscriptions* (i.e., filters which X has subscribed to and not unsubscribed to yet).
2. a set P_X of *published notifications* (i.e., the subset of $\mathcal{N}$ containing all notifications X has previously published).
3. a multiset D_X of *delivered notifications* (i.e., the subset of $\mathcal{N}$ containing all notifications which have previously been delivered to X. $\#(D_X, n)$ gives the number of occurrences of n in D_X.

Table 2.3. Changes of the state variables caused by interface operations

$pub(X, n)$	$P'_X = P_X \cup \{n\}$
$sub(Y, F)$	$S'_Y = S_Y \cup \{F\}$
$unsub(Y, F)$	$S'_Y = S_Y \setminus \{F\}$
$notify(Y, n)$	$D'_Y = D_Y \cup \{n\}$

The sets are initially empty, and they are updated faithfully according to the operations that occur at the system's interface. When X publishes a notification n, it is added to P_X. Whenever X subscribes to F, F is added to S_X, and whenever X unsubscribes to F, F is removed from S_X. Hence, multiple (un)subscriptions to the same filter are idempotent. For example, if a component X subscribes to a filter F multiple times and then unsubscribes to this filter once, then F is no longer in S_X afterwards. The state changes caused by the interface operation are specified in Table 2.3. With a prime we indicate the state of a variable after the execution of an interface operation (e.g., S'_Y). Beyond the changes above caused by the interface operations, the specification variables are not changed.

We have now specified how the state of the system is changed by the interface operations. Note that only the *notify* operation is raised by the event notification service, while all other operations are raised by components. To specify the correct behavior of the system, we must thus define in what situations the notification service must or must not execute the *notify* operations in reaction to the other operations. We will specify the intended behavior of the notification service by giving a set of temporal formulas.

Before we present the full specification in the next section, we give some examples to better understand the semantics of temporal formulas in the context of event systems. Note that free variables are — if nothing else is said — assumed to be implicitly universally quantified.

$$\Diamond notify(X, n)$$

specifies all traces in which component X is eventually notified about n.

$$\Box \neg unsub(X, F)$$

specifies all traces in which X never unsubscribes to F.

$$\Box \big[notify(X, n) \Rightarrow \bigcirc \Box \neg notify(X, n) \big]$$

specifies all traces in which, if X is notified about n, X is never notified about n again.

$$\Box \big[notify(X, n) \Rightarrow n \in N(S_X) \big]$$

specifies all traces in which the fact that X is notified about n implies that X currently has a subscription that matches n. It is important to keep in mind that the temporal operators determine the place in the trace to which the imposed conditions are applied. As a last example,

$$\Box \big[notify(Y, n) \Rightarrow n \in \bigcup_{X \in \mathcal{C}} P_X \big]$$

requires that the fact that Y is notified about n implies that there is a component X for that n is in P_X. This implies X has published n before.

Trace-Based Specification

In the following, a specification of simple event systems is presented that relies on the trace-based semantics introduced above (cf. [140, 144]). It conforms to the following informal requirements: A component receives

(a) only notifications it is currently subscribed to
(b) only notifications that have previously been published
(c) a notification at most once
(d) all future notifications matching one of its active subscriptions

While properties (a) to (c) are relatively easy to express, the exact meaning of property (d) requires the most attention.

Definition 2.5 (Simple Event System). *A simple event system is a system that exhibits only traces satisfying the following requirements:*

- *(Safety)*

$$\Box\Big[notify(Y, n) \ \Rightarrow \ \big[n \in N(S_Y)\big]$$
$$\wedge \big[n \in \cup_{X \in e} P_X\big] \qquad\qquad (2.1)$$
$$\wedge \big[\bigcirc\Box\neg notify(Y, n)\big]\Big]$$

- *(Liveness)*

$$\Box\Big[\Box(F \in S_Y) \Rightarrow \big[\Diamond\Box(pub(X, n) \wedge n \in N(F) \ \Rightarrow \ \Diamond notify(Y, n))\big]\Big]$$
$$(2.2)$$

The specification consists of a safety and a liveness condition [228]. A *safety* condition demands that "something irremediably bad" will never happen, while a *liveness* condition requires that "something good" will eventually happen.[3] It has been shown that all properties on traces can be expressed as the intersection of safety and liveness conditions [10, 168, 169].

Here, the safety condition states that a notification should never be delivered to a consumer more than once, that a delivered notification must have been published by a component in the past, and that a notification should only be delivered to a component if it matches one of the component's active subscriptions at the time of delivery; entailing requirements (a) to (c) from the beginning of this section.

The liveness condition is more difficult to understand. It describes precisely under which conditions a notification must be delivered. The condition can be rephrased as follows: When a component Y subscribes to a filter F and does not issue an unsubscription for this filter, then, from some time on, every notification n that is published by some component X thereafter and matches the filter will be delivered to the subscribing component. The liveness condition can be regarded as a precise formulation of requirement (d). Note that no delivery order (e.g., causal order) is imposed on notifications because ordering is a highly implementation-dependent and application-specific issue, and hence is left out of consideration when defining the semantics of simple event systems. Specifying ordering requirements is discussed in Sect. 2.5.3.

Intuitively, the liveness requirement states that any *finite* processing delay of a subscription is acceptable. By abstracting away from real time, a concise and unambiguous characterization system behavior is obtained. For example, if a component has subscribed to a filter F and later unsubscribes to it, the system does not have to notify the component about *any* notifications that match F and are published in the meantime; it may nevertheless do so. Liveness requires delivery if the component continuously remains subscribed to F. Because the system cannot tell the future, it must at some point start to deliver notifications until the component unsubscribes to F.

---

[3] For a formal definition of safety and liveness refer to Broy and Olderog [53].

Furthermore, the definition of liveness does not directly relate subscribing and publishing operations to each other, because they are causally independent and no semantics is implied here. As an advantage future extensions can build on this definition to introduce real-time requirements that prevent old notifications from being delivered to new subscriptions, or caching strategies that allow for a defined history of notifications to be delivered to newly issued subscriptions.

A system that satisfies only the safety condition is trivial to implement. Any system that never invokes a *notify* operation satisfies the imposed conditions. Similarly, it is easy to implement a system which satisfies only the liveness condition. Any system that delivers every published notification to all components fulfills this condition. The challenge is to implement a system that satisfies *both* requirements.

2.5.3 A Simple Event System With Ordering Requirements

A simple event system (cf. Def. 2.5) may deliver notifications in an arbitrary order. For many applications, however, it is important that certain notification ordering guarantees are given by the event system. In the following, we introduce FIFO-producer ordering, causal ordering, and total ordering as additional safety properties. These properties impose the desired ordering independent of other safety properties (e.g., those of Def. 2.5). Total ordering will probably not be used in most event systems because enforcing a total order seriously affects the scalability of a system. Note that causal ordering implies FIFO-producer ordering, but total ordering is orthogonal to both FIFO-producer and causal ordering. Hence, for example, total order and causal ordering can both be required to hold. All ordering properties are of the form $\Diamond A \Rightarrow \neg \Diamond B$ with appropriate temporal formulas inserted for A and B. The equation means that if A occurs in a trace, then B should not occur. For FIFO-producer and total ordering, the given property must hold for all ordered pairs of notifications (n_1, n_2), where $n_1 \neq n_2$. This means that for two concrete notifications a and b the given property must hold for (a, b) *and* (b, a). For causal ordering, the given property must hold for all ordered k-tuples $(n_1, \ldots, n_k)$, where $\forall (1 \geq i, j \leq k). i \neq j \Rightarrow n_i \neq n_j$ for all $k \geq 2$.

Definition 2.6. *An event system respects* FIFO-producer ordering *iff it only exhibits traces satisfying the following requirements:*

- *(Safety FIFO)*

$$
\begin{aligned}
n_1 \neq n_2 \;\wedge \\
\Diamond \big[pub(C_1, n_1) \wedge \Diamond pub(C_1, n_2) \big] \qquad\qquad (2.3) \\
\Rightarrow \neg \Diamond \big[notify(C_2, n_2) \wedge \Diamond notify(C_2, n_1) \big].
\end{aligned}
$$

Equation 2.3 states that the notifications that are published by a component C_1 should not be delivered to a component C_2 in an order different from the order in which they were published.

Definition 2.7. *An event system respects* causal *ordering iff it only exhibits traces satisfying the following requirements for every $k \geq 2$:*

- *(Safety Causal)*

$$\forall(1 \geq i, j \leq k).\, i \neq j \Rightarrow n_i \neq n_j \,\wedge$$
$$\Diamond\big[pub(C_1, n_1)\,\wedge$$
$$\Diamond\big[notify(C_2, n_1) \wedge \Diamond\big[pub(C_2, n_2)\,\wedge$$
$$\cdots \tag{2.4}$$
$$\Diamond\big[notify(C_k, n_{k-1}) \wedge \Diamond pub(C_k, n_k)\big]\ldots\big]\big]\big]$$
$$\Rightarrow \neg\Diamond\big[notify(Y, n_k) \wedge \Diamond notify(Y, n_1)\big].$$

Equation 2.4 states that if there is a sequence of components $C_1, \ldots, C_k$ such that each component C_i publishes a notification n_i that is notified to component C_{i+1} if $i < k$ then a component Y should not be notified about n_1 after it was notified about n_k.

Definition 2.8. *An event system respects* total *ordering iff it only exhibits traces satisfying the following requirements:*

- *(Safety Total)*

$$n_1 \neq n_2 \,\wedge$$
$$\Diamond\big[notify(C_1, n_1) \wedge \Diamond notify(C_1, n_2)\big] \tag{2.5}$$
$$\Rightarrow \neg\Diamond\big[notify(C_2, n_2) \wedge \Diamond notify(C_2, n_1)\big].$$

Equation 2.5 states that if a component C_1 is notified about n_1 and eventually notified about n_2, then a component C_2 should not be notified about n_1 after it was notified about n_2.

2.5.4 Simple Event System With Advertisements

Advertisements are filters issued by producers to indicate their intention to publish certain kinds of notifications in the future. Some implementations of event systems use advertisements to optimize content-based routing [65]. Advertisements can also be used to control the notifications a producer publishes, for example, to enforce security policies [34]: if a notification is published by a component that does not match any of its active (and authorized) advertisements, it should be discarded and not delivered to any component. Moreover, issued advertisements can be used by components to find out what notifications currently are potentially published in the system.

Advertisements are easily integrated into the formal model of event systems presented here: We introduce two more interface operations $adv(F)$ and $unadv(F)$ (see Table 2.4) that are used by components to issue and revoke

Table 2.4. Additional interface operations for advertisements

$adv(X, F)$	Component X advertises filter F
$unadv(X, F)$	Component X unadvertises filter F

Table 2.5. Changes of the state variables caused by the additional interface operations for advertisements

$adv(X, F)$	$A'_X = A_X \cup \{F\}$
$unsub(X, F)$	$A'_X = A_X \setminus \{F\}$

advertisements, respectively, and a further state variable A_X, which is the set containing all active advertisements of a component X (i.e., all filters which X has advertised and not yet unadvertised). The state changes caused by the two new interface operations are specified in Table 2.5. Again, A_X is not changed besides the given changes above.

In the context of content-based routing, advertisements are used to restrict the forwarding of subscriptions into those subnets where matching notifications can be produced. This also means that in reaction to a new advertisement, it might be necessary to forward some new subscriptions into a subnet. Similar to new subscriptions, this change should intuitively take effect immediately to ensure that a notification published right after a new advertisement has been issued is delivered to all interested consumers. Again, this is not sensible in a loosely coupled distributed system. Hence, we allow a finite delay of advertisement processing, too.

Definition 2.9. *A simple event system with advertisements is a system which exhibits only traces satisfying the following requirements:*

- *(Safety)*

$$\Box\Big[\big[notify(Y, n) \Rightarrow \bigcirc\Box\neg notify(Y, n)\big] \wedge$$
$$\big[notify(Y, n) \Rightarrow n \in \bigcup_{X \in \mathcal{C}} P_X \cap N(S_Y)\big] \wedge \qquad (2.6)$$
$$\big[pub(X, n) \wedge n \notin N(A_X) \Rightarrow \Box\neg notify(Y, n)\big]\Big]$$

- *(Liveness)*

$$\Box\Big[\big[\Box(F \in S_Y) \wedge \Box(G \in A_X)\big]$$
$$\Rightarrow \big[\Diamond\Box\big(pub(X, n) \wedge n \in N(F) \cap N(G) \qquad (2.7)$$
$$\Rightarrow \Diamond notify(Y, n)\big)\big]\Big]$$

The safety condition has been *strengthened* such that if a notification is published that does not match any of the active advertisements of the publishing component, the notification should not be delivered to any component.

The liveness condition has been *weakened* and can be rephrased as follows: If a component Y is always subscribed to F and a component X always advertises G, then there exists a future time where a notification n published by X that matches F *and* G will lead to a delivery of n to Y. The specification without advertisements can be seen as a special case of those with advertisements if it is implicitly assumed that every component initially advertises and never unadvertises a filter that matches any notification.

2.6 Further Reading

The field of event-based communication and computation can be approached in different ways. There are standards, products, books, and articles from the distributed systems perspective as well as from other related areas such as active databases and tuple spaces.

The book by Luckham [242] gives a general introduction to event-based computing, taking a viewpoint similar to ours. It concentrates on the detection of event pattern as a fundamental way of implementing distributed applications. Our goal is a broader treatment of event-based systems, spanning over multiple areas including algorithmic and practical engineering concerns.

Standardization efforts try to establish a common API as basis for notification service implementations and use. The Object Management Group (OMG) included standard services into the CORBA specification [283]. The CORBA Notification Service offers channels, which publishers and subscribers choose to dissemination and receive notifications [287]. Consumers may register additional filters to reduce the amount of messages. The Notification Services subsumes and obsoletes the CORBA Event Service [280].

The Java Message Service (JMS) is an API specification as part of the Java 2 Enterprise Edition (J2EE) [364, 365]. JMS coined the term *topic-based subscription*, which stands for message grouping according to abstract topics plus content-based filtering on a set of header fields and properties. JMS is becoming the dominant messaging API, and lots of commercial, academic, and open source notification services are implementing this API. A more detailed discussion of products and prototypes is given in Chap. 9.

The field of software architecture is concerned with the overall organization of a software systems [165]. It corresponds to the coordination paradigm, since both deal with the high-level interaction of system components. The architectural point of view focuses more on the static, immutable characteristics of these constellations. Architecture definition languages (ADLs)[4] are employed to describe the high-level conceptual architecture consisting of components, connectors, and specific configurations [256] of these. Typical, well-understood arrangements of connectors and configurations are identified as *architectural styles* [3], the patterns of software architecture, and events and implicit invocation is one of them. The event-based architectural style comprises exactly

---

[4] Also: architecture description languages

the concepts given in Sect. 2.1, featuring the independence of producer and consumer components [68, 165]. In fact, Garlan et al. [166] identified early the prominent importance of using events for the construction of flexible software architectures.

As mentioned above, Garlan et al. [166] emphasize the importance of events for flexible software systems, which is corroborated by [355] and others. One of the first contributions is the Field environment [325], an early work on tool integration that is built around a centralized server that distributes messages. Messages sent to the server are selectively rebroadcasted to receivers that have registered patterns matching the message.

The InfoBus [84] is a small Java API that facilitates communication between several JavaBeans or cooperating applets on a Web page. Multiple instances of InfoBus might be manually connected with bridges, providing a limited means of structuring, but without any inherent interfaces or composition support. Matching of messages is done by names, i.e., string matching. Besides being limited to one virtual machine, it is a tool for connecting components, not for composing new ones.

3

Content-Based Models and Matching

3.1 Content-Based Data and Filter Models

This section discusses some important content-based data models in conjunction with corresponding filter models. Informally, a *data model* defines how the content of notifications is structured, while a *filter model* defines how subscriptions can be specified, i.e., how notifications can be selected by applying filters that evaluate predicates over the content of notifications. The filter model always depends on the underlying data model, and there can be more than one filter model for a given data model. The data/filter model has to be chosen carefully because it has a large impact on the expressiveness and the scalability of a content-based notification service. In the following, we discuss tuples, structured records, semistructured records, and objects.

3.1.1 Tuples

In tuple-oriented models a notification is a *tuple*, i.e., an *ordered* set of *attributes*. All approaches using tuples deploy some sort of templates as subscription mechanisms. Similarly, to a query-by-example mask, a *template* specifies matching notifications by a partial tuple which can contain wildcards. The attributes in the notification are matched to the attributes in the template according to their position. For example, the notification (*StockQuote*, *"Foo Inc."*, 45) is matched by the subscription template (*StockQuote*, *"Foo Inc."*, *). "Matching by position" is inflexible because attributes cannot be optional. Tuples in conjunction with templates were first proposed by Gelernter in work onLinda Tuple spaces [174], which use typed attributes. The original version of Linda, however, did not support a subscription mechanism, but newer approaches based on Tuple spaces, e.g., JavaSpaces [366], do. Also, some notification services are built upon tuples: JEDI models a notification as a tuple of strings [91] in which the first string corresponds to the notification name, while the others are normal attributes. JEDI supports the equality and the prefix operator for matching. Bates et al. [32] define notifications as instances of classes. An instance consists of a tuple of typed attributes derived from a class definition. Here, a template either specifies the exact value of an attribute or it does not care about the value. Concluding, tuples with templates provide a simple model that is not flexible enough because attributes of notifications and templates are matched to each other according to their position. This disadvantage is diminished by record-oriented models which use "matching by attribute names." However, "matching by position" is more efficient.

3.1.2 Structured Records

In this section structured records are discussed in detail. In a record-oriented model a notification consists of a *named* set of attributes. Record-oriented models can be divided into two categories, which are structured records and semistructured records, respectively. Roughly speaking, the models can be distinguished by the fact that in structured records attribute names are unique, while in the semistructured models several attributes with the same name can exist. In this section, structured records are discussed; semistructured records are discussed in Sect. 3.1.3.

Many systems model notifications similarly to structured records consisting of a set of name/value pairs called attributes. Examples are SIENA [65], Gryphon [6, 26], REBECA [136], JMS [364], and the CORBA Notification Service [279]. In this model filters address attributes by their unique names and impose constraints on the values of the respective attributes. In most models a constraint is assumed to evaluate to *false* if the addressed attribute is not contained in the notification. Therefore, each constraint implicitly defines an existential quantifier over the notification. Besides *flat records* in which values are atomic types, *hierarchical records* in which attributes may be nested can also be supported easily by using a dotted naming scheme (e.g., *Position*.x).

Some systems (e.g., SIENA) restrict constraints to depend on a single attribute (e.g., $\{x = 1\}$). This class of constraints is called *attribute filters*. Other systems, such as ELVIN, allow constraints to evaluate multiple attributes which are combined by operators (e.g., $\{x + y = 5\}$). In general, multiple constraints can be combined to form filters by Boolean operators (e.g., $\{y < 3 \wedge x = 4\}$). SIENA and REBECA restrict filters to be conjunctions of attribute filters. On one hand, this restriction reduces the expressiveness of the filter model, but on the other hand it enables routing optimizations like covering (cf. Chap. 4) to be applied efficiently. The limitation is also not as serious as it seems first. For example, a filter that is defined by an arbitrary Boolean expression can always be converted to and treated as a collection of conjunctive filters.

Although records and tuples seem to be similar at a first glance, records are clearly more powerful because they allow for optional attributes in the notifications. They also avoid unnecessary "don't care" constraints in the templates, and enable the easy addition of new attributes without affecting existing filters.

Data Model

A notification is a message that contains information about an event that
has occurred. Formally, a *notification* n is a nonempty set of *attributes*
$\{a_1, \ldots, a_n\}$, where each a_i is a *name/value pair* (n_i, V_i) with name n_i and
value v_i. It is assumed that names are unique, i.e., $i \neq j \Rightarrow n_i \neq n_j$, and that
there exists a function that uniquely maps each n_i to a type T_j that is the
type of the corresponding value v_i.

In the following we distinguish between *simple values* that are a single
element of the domain of T_j, i.e., $v_i \in dom(T_j)$, and *multi values* that are
a finite subset of the domain, i.e., $v_i \subseteq dom(T_j)$. An example of a simple
notification is $\{(type, StockQuote), (name, \text{``Infineon''}), (price, 45.0)\}$.

Filter Model

A *filter* F is a stateless Boolean function that is applied to a notification,
i.e., $F(n) \rightarrow \{true, false\}$. A notification *matches* F if $F(n)$ evaluates to *true*.
Consequently, the set of matching notifications $N(F)$ is defined as $\{n \mid F(n) =
true\}$. Two filters F_1 and F_2 are *identical*, written $F_1 \equiv F_2$, iff $N(F_1) = N(F_2)$.
Moreover, they are *overlapping*, denoted by $F_1 \sqcap F_2$, iff $N(F_1) \cap N(F_2) \neq \emptyset$.
Otherwise they are *disjoint*, denoted by $F_1 \not\sqcap F_2$.

A filter is usually given as a Boolean expression that consists of predicates
that are combined by Boolean operators (e.g., *and*, *or*, *not*). A filter consisting
of a single atomic predicate is a *simple filter* or *constraint*. Filters that are
derived from simple filters by combining them with Boolean operators are
compound filters. A compound filter that is a conjunction of simple filters
is called a *conjunctive filter*. In the model proposed filters are restricted to
be conjunctive filters. It is sufficient to consider conjunctive filters because a
compound filter can always be broken up into a set of conjunctive filters that
are interpreted disjunctively and can be handled independently.

An *attribute filter* is a simple filter that imposes a constraint on the value
of a single attribute (e.g., $\{name = \text{``Foo Inc.''}\}$). It is defined as a triple
$A_i = (n_i, Op_i, C_i)$, where n_i is an attribute name, Op_i is a test operator
and C_i is a set of constants that may be empty. The name n_i determines to
which attribute the constraint applies. If the notification does not contain an
attribute with name n_i then A_i evaluates to *false*. Therefore, each constraint
implicitly defines an existential quantifier over the notification. Otherwise, the
operator Op_i is evaluated using the value of the addressed attribute and the
specified set of constants C_i. It is assumed that the types of operands are
compatible with the used operator. The outcome of A_i is defined as the result
of Op_i that evaluates either to *true* or *false*. Furthermore, an attribute filter is
provided that simply checks whether a given attribute is contained in n. For
the sake of simplicity the more readable notation $\{price > 10\}$ is used instead
of $\{(price, >, \{10\})\}$. In contrast to most other work (e.g.,)SIENA, constraints
that depend on more than one constant are considered in this chapter. This

enables more operators and enhances the expressiveness of the filtering model and can be done without affecting scalability.

By $L_A(A_i) \subseteq dom(T_k)$ the set of all values is denoted that cause an attribute filter to match an attribute, i.e., $\{v_i \mid Op_i(v_i, C_i) = true\}$. It is assumed that $L_A(A_i) \neq \emptyset$. An attribute filter A_1 *covers* an attribute filter A_2, written $A_1 \sqsupseteq A_2$, iff $n_1 = n_2 \wedge L_A(A_1) \supseteq L_A(A_2)$. For example, $\{price > 10\}$ covers $\{price \in [20, 30]\}$. A_1 and A_2 are *identical*, denoted by $A_1 \equiv A_2$, iff $n_1 = n_2 \wedge L_A(A_1) = L_A(A_2)$. A_1 and A_2 are *overlapping* iff $n_1 = n_2 \wedge L_A(A_1) \cap L_A(A_2) \neq \emptyset$, denoted by $A_1 \sqcap A_2$. Otherwise they are *disjoint*, denoted by $A_1 \not\sqcap A_2$. For example, $\{price > 10\}$ and $\{price < 20\}$ are overlapping, while $\{price < 10\}$ and $\{price > 20\}$ are disjoint.

In the described model a *filter* is defined as a conjunction of attribute filters, i.e., $F = A_1 \wedge \ldots \wedge A_n$. To enable efficient evaluation of routing optimizations like covering and merging, at most one attribute filter for each attribute is allowed. A notification n *matches a filter F* iff it satisfies all attribute filters of F. Moreover, a filter with an empty set of attribute filters matches any notification. An example for a conjunctive filter consisting of attribute filters is $\{(type = StockQuote), (name = \text{``Foo Inc.''}), (price \notin [30, 40])\}$.

The limitation to at most one attribute filter for each attribute is not as serious as it seems at first glance because the proposed model provides complex data types as attribute values and an extensible set of constraints that can be imposed. Moreover, it is often possible to merge several conjunctive constraints imposed on a single attribute into a single constraint on the same attribute. Especially suited for this kind of merging are constraints which are either contradicting (if they are conjuncted) or can be replaced by a single constraint of the same type. Such types of constraints and their corresponding attribute filters are called *conjunction-complete*. For example, interval constraints and constraints testing whether a point is in a given rectangle in a two-dimensional plane are conjunction-complete. As an example, $\{x \in [3, 7] \wedge x \in [5, 8]\}$ can be substituted by $\{x \in [5, 7]\}$. If a constraint type is not conjunction-complete it is often possible to substitute a set of such constraints by a single constraint of a more general type. For example, a set of ordering constraints defined on a totally ordered set (e.g., integer numbers) are either contradictory or can be replaced by a single interval constraint. As an example, $\{x \geq 3 \wedge x \leq 5\}$ can be merged to $\{x \in [3, 5]\}$.

Subscriptions and *advertisements* are simply filters that are issued by consumers and producers of notifications, respectively. There is no difference in their model, and hence, subscriptions and advertisements are the exact dual of each other. This is in contrast to SIENA, where subscriptions and advertisements are not exactly complementary, raising a number of problems.

Generic Constraints and Types

Earlier work dealing with content-based notification selection mechanisms often tightly integrated the constraints that can be put on values and the types of values supported by the matching and the routing algorithms [6, 26]. An exception is SIENA, where matching and routing algorithms are separated from constraints. However, SIENA only supports a fixed set of constraints on some predefined primitive types.

We propose to use a collection of abstract attribute filter classes. Each of these classes offers a generic implementation of the methods needed by the matching and the routing algorithms (e.g., a covering and a matching test) and imposes a certain type of constraint on an attribute that can be used with values of all types that implement the operators needed. The appropriate implementation of the operators is called by the constraint class at runtime using polymorphism. This enables new constraints and types to be defined and to be supported without requiring changes to the routing and or to the matching algorithms. Note that although an object-oriented approach is suggested, it is not mandatory to use it.

For example, a constraint class can realize comparison constraints on totally ordered sets. This class can be used to impose comparison constraints on all kinds of ordered values (e.g., integer numbers). Consider a type "person" that consists of first and second name, the date of birth, and the place of birth. This type is easily supported by providing implementations for the comparison operators which are called by the constraint class to provide the covering and matching methods using polymorphism.

In the following subsections, some generic attribute constraints are presented that cover a wide range of practically relevant constraints, but more important, they illustrate the feasibility of the approach. Of course, this collection is not exhaustive, but other constraints can be integrated easily. For example, intervals could be used as values. In this case the same operators as for set constraints can be used because intervals are essentially sets. The investigation of a subset of regular expressions seems to be promising, too. Most paragraphs also present a table that gives an overview of covering implication dealing with the discussed type of constraint. The meaning of a single row in the Tables 3.1 through 3.7 is: Given A_1 and A_2 as specified in column 1 and 2, $A_1 \sqsupseteq A_2$ iff the condition in column 3 is satisfied. In order to test whether a filter *covers* another, covering must hold for all attributes, as will be shown later.

General Constraints

Two general constraints are considered that can be imposed on all attributes regardless of the type of their value: *exists(n)* tests whether an attribute with name n is contained in a given notification, i.e., whether $\exists A_i. n_i = n$. The *exists* constraint covers all other constraints that can be imposed on an attribute.

Constraints on the Type of Notifications

Most work on notification services has a notion of types or classes of notifications. Usually, the type of a notification is specified by a textual string that can be tested for equality and prefix. If a dot notation is used, a type hierarchy with single inheritance can be supported, allowing for the automatic propagation of interest in subclasses [32]. Unfortunately, multiple inheritance cannot be supported by a dotted naming scheme. In contrast to that, a direct support of notification types has a number of advantages. Such an approach can enable multiple inheritance and achieve a better programming language integration [120]. Moreover, type inclusion tests can be evaluated more efficiently than the corresponding string operation (i.e., whether the string starts with a given prefix) [388].

Consequently, a separate constraint that evaluates to *true* if n is an instance of type T and *false* otherwise, written n *instanceof* T, is defined. A constraint n *instanceof* T_1 covers a constraint n *instanceof* T_2 iff T_1 is either the same type or a supertype of T_2 (Table 3.1). It is assumed that the set of attributes that can be contained in a notification of type T is a superset of the union of all attribute names of all supertypes of T.

Table 3.1. Covering among notification types

A_1	A_2	$A_1 \sqsupseteq A_2$ iff
n *instanceof* T_1	n *instanceof* T_2	$T_1 = T_2 \lor T_1$ *supertype of* T_2

Equality and Inequality Constraints on Simple Values

The simplest constraints that can be imposed on a value are tests for equality and inequality. Covering implications among these tests can always be reduced to a simple comparison of their respective constants (Table 3.2).

Table 3.2. Covering among (in)equality constraints on simple values

A_1	A_2	$A_1 \sqsupseteq A_2$ iff
$x = c_1$	$x = c_2$	$c_1 = c_2$
$x \neq c_1$	$x = c_2$	$c_1 \neq c_2$
	$x \neq c_2$	$c_1 = c_2$

Comparison Constraints on Simple Values

Another common class of constraints are comparisons on values for which the domain and the comparison operators define a totally ordered set (e.g., integers with the usual comparison operators). Again, covering among these tests can be reduced to a simple comparison of their respective constants. Table 3.3 depicts covering implications of inequality and greater than; for brevity the other comparison operators are omitted.

Table 3.3. Covering among comparison constraint on simple values

A_1	A_2	$A_1 \sqsupseteq A_2$ iff
	$x < c_2$	$c_1 \geq c_2$
	$x \leq c_2$	$c_1 > c_2$
$x \neq c_1$	$x = c_2$	$c_1 \neq c_2$
	$x \geq c_2$	$c_1 < c_2$
	$x > c_2$	$c_1 \leq c_2$
	$x = c_2$	$c_1 < c_2$
$x > c_1$	$x > c_2$	$c_1 \leq c_2$
	$x \geq c_2$	$c_1 < c_2$

Interval Constraints on Simple Values

Interval constraints test whether a value x is within a given interval I or not, i.e., $x \in I$ and $x \notin I$, respectively, where I is a closed interval $[c_1, c_2]$ with $c_1 \leq c_2$. Here, computing coverage involves two comparisons (Table 3.4).

Table 3.4. Covering among interval constraints on simple values

A_1	A_2	$A_1 \sqsupseteq A_2$ iff
$x \in I_1$	$x \in I_2$	$I_1 \supseteq I_2$
$x \notin I_1$	$x \notin I_2$	$I_1 \subseteq I_2$

Constraints on Strings

Constraints on strings can be used to realize subjects. In addition to the comparison operators based on the lexical order, a prefix, a substring, and a postfix operator are defined. *s hasPrefix S* and *s hasPostfix S* mean that *s* has the prefix and the postfix *S*, respectively. *s containsSubstring S_1* means that *s* contains the substring S_1. Computing coverage among them requires a single test (Table 3.5).

Table 3.5. Covering among constraints on strings

A_1	A_2	$A_1 \sqsupseteq A_2$ iff
s hasPrefix S_1	*s hasPrefix S_2*	*S_2 hasPrefix S_1*
s hasPostfix S_1	*s hasPostfix S_2*	*S_2 hasPostfix S_1*
s hasSubstring S_1	*s hasSubstring S_2*	*S_2 hasSubstring S_1*

Set Constraints on Simple Values

Set constraints on simple values test whether or not a value is a member of a given set. For computing coverage among two of these constraints, a single set inclusion test is sufficient (Table 3.6). Its complexity depends on the characteristics of the underlying set. Set constraints can be combined with comparison constraints if the domain of the value is a totally ordered set.

Table 3.6. Covering among set constraints on simple values

A_1	A_2	$A_1 \sqsupseteq A_2$ iff
$x \in M_1$	$x \in M_2$	$M_1 \supseteq M_2$
$x \notin M_1$	$x \notin M_2$	$M_1 \subseteq M_2$

Set Constraints on Multi Values

The idea of multi values is to allow a value to be a set of elements. This enables set-oriented operators which are defined on a multi value $X = \{v_1, \ldots, v_n\}$. For example, the following common operators can be defined:

$$X \text{ subset } M \Leftrightarrow X \subseteq M$$
$$X \text{ superset } M \Leftrightarrow X \supseteq M$$
$$X \text{ contains } a_1 \Leftrightarrow a_1 \in X$$
$$X \text{ notcontains } a_1 \Leftrightarrow a_1 \notin X$$
$$X \text{ disjunct } M \Leftrightarrow X \cap M = \emptyset$$
$$X \text{ overlaps } M \Leftrightarrow X \cap M \neq \emptyset$$

To determine covering with respect to these constraints either the evaluation of a set inclusion test or of a set membership test is needed (Table 3.7).

Table 3.7. Covering among set constraints on multi values

A_1	A_2	$A_1 \sqsupseteq A_2$ iff
X subset M_1	X subset M_2	M_1 superset M_2
X contains a_1	X superset M_2	$a_1 \in M_2$
X superset M_1	X superset M_2	M_1 subset M_2
X notContains a_1	X disjunct M_2	$a_1 \in M_2$
X disjunct M_1	X disjunct M_2	M_1 subset M_2
X overlaps M_1	X overlaps M_2	M_1 superset M_2

Support for Routing Optimizations

For routing algorithm such as identity-based, covering-based, or merging-based routing (cf. Chap. 4) as well as for enabling the use of advertisement, some routing optimization must be efficiently computable.

Identity of Conjunctive Filters

In the following it is shown how identity of conjunctive filters can be reduced to the respective attribute filters. An identity test among filters is necessary to implement identity-based routing.

Lemma 3.1. *Given two filters $F_1 = A_1^1 \wedge \ldots \wedge A_n^1$ and $F_2 = A_1^2 \wedge \ldots \wedge A_m^2$ that are conjunctions of attribute filters, the following holds: the fact that F_1 and F_2 contain the same number of attribute filters and that $\forall A_i^1 \exists A_j^2. A_i^1 \equiv A_j^2$ implies that F_1 and F_2 are identical.*

Proof. The proof is rather trivial. A notification that matches F_1 satisfies all attribute filters A_i^1. For each of these A_i^1 there is an identical A_j^2. Hence, A_j^2 is matched, too. As F_1 and F_2 contain the same number of attribute filters, this implies that all attribute filters of F_2 are matched, too. Therefore, F_2 is also matched. As the same argumentation can be applied to notifications that match F_2, this implies that F_1 and F_2 match identical sets of notifications, i.e., they are identical. $\square$

It is necessary to restrict filters to contain at most one attribute filter for each attribute in order to strengthen Lemma 3.1 to an equivalence. As a simple example, $\{x > 5 \wedge x < 5\}$ is identical to $\{x \neq 5\}$, although neither $\{x > 5\} \equiv \{x \neq 5\}$ nor $\{x < 5\} \equiv \{x \neq 5\}$.

Lemma 3.2. *Given two filters $F_1 = A_1^1 \wedge \ldots \wedge A_n^1$ and $F_2 = A_1^2 \wedge \ldots \wedge A_m^2$ that are conjunctions of attribute filters with at most one attribute filter for each attribute, the following holds: $F_1 \equiv F_2$ implies $\forall A_i^1 \exists A_j^2. A_i^1 \equiv A_j^2$.*

Proof. The proof is by contradiction. We assume that

1. $F_1 \equiv F_2$
2. $\forall A_i^1 \exists A_j^2. A_i^1 \equiv A_j^2$ does not hold

and prove that this cannot hold.

The second assumption implies that there is an A_i^1 for which no identical A_j^2 exists. This means that either no attribute filter with the same name is contained in F_2 or that $L(A_i^1) \neq L(A_j^2)$. In the first case, a notification can be constructed that does not contain the respective attribute and which matches F_2 but does not match F_1. Hence, F_1 and F_2 cannot be identical and the first assumption is violated. In the second case, a notification can be constructed, where the value of the respective attribute is in $L(A_i^1)$ but not in $L(A_j^2)$ if $L(A_i^1) \supset L(A_j^2)$. This notification matches F_1 but not F_2. The other way around, a notification can be constructed, where the value of the respective attribute is in $L(A_j^2)$ but not in $L(A_2^j)$ if $L(A_i^1) \subset L(A_j^2)$. This notification matches F_2 but not F_1. At least one of these two cases needs to occur because $L(A_i^1) \neq L(A_j^2)$. Hence, F_1 and F_2 cannot be identical and the first assumption is violated. The above cases cover all possible cases.   □

Lemma 3.3. *Given two filters $F_1 = A_1^1 \wedge \ldots \wedge A_n^1$ and $F_2 = A_1^2 \wedge \ldots \wedge A_m^2$ that are conjunctions of attribute filters with at most one attribute filter for each attribute, the following holds: $F_1 \equiv F_2$ implies that F_1 and F_2 contain the same number of attribute filters.*

Proof. By Lemma 3.2 and the fact the identity relation among filters is symmetrical.   □

Corollary 3.1. *Two filters $F_1 = A_1^1 \wedge \ldots \wedge A_n^1$ and $F_2 = A_1^2 \wedge \ldots \wedge A_m^2$ that are conjunctions of attribute filters with at most one attribute filter for each attribute are identical iff they contain the same number of attribute filters and $\forall A_i^1 \exists A_j^2. \, A_i^1 \equiv A_j^2$.*

Proof. By Lemmas 3.1, 3.2, and 3.3.   □

The above corollary essentially states that two filters are identical iff they constrain the same attributes and iff the attribute filters of each constrained attribute are pairwise identical (Fig. 3.1).

$$
\begin{array}{ccc}
F_1 = & \{x \geq 2\} \wedge & \{y > 5\} \\
| & | & | \\
\equiv & \equiv & \equiv \\
| & | & | \\
F_2 = & \{x \geq 2\} \wedge & \{y > 5\}
\end{array}
$$

Fig. 3.1. Identity of filters consisting of attribute filters

Covering of Conjunctive Filters

In the following it is shown how covering of conjunctive filters can be reduced to the respective attribute filters. A covering test among filters is necessary to implement covering-based routing.

Lemma 3.4. *Given two filters $F_1 = A_1^1 \wedge \ldots \wedge A_n^1$ and $F_2 = A_1^2 \wedge \ldots \wedge A_m^2$ that are conjunctions of attribute filters, the following holds: $\forall i \exists j. \, A_i^1 \sqsupseteq A_j^2$ implies $F_1 \sqsupseteq F_2$.*

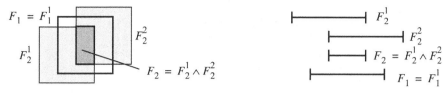

Fig. 3.2. $F_1 \sqsupseteq F_2$ although neither $F_1^1 \sqsupseteq F_2^1$ nor $F_1^1 \sqsupseteq F_2^2$ (two examples)

Proof. Assume $\forall i \exists j.\, A_i^1 \sqsupseteq A_j^2$. Prove $F_1 \sqsupseteq F_2$. If an arbitrary notification n is matched by F_2 then n satisfies all A_j^2. This fact together with the assumption implies that n also satisfies all A_1^i. Therefore, n is matched by F_1, too. Hence, $F_1 \sqsupseteq F_2$. □

If several attribute filters can be imposed on the same attribute then $\forall i \exists j.A_i^1 \sqsupseteq A_j^2$ is not a necessary condition for $F_1 \sqsupseteq F_2$ (Fig. 3.2). For example, $\{x \in [5,8]\}$ covers $\{x \in [4,7] \wedge x \in [6,9]\}$, although $\{x \in [5,8]\}$ covers neither $\{x \in [4,7]$ nor $\{x \in [6,9]\}$. If conjunctive filters are restricted to have at most one attribute filter for each attribute, then Lemma 3.4 can be strengthened to an equivalence:

Lemma 3.5. *Given two filters $F_1 = A_1^1 \wedge \ldots \wedge A_n^1$ and $F_2 = A_1^2 \wedge \ldots \wedge A_m^2$ that are conjunctions of attribute filters with at most one attribute filter for each attribute, the following holds: $F_1 \sqsupseteq F_2$ implies $\forall i \exists j.\, A_i^1 \sqsupseteq A_j^2$.*

Proof. Assume $\neg(\forall i \exists j.\, A_i^1 \sqsupseteq A_j^2)$. Prove $\neg(F_1 \sqsupseteq F_2)$. A notification n is constructed that matches F_2 but not F_1 to prove that F_1 does not cover F_2. The assumption implies that there is at least one A_k^1 that does not cover any A_j^2. If there exists an A_l^2 that constrains the same attribute as such an A_k^1 then choose for this attribute a value that matches A_l^2 but not A_k^1. Such a value exists because $L_A(A_k^1) \neq \emptyset$ and $A_k^1 \not\sqsupseteq A_l^2$. Add name/value pairs for all other attributes that are constrained in F_2 such that they are matched by the appropriate attribute filters of F_2. The constructed notification matches F_2 but not F_1. Therefore, F_1 does not cover F_2. □

Corollary 3.2. *Given two filters $F_1 = A_1^1 \wedge \ldots \wedge A_n^1$ and $F_2 = A_1^2 \wedge \ldots \wedge A_m^2$ that are conjunctions of attribute filters with at most one attribute filter per attribute, the following holds: $F_1 \sqsupseteq F_2$ is equivalent to $\forall i \exists j.\, A_i^1 \sqsupseteq A_j^2$.*

Proof. By Lemmas 3.4 and 3.5. □

The above corollary essentially states that a filter F_1 covers a filter F_2 iff for each attribute filter in F_1 there is an attribute filter in F_2 that is covered by the former (Fig. 3.3).

$$F_1 = \{x \geq 2\} \wedge \{y > 5\}$$

$$F_2 = \{x = 4\} \wedge \{y = 7\} \wedge \{z \in [3,5]\}$$

Fig. 3.3. Covering of filters consisting of attribute filters

Overlapping of Conjunctive Filters

In the following it is shown how overlapping of conjunctive filters can be reduced to the respective attribute filters. An overlapping test among filters is necessary to use advertisements for routing optimizations.

Lemma 3.6. *Given two filters $F_1 = A_1^1 \wedge \ldots \wedge A_n^1$ and $F_2 = A_1^2 \wedge \ldots \wedge A_m^2$ that are conjunctions of attribute filters, $\exists A_i^1, A_j^2. (n_i^1 = n_j^2 \wedge L_A(A_i^1) \cap L_A(A_j^2) = \emptyset)$ implies that F_1 and F_2 are disjoint.*

Proof. Proof: Suppose that F_1 and F_2 contain attribute filters A_i^1 and A_j^2 such that $(n_i^1 = n_j^2 \wedge L_A(A_i^1) \cap L_A(A_j^2) = \emptyset)$. This means that both filters require the existence of an attribute with name n_i^1 and that the value of this attribute must match $L_A(A_i^1)$ in order to make a notification match F_1 and $L_A(A_j^2)$ in order to match F_2. As $L_A(A_i^1)$ and are $L_A(A_j^2)$ disjoint, this implies that a given notification can be matched either by F_1 or by F_2. Hence, F_1 and F_2 are disjoint.   □

It is necessary to restrict filters to contain at most one attribute filter for each attribute in order to strengthen Lemma 3.6 to an equivalence. As a simple example, $\{x \in \{3,5\} \wedge x \in \{4,5\}\}$ is disjoint with $\{x \in \{3,5\} \wedge x \in \{3,4\}\}$ although there are no disjoint attribute filters.

Lemma 3.7. *Given two filters $F_1 = A_1^1 \wedge \ldots \wedge A_n^1$ and $F_2 = A_1^2 \wedge \ldots \wedge A_m^2$ that are conjunctions of attribute filters with at most one attribute filter for each attribute, the fact that F_1 and F_2 are disjoint implies that $\exists A_i^1, A_j^2. (n_i^1 = n_j^2 \wedge L_A(A_i^1) \cap L_A(A_j^2) = \emptyset)$.*

Proof. Proof: The proof is by contradiction. Suppose that F_1 and F_2 are disjoint and that there are no A_i^1, A_j^2 such that $n_i^1 = n_j^2 \wedge L_A(A_i^1) \cap L_A(A_j^2) = \emptyset$. We construct a notification that matches F_1 and F_2 to imply a contradiction in following way: For each attribute that is constrained in F_1 or F_2 add an attribute whose value satisfies the attribute filters contained in F_1 and F_2 regarding this attribute. This value must exist because there are no A_i^1, A_j^2 such that $n_i^1 = n_j^2 \wedge L_A(A_i^1) \cap L_A(A_j^2) = \emptyset$. Hence, the constructed notification matches F_1 and F_2, and therefore F_1 and F_2 are not disjoint.   □

Corollary 3.3. *Two filters* $F_1 = A_1^1 \wedge \ldots \wedge A_n^1$ *and* $F_2 = A_1^2 \wedge \ldots \wedge A_m^2$ *that are conjunctions of attribute filters with at most one attribute filter for each attribute are disjoint, i.e., not overlapping, iff* $\exists A_i^1, A_j^2 . \left(n_i^1 = n_j^2 \wedge L_A(A_i^1) \cap L_A(A_j^2) = \emptyset \right)$.

Proof. By Lemmas 3.6 and 3.7. $\square$

$$
\begin{array}{ccc}
F_1 = \{x \geq 2\} & \wedge & \{y > 5\} \\
| & | & | \\
\not\sqsubseteq & \not\sqsubseteq & \sqcap \\
| & | & | \\
F_2 = \{x < 1\} & \wedge & \{y < 7\}
\end{array}
$$

Fig. 3.4. Disjoint filters consisting of attribute filters

$$
\begin{array}{ccc}
F_1 = \{x \geq 2\} & \wedge & \{y > 5\} \\
| & | & | \\
\sqcap & \sqcap & \sqcap \\
| & | & | \\
F_2 = \{x < 5\} & \wedge & \{y < 7\}
\end{array}
$$

Fig. 3.5. Overlapping filters consisting of attribute filters

The above corollary essentially states that two filters are disjoint iff for an attribute that is constrained in both filters the corresponding attribute filters are disjoint (Fig. 3.4). Hence, two filters are overlapping iff no such attribute filters exist (Fig. 3.5).

Merging of Conjunctive Filters

Merging-based routing algorithms use abstract merging operations. In this section merging of conjunctive filters is discussed. The aim of filter merging is to determine a filter that is a merger of a set of filters. Merging of filters can be used to drastically reduce the number of subscriptions and advertisements that have to be stored by the brokers.

Perfect Merging

A set of conjunctive filters with at most one attribute filter for each attribute can be perfectly merged into a single conjunctive filter if, for all except a single attribute, their corresponding attribute filters are identical and if the attribute filters of the distinguishing attribute can be merged into a single attribute filter. For example, the two filters $F_1 = \{x = 5 \wedge y \in \{2, 3\}\}$ and $F_2 = \{x = 5 \wedge y \in \{4, 5\}\}$ can be merged to $F = \{x = 5 \wedge y \in \{2, 3, 4, 5\}\}$. Moreover, a set of attribute filters imposed on the same attribute with name n can be merged to an $exists(n)$ test if at least one of them is satisfied by any value. Note that an existence test is equivalent to no constraint if the attribute is mandatory for the corresponding type of notification.

An algorithm that determines the possibly empty set of filters which are candidates to be merged with a given filter is depicted later. From the set of merging candidates the set of attribute filters to be merged can easily be extracted. This set is used as input of a merging algorithm which has a specialized implementation for each type of constraint. In the general case purely algebraic merging techniques have exponential time complexity. Alternatively, a predicate proximity graph can be used to implement a greedy algorithm [218]. For many practical cases (e.g., set operators) efficient algorithms exist. Only in rare cases is it necessary to use an exhaustive combinatorial or a suboptimal greedy algorithm.

The characteristics of the constraints that are used to define attribute filters are important for merging. Constraints which only exist in a normal and a negated form can be directly merged by using some basic laws of Boolean algebra. For example, the filters $F_1 = (y = 3 \wedge x = 5)$ and $F_1 = (y = 3 \wedge x \neq 5)$ can be merged to $F = (y = 3 \wedge \exists x)$. In general, constraints are not restricted to be the negated form of each other, and hence better merging can be achieved by taking the specific characteristics of the imposed constraints into account.

A class of constraints that is *complete under disjunction* allows a set of constraints of this class to be merged into a single constraint of the same class. Examples for disjunction-complete constraints are *set inclusions* (e.g., $x \in \{2, 3, 7\}$) and *set exclusions* (e.g., $x \notin \{2, 3, 7\}$) while *comparison constraints* (e.g., $x < 4$) are not disjunction-complete. If a constraint class is not disjunction-complete it may still be possible to carry out merging if a specific *merging condition* is met. For example, a set of *interval tests* (e.g., $x \in [2, 4]$ and $x \in [3, 5]$) can be merged into a single interval test (here, $x \in [2, 5]$) if the intervals form a connected set. Otherwise, merging may be possible if a more general constraint is considered as merging result. For example, two comparison constraints (e.g., $x < 4$ and $x > 7$) can be merged to an interval test (here, $x \notin [4, 7]$).

Merging on the level of attribute filters is implemented by each generic attribute filter class. Table 3.8 presents some perfect merging rules. The meaning of a single row is that A_1 and A_2 can be perfectly merged to the indicated merger (column 4) if the given merging condition (column 3) holds. The first two rules can also be applied to equality and inequality tests because $x = a_1 \Leftrightarrow x \in \{a_1\}$ and $x \neq a_1 \Leftrightarrow x \notin \{a_1\}$.

Table 3.8. Perfect merging rules for attribute filters

A_1	A_2	Condition	$A_1 \cup A_2$
$x \in M_1$	$x \in M_2$	-	$x \in M_1 \cup M_2$
$x \notin M_1$	$x \notin M_2$	$M_1 \cap M_2 = \emptyset$	$\exists x$
		$M_1 \cap M_2 \neq \emptyset$	$x \notin M_1 \cap M_2$
X overlaps M_1	X overlaps M_2	-	X overlaps $M_1 \cup M_2$
X disjunct M_1	X disjunct M_2	$M_1 \cap M_2 = \emptyset$	$\exists X$
		$M_1 \cap M_2 \neq \emptyset$	X disjunct $M_1 \cap M_2$
$x = a_1$	$x \neq a_1$	$a_1 = a_2$	$\exists x$
$x < a_1$	$x > a_2$	$a_1 > a_2$	$\exists x$
	$x \geq a_2$	$a_1 \geq a_2$	
$x \leq a_1$	$x > a_2$	$a_1 \geq a_2$	$\exists x$
	$x \geq a_2$		

Imperfect Merging

At a first glance, imperfect merging seems to be less promising, but in situations in which perfect merging is either too complex or not computable it is a good compromise. Clearly, there exists a trade-off between filtering overhead and network resource consumption. Imperfect merging may result in notifications being forwarded that do not match any of the original subscriptions, but on the other hand, it reduces the number of subscriptions and advertisements that must be dealt with.

In order to use imperfect merging, heuristics are necessary that define in what situations and to what degree imperfect merging should be carried out. For example, filters that differ in few attribute filters could be merged imperfectly by imposing on each attribute a constraint that covers all original constraints. In order to decide whether two given filters should be merged a heuristic that allows the amount of introduced imperfection to be estimated is needed. This could also be accomplished by explicitly replacing an attribute filter with another that only tests for the existence of the given attribute or by simply dropping the attribute filter. Statistical online evaluation of filter selectivity would be also a good basis for merging decisions that enables adaptive filtering strategies. Imperfect merging requires further investigation.

Algorithms

In this section algorithms are presented that are superior to the naïve algorithms (cf. Sect. 3.2.1). The presented algorithms use the generic approach presented in the previous section: Each generic constraint class (e.g., constraints on ordered values) offers specialized indexing data structures to efficiently manage constraints on attributes. For example, hashing is used for equality tests. In the following, algorithms for matching, covering, and for detecting merging candidates are described that are all based on the predicate counting algorithm (cf. Sect. 3.2.2). Algorithms for detecting identity and overlapping among filters can be derived similarly.

Matching Algorithm

The naïve algorithm separately matches a given notification against all filters to determine the set of matched filters. This implies that the same attribute filter may be evaluated many times. More advanced algorithms avoid this. Some of these require a costly compilation step (e.g., [181]) that makes them less suitable for publish/subscribe systems in which subscriptions change dynamically. In contrast to that, the algorithm presented here allows filters to be added or removed at any time. The algorithm is based on the idea of *predicate counting* [305, 404] and makes use of our generic approach. The algorithm is depicted in Fig. 3.6. It determines all filters that match a given notification.

```
1 Matching Algorithm
   Input: notification n, set of filters F
   Output: the set M of all filters in F that match n.
   {
       <For each filter in F a counter is initialized to zero.>
6      for <each Aᵢ contained in n> {
           for <each filter S in F that has a constraint on Aᵢ that
                 is satisfied by the value of the corresponding
                 attribute of n> {
               <Increment the counter of S>
11         }
       }
       M:=<all filters in F whose counter is equal to their
             number of attribute filters>
   }
```

Fig. 3.6. Matching algorithm based on counting satisfied attribute filters

Covering Algorithm

Covering-based routing is built upon two tests: a first test that determines all filters that cover a given filter, and a second one determines all filters that are covered by a given filter. The naïve implementation simply tests each filter against all others sequentially. The algorithms presented here are more efficient. They are derived from the matching algorithm presented above (Figs. 3.7 and 3.8).

```
    Covering Algorithm I
    Input: filter F₁, set of filters F
    Output: the set C of all filters in F that cover F₁.
    {
5      <For each filter in F a counter is initialized to zero.>
       for <each Aᵢ contained in F₁> {
          for <each filter S in F that has a constraint Aⱼ that
                covers Aᵢ> {
            <Increment the counter of S>
10        }
       }
       C:=<all filters in F whose counter is equal to their
             number of attribute filters>
    }
```

Fig. 3.7. Covering algorithm that determines all *covering* filters

```
1  Covering Algorithm II
   Input: filter F₁, set of filters F
   Output: the set C of all filters in F that are covered by F₁.
   {
      <For each filter in F a counter is initialized to zero.>
6     for <each Aᵢ contained in F₁> {
         for <each filter S in F that has a constraint Aⱼ that
            is covered by Aᵢ> {
           <Increment the counter of S>
         }
11    }
      C:=<all filters in F whose counter is equal to the
            number of attribute filters of F₁>
   }
```

Fig. 3.8. Covering algorithm that determines all *covered* filters

Merging Algorithm

We present an algorithm that determines all possible merging candidates. These are those filters that are identical to a given filter in all but a single attribute. The algorithm avoids testing all filters against all others. It counts the number of identical attribute filters to find merging candidates (Fig. 3.9).

The further handling of the set of merging candidates depends on the constraints involved. For all constraints discussed (e.g., set constraints on simple values) there exists an efficient algorithm which outputs a single merged filter and a set of filters not included in the merger. For other constraints, an optimal algorithm requires exponential time complexity [87]. In this case the

use of greedy algorithms or heuristics (e.g., using a predicate proximity graph)
seems to be promising.

```
 1 Merging Algorithm
   Input: filter F₁, set of filters F
   Output: set M of all merging candidates
   {
       <For each filter in F a counter is initialized to zero.>
 6     for <each Aᵢ contained in F₁> {
           for <each filter S in F that has a constraint Aⱼ that
           is identical to Aᵢ> {
               <Increment the counter of S>
           }
11     }
       M:=<all filters in F whose counter is one smaller than or
               equal to their number of attribute filters>
   }
```

Fig. 3.9. Merging algorithm based on counting identical attribute filters

3.1.3 Semistructured Records

In the previous section structured records have been discussed in detail. In this
section a model for semistructured records is presented. The structured and
the semistructured model are mainly distinguished by the following fact: In
the structured model attribute names are unique, and hence an attribute name
uniquely addresses a single attribute. On the contrary, in the semistructured
model sibling attributes can have the same name, and therefore names address
sets of attributes.

In the following, a model for semistructured records is presented in which
notifications are essentially XML [399] documents. The filtering mechanisms
are similar to but less powerful than XPath [398]. After the model has been
introduced, how routing optimizations can be achieved is discussed.

According to Bunemann [55] semistructured data can be characterized as
some kind of graphlike or treelike structure that is often called *self-describing*
because the schema of the data is contained in the data itself. At the moment,
the most prominent semistructured data model is XML [399]. Similarly, to
structured records, a semistructured record is a set of nested attributes, but
in contrast to structured records, in semistructured records sibling attributes
can have the same name. In consequence, a single attribute can no longer be
uniquely addressed by its name alone. Instead, names (e.g., *car.price*), which
are usually called *paths* in this context, select sets of attributes. Therefore,
filtering strategies assuming that a single attribute is addressed by a given

name cannot directly be used in this scenario. One way to approach this problem is to use path expressions (e.g., XPath [398]), which select a set of attributes and impose constraints on the selected attributes.

Clearly, the semistructured model is more powerful than structured records, but work in this area related to content-based routing is still in its early stages. Lately, using XML and path expressions has gained increased attention. Nguyen et al. [271] and Chen et al. [77] described approaches for XML continuous queries. Altinel and Franklin [12] presented an efficient method for filtering XML documents using XPath expressions. All this work concentrates on efficient local matching and does not deal with distributed content-based routing. First ideas on how to support routing optimizations like covering and merging for semistructured records was presented by Mühl and Fiege [264]. These ideas are discussed later in this section.

Data Model

In the semistructured data model a *notification* is a well-formed XML document [399] and consists of a set of elements that are arranged in a hierarchy with a single root element uniquely named "notification". Each *element* consists of a set of attributes whose names must be distinct and a set of subordinate *child elements*, which are named but whose names must not necessarily be distinct. An *attribute A* is a pair (n_i, v_i) with name n_i and value v_i. Names of attributes must be unique with respect to elements. A simple notification that describes an auction is shown in Fig 3.10. In this example, the element *auction* has two subelements that are named *item*. Furthermore, the element *cpu* contains an attribute *clock* whose value is 800. Note that XML documents can contain free text between the opening and the closing tag of an element. Here, this text is simply ignored.

Filter Model

In the semistructured filter model a filter is a conjunction of path filters. Each of the path filters selects a subset of the elements in a notification by an element selector and places constraints on the attributes of the selected elements by an element filter, which consists of a set of attribute filters. In the following, this model is described in full detail.

An *element selector* selects a subset of the elements of a notification and is specified by an attribute path. It is distinguished between absolute and abbreviated paths. An *absolute path* is a slash-separated string that starts with a single slash (e.g., */notification/auction*). An *abbreviated path* is a slash-separated string that starts with two slashes (e.g., *//cpu*). An absolute (abbreviated) path selects all elements whose path is equal to (ends with) the given path. For example, *//item* selects both *item* elements of the notification in Fig. 3.10.

```
 1  <notification>
        <auction
            endtime="05/18/02 22:17:42"
            minprice="50">
            <seller
 6            name="Smith"
              id="1234"/>
            <item>
                <board
                  manufacturer="Elitegroup"
11                type="K7S5"
                  socket="Socket A"/>
            </item>
            <item>
                <cpu
16                manufacturer="AMD"
                  type="Athlon"
                  socket="Socket A"
                  clock="800"/>
            </item>
21      </auction>
    </notification>
```

Fig. 3.10. A simple notification

An *attribute filter* is a pair $A = (n, Q)$ consisting of a name n (e.g., *manufacturer*) and a constraint Q (e.g., $=$ *"AMD"*). An element *matches an attribute filter* if the element contains an attribute with name n whose value v satisfies Q, e.g. (*manufacturer*, *"AMD"*). This means that an attribute filter evaluates to false if the element does not contain an attribute with name n. Therefore, an attribute filter implicitly defines an existential quantifier over an element.

An *element filter* C is a conjunction of a nonempty set A of attribute filters $\{A_1, \ldots, A_i\}$, i.e., $C = \wedge_i A_i$. Hence, an element *matches an element filter* iff all attribute filters are satisfied. An example of an element filter based on the syntax of XPath is [@*manufacturer* $=$ *"AMD"* $\wedge$ @*clock* $\geq$ 700]. Note that in this notation attribute names are prefixed by an "@".

A *path filter* $P = (S, C)$ consists of an element selector S and an element filter C. A notification n *matches a path filter* P if at least one element of n is selected by S that matches C. It is possible to extend this model in such a way that an interval constraint can be imposed on both the number of elements that match an element filter and the number of elements that must not match. These extensions are not discussed for brevity. An example of a complete path filter based on an absolute path is: /*notification*/*auction*/*item*/*cpu*[@*manufacturer* $=$ *"AMD"* $\wedge$ @*clock* $\geq$ 700].

A *filter* F is a conjunction of path filters $\{P_1, \ldots, P_n\}$. Hence, a notification *matches a filter* if all path filters are satisfied. The set of all notifications that match a given filter F is $N(F)$.

Covering

This section discusses how covering among filters can be detected in the semistructured model. Similar results can easily be obtained for identity and overlapping, too. These are not discussed for brevity.

Let $L_A(A)$ be the set of all values that cause an attribute filter A to match an attribute. An attribute filter $A_1 = (n_1, Q_1)$ *covers* an attribute filter $A_2 = (n_2, Q_2)$, denoted by $A_1 \sqsupseteq A_2$, iff $n_1 = n_2 \wedge L_A(A_1) \supseteq L_A(A_2)$. For example, $[@clock \geq 600]$ covers $[@clock \geq 700]$.

Let $L_E(C)$ be the set of all elements that match an element filter C. An element filter C_1 *covers* an element filter C_2, denoted by $C_1 \sqsupseteq C_2$, iff $L_E(C_1)$ is a superset of $L_E(C_2)$. For example, $[@clock \geq 600]$ covers $[@manufacturer = \text{``}AMD\text{''} \wedge @clock \geq 700]$. Furthermore, $C_1$ is *disjoint* with $C_2$ with respect to the constrained attributes if there exists no attribute that is constrained in both element filters. For example, $[@minprice < 100]$ is disjoint with $[@name = \text{``}Pu\text{''}]$ with respect to their constrained attributes.

Corollary 3.4. *Given two element filters C_1 and C_2, neither of which contains two attribute filters with the same name, the following holds: $C_1 \sqsupseteq C_2$ is equivalent to $\forall j \exists i. A_i^1 \sqsupseteq A_j^2$.*

Let $L_S(S)$ be the set of all elements that are selected by an element selector S. An element selector S_1 *covers* an element selector S_2, denoted by $S_1 \sqsupseteq S_2$, iff $L_S(S_1) \supseteq L_S(S_2)$. S_1 is *disjoint* with S_2, iff $L_S(S_1) \cap L_S(S_2) = \emptyset$.

In the model presented here, an absolute path covers another absolute path iff both are identical. An absolute path only covers an abbreviated path iff the former is /notification and the latter is //notification, as the root element has a unique name. An abbreviated path covers another (abbreviated or absolute) path iff the former is a suffix of the latter (without the leading // or /). For example, //cpu covers //item/cpu because the former path selects all elements named *cpu*, while the latter only selects those elements named *cpu* which are a subelement of an element with name *item*.

Let $L_P(P)$ be the set of all elements that match a path filter P. A path filter $P_1 = (S_1, C_1)$ *covers* another path filter $P_2 = (S_2, C_2)$, written $P_1 \sqsupseteq P_2$, iff $L_P(P_1) \supseteq L_P(P_2)$. For example, the path filter //cpu[@manufacturer = ``AMD''] covers //cpu[@manufacturer = ``AMD'' \wedge @clock \geq 700]. P_1 is *disjoint* with P_2, iff either S_1 is disjoint with S_2 or if C_1 is disjoint with C_2 with respect to their constrained attributes.

Corollary 3.5. *Given two path filters $P_1 = (S_1, C_1)$ and $P_2 = (S_2, C_2)$, the following holds: $P_1 \sqsupseteq P_2$ is equivalent to $S_1 \sqsupseteq S_2 \wedge C_1 \sqsupseteq C_2$.*

A filter F_1 *covers* a filter F_2, denoted by $F_1 \sqsupseteq F_2$, iff $N(F_1) \supseteq N(F_2)$.

Corollary 3.6. *Given two filters $F_1 = P_1^1 \wedge \ldots \wedge P_n^1$ and $F_2 = P_1^2 \wedge \ldots \wedge P_m^2$ which are conjunctions of disjoint path filters the following holds: $F_1 \sqsupseteq F_2$ is equivalent to $\forall i \exists j.\ P_i^1 \sqsupseteq P_j^2$.*

For example, the filter $\{//cpu[@type = \text{``Athlon''}]\}$ covers $\{//seller[@name = \text{``Pu''}] \wedge //cpu[@type = \text{``Athlon''} \wedge @clock \geq 600]\}$.

3.1.4 Objects

Using objects as notifications is widely used in GUIs (e.g., Java AWT [358]) and visual components (e.g, JavaBeans [359]). The Java Distributed Event Specification [361], which is built upon Java RMI, also uses objects. The difference between this approach and a notification service is that consumers must directly register with the source of an event. Eugster and Guerraoui [124] present how to use structural reflection for content-based filtering of notifications. The object-oriented model is most flexible and powerful, but routing optimizations like covering and merging are difficult to achieve if filters can contain arbitrary code. Mühl and Fiege [264] have presented first ideas on how to support routing optimizations like covering and merging for objects. These ideas are discussed later in this section.

A purely object-oriented approach models notifications and filters as objects. A clear advantage of such a model is that it can easily be integrated with object-oriented programming languages. In contrast to that, models that are based on, e.g., name/value pairs, can only operate on serialized instances of objects violating object encapsulation. Unfortunately, routing optimizations, and in particular, covering and merging, are difficult to achieve if filters can contain arbitrary code. In this section three scenarios for which covering and merging can be supported are described.

Calling Methods on Attribute Objects

Regardless of whether the data models depend on structured or on semistructured records, it is possible to embed objects in notifications. In this case public members can be accessed and public inspector methods can be invoked on the embedded object after it has been instantiated. The returned member or the return value of the inspector method can either be a Boolean value that is directly interpreted as result of the attribute filter or a value that is used in order to evaluate the actual constraint.

For example, suppose that an instance of a class **StockQuote** has been embedded in a notification as an attribute with name *quote*. Then an attribute filter that evaluates this attribute could be specified like this: $\{quote.id() = \text{``IBM''}\}$. For example, this filter covers $\{quote.isRealTime() \wedge quote.id() = \text{``IBM''} \wedge quote.Price() > 45.0\}$. Moreover, it could be merged with a filter $\{quote.id() = \text{``MSFT''}\}$ to a filter $\{quote.id() \in \{\text{``IBM''}, \text{``MSFT''}\}\}$.

As stated in [121, 124], structural reflection (e.g., supported by Java) can be used to invoke the specified methods. Unfortunately, the model does not allow us to detect all covering relations among filters. For example, a filter $\{quote.Volume() > 10,000\}$ covers a filter $\{quote.Price() > 100 \wedge quote.Quantity() > 100\}$ because the volume is defined as the product price multiplied by the quantity.

Filtering on Notification Classes

Here, notifications are objects and consequently they are an instance of some class. Hence, class filters can be used that evaluate the class of a notification: A notification matches a filter if it is assignable to the specified class. It is also possible to support covering and merging. A class filter covers another class filter if an instance of the latter class can be assigned to an instance of the former one. A set of class filters can be merged perfectly if they either contain a class which covers all other classes or if they represent all direct subclasses of their common superclass. Figure 3.11 shows the implementation of a `ClassFilter` in Java. The integration with content-based filtering can be achieved by supporting filters that are conjunctions of a class filter and a specialized filter object whose match method is invoked if the class filter returned true.

Specialized Filter Objects

Another possibility is to use specialized filter objects, an approach that can also be combined with class filters. Such a filter implements a `match` method that evaluates whether a notification matches this filter instance or not. Moreover, it can also implement methods for covering and merging. Figure 3.12 shows the implementation of a `QuoteFilter` in Java. Note that the filters can also be built upon a more generic filter library, which offers, for example, set-oriented filters.

3.2 Matching Algorithms

Matching is probably the most fundamental functionality in a publish/subscribe system. A matching algorithm determines the filters, and thus the recipients, that are matched by a given notification. In this chapter several common approaches are discussed, including brute force, predicate counting, decision trees, binary decision diagrams, and efficient XML matching.

One must carefully distinguish between *notification matching* and *notification forwarding*. While matching aims at determining all filters that match a given notification, notification forwarding aims at determining all destinations for which a filter exists that matches a given notification. This means that for

```
  class ClassFilter {
    protected Class class;
3
    public boolean covers(ClassFilter filter) {
      return class.isAssigneableFrom(filter.class);
    }

8   public static ClassFilter merge(ClassFilterSet filters) {
      Class superClass=filters.getCommonSuperClass();
      if (superclass!=null) {
        if (filters.contain(superClass))
          return new ClassFilter(superClass);
13       if (filters.containAllSubclasses(superClass))
          return new AllSubclassesFilter(superClass);
      }
      return null;
    }
18
    public boolean match(Notification n){
      return class.isInstance(n);
    }
  }
```

Fig. 3.11. Implementation of a ClassFilter in Java

```
public class QuoteFilter {

3   public boolean covers(QuoteFilter qf){
      return getSymbolSet().isSuperSet(qf.getSymbolSet());
    }

    public static QuoteFilter merge(QuoteFilter[] qf){
8     return new QuoteFilter(QuoteFilter.
          unionOfSymbolSets(qf));
    }

    public boolean match(Event e) {
13     if (!(e instanceof QuoteEvent))
        return false;
      return (qf.getSymbolSet().contains(
          ((QuoteEvent)e).getSymbol()));
    }
18 }
```

Fig. 3.12. Implementation of a QuoteFilter in Java

the latter it may not be necessary to determine all matching filters. However, most algorithms do not exploit this difference. They determine all matching filters and derive the set of destinations by "or-ing" the individual destination of each filter. In the following, we concentrate on notification matching.

3.2.1 Brute Force

This is the simplest algorithm. It tests the given notification sequentially against all filters. The main advantage of this algorithm is that it can be used for all kind of filters; for example, it does not presume that filters are conjunctive filters. Moreover, it does not require some kind of preprocessing as other algorithms do. The main disadvantage of this naïve algorithm is its degraded performance. This is because the same predicate is evaluated many times if it is part of many filters. Moreover, the dependencies among predicates are not exploited. For example, the algorithm does not exploit that if the predicate $\{x = 5\}$ is matched, the predicate $\{x = j\}$ for any $j \neq 5$ cannot be matched.

3.2.2 Counting Algorithm

Yan and Garcia-Molina have proposed to use the counting algorithm for document matching [404]. This algorithm separates filter matching from predicate matching. This way, the algorithm avoids evaluating predicates more than once. In the following, we depict the algorithm for conjunctive filters consisting of attribute filters.

For each filter there is a counter that is initialized to 0. Then, all matching attribute filters are determined. For each matching attribute filter, the counters of those filters are incremented which contain the attribute filter as conjunctive term. After all matching attribute filters have been processed, those filters whose counter equals the number of predicates this filter consists of match the given notification.

The simplest strategy to find all matching predicates is to sequentially test each attribute filter as to whether or not it is matched by the given notification. A more advanced strategy is to use multilevel index structures that depend on the type of constraint (e.g., a hash table can be used for equality tests). The first level of the index (the attribute name index) is used to look up all attribute filters constraining an attribute by its name. The second level (the operator index) is used to look up all of those constraints that use a given operator (e.g., equivalence or greater than). The third level (the value index), finally, allow to find all of those attributes for the respective attribute and operator that are satisfied. In this way all matching attribute filters can be found without testing all attribute filters for satisfaction.

Figure 3.13 shows a simple example, where a notification is matched against three filters F_1, F_2, and F_3. From these filters only F_1 is matched by the notification.

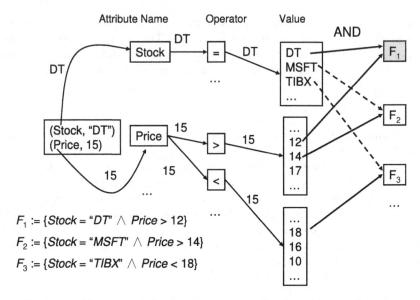

$F_1 := \{Stock = \text{"DT"} \land Price > 12\}$

$F_2 := \{Stock = \text{"MSFT"} \land Price > 14\}$

$F_3 := \{Stock = \text{"TIBX"} \land Price < 18\}$

Fig. 3.13. Using a multilevel index structure for the counting algorithm

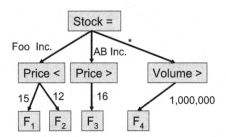

Fig. 3.14. An exemplary decision tree

3.2.3 Decision Trees

Aguilera et al. [6] have proposed using decision trees for matching in publish/-subscribe systems. A *decision tree* arranges tests, test results, and filters in a tree; usually conjunctive filters consisting of attribute filters are assumed. In the tree, nonleaf nodes are tests (e.g., *price* <), while leaf nodes represent filters. Finally, edges are test constants (e.g., 10). The decision tree is usually traversed in depth-first order. The traversal follows an edge if the notification matches the attribute filter that is formed by the test and the test constants (e.g., *price* < 10). The filters that are reached, match the given notification. Figure 3.14 shows an exemplary decision tree. The tree contains the filters $F_1 = \{Stock = \text{"Foo Inc"} \land Price < 15\}$, $F_2 = \{Stock =$

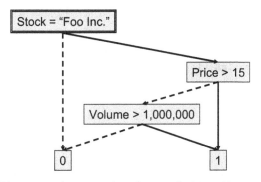

Fig. 3.15. An exemplary binary decision diagram

"Foo Inc." $\land$ Price $< 12\}$, $F_3 = \{Stock = $ "AB Inc." $\land$ Price $> 16\}$, and
$F_4 = \{Volume > 1,000,000\}$.

3.2.4 Binary Decision Diagrams

Campailla et al. [58] suggested using *binary decision diagrams (BDDs)* for
matching in publish/subscribe systems. BDDs are not restricted to conjunctive filters. They can be used to express arbitrary Boolean functions. In the
following, we describe the basics of BDDs and how they can be used in pub-
lish/subscribe systems.

BDDs are directed acyclic graphs. In a BDD, there are two terminal nodes
(i.e., nodes without outgoing edges) with the labels 1 and 0. These stand for
the predicates *true* and *false*, respectively. Each nonterminal node corresponds
to a predicate (e.g., *price* < 10) and has two outgoing edges, the *low edge*
and the *high edge*. A subset of the nodes is marked as *output nodes*; each
output node represents a filter. Figure 3.15 shows a simple BDD with a single
output node. The solid lines are the high edges while the dashed lines are
the low edges. The filter that corresponds to the output node is $\{Stock = $
"Foo Inc." $\land$ (*price* > 15 $\lor$ *Volume* > 1,000,000)$\}$.

A filter is evaluated by traversing the BDD starting from the given output
node (Fig. 3.16). While traversing the BDD, the high edge is followed if the
predicate corresponding to the visited node is fulfilled by the given notification; the low edge is followed otherwise. A notification matches a filter if finally
the node 1 is reached; if 0 is reached, the notification does not match. For example, the notifications $\{\{Stock, $ "Foo Inc."$\}, \{Price, 16\}, \{Volume, 10,000\}\}$
and $\{\{Stock, $ "Foo Inc."$\}, \{Price, 14\}, \{Volume, 1,000,000\}\}$ match the BDD
shown in Fig. 3.15.

Evaluating all filters separately can be avoided by using *ordered binary
decision diagrams (OBDDs)*. In a OBDD, the nodes are numbered such that
for every path, the numbers of the visited nodes are strictly monotonically
increasing. This means that the nodes 0 and 1 are numbered by n and $n - 1$,

```
  v := <output node of filter>;
2 while <v is not a terminal node> do
     if eval[v] then
        v := high[v];
     else
        v := low[v];
7    endif
  endwhile
  matched := label[v];
```

Fig. 3.16. Evaluating a filter using a binary decision diagram

```
1 for v := n downto 1 do
     if <v is terminal node> then
        value[v] := label[v];
     else
        a := eval[v];
6       value[v] := a and value[high[v]] or
                    not a and value[low[v]];
     endif
  endfor
```

Fig. 3.17. Evaluating an ordered binary decision diagram

respectively. OBDDs are evaluated bottom-up by visiting the nodes in decreasing order starting by node n. If the visited node is a terminal node, a value of 1 is assigned if node 1 is visited and 0, otherwise. If a nonterminal node v is visited it is assigned the value $p(v) \land low(v) \lor \neg p(v) \land high(v)$, where $p(v)$ is the result of the predicate corresponding to node v, and $low(v)$ and $high(v)$ are the values assigned to the node to which the low and the high edge originating at v are leading, respectively. A filter is matched, if to its output node 1 is assigned; otherwise it is not matched. The algorithm is shown in Fig. 3.17.

A *reduced ordered binary decision diagram (ROBDD)* is an OBDD from which redundant nodes and isomorphic subgraphs are removed. It is known from the research on Boolean function minimization that ROBDDs exhibit exponential grow for some Boolean functions (e.g.,the chessboard function). The predicate numbering has a large effect on the size of the ROBDD, too. While some functions require exponential size only for a subset of the potential predicate orderings, other functions require exponential size for all possible variable orderings. Finding the optimal ordering is known to be NP-hard. BDDs can easily be logically combined. For example, the BDD of a negated function is the BDD of the function, where the nodes 0 and 1 are swapped. BDDs can also "or-ed" and "and-ed" together.

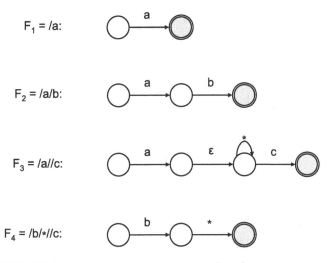

Fig. 3.18. XPath Queries and their corresponding finite state automaton

3.2.5 Efficient XML Matching

As XML becomes more popular, using XML as a data model for publish/-subscribe systems is also gaining increased attention. In the area of XML processing, XPath [398] is often used to select parts of an XML document that match a *path expression*. This approach can also be used to test whether a document contains a matching part. A path expression searches for elements and attributes in an XML document that satisfy the given condition. Because XPath allows for very complex queries, implementing efficient matching for XPath filters is challenging. In the literature, XFilter and YFilter have been proposed to facilitate XPath for matching XML documentsr. Both approaches are based on *finite state machines (FSMs)*. Recent approaches [183] are based on a constructing a *deterministic finite automaton (DFA)* from the given NFA. In the following, we give an overview of XFilter and its successor YFilter. Altinel and Franklin [12] have proposed XFilter, which was the first FSM-based approach. XFilter translates each XPath query into a separate FSM (Fig. 3.18) and uses a novel indexing mechanism to allow all of the FSMs to be executed simultaneously during the processing of a document. When a document arrives, it is processed by an event-based XML parser (e.g., based on the SAX interface). The events raised (e.g., an element is opened or an element is closed) during parsing are used to drive the FSMs through their various transitions. A query is said to match a document if during parsing, an accepting state for that query is reached. The approach of XFilter to use one FSM per XPath query has the disadvantage that commonalities among queries are not exploited.

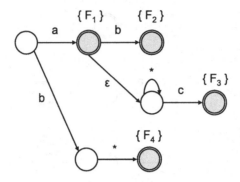

Fig. 3.19. Combined nondeterministic finite state automaton

YFilter, which can be seen as the successor of XFilter, was proposed by Diao et al. [110, 111, 112]. YFilter combines all path expressions into a single *nondeterministic finite automaton (NFA)* (Fig. 3.19), where the common prefixes among path expressions are shared, i.e., represented only once. This NFA-based approach can be extended to also process predicates attached to path expressions. The authors have developed two alternatives to combining the NFA execution and predicate evaluation. One approach evaluates predicates as early as their addressed elements are matched, while the other delays predicate evaluation until the corresponding path expression has been entirely matched.

3.3 Further Reading

Approximate Matching

In this chapter we assumed the Boolean filter model [404]. Either the notification exactly matches the filter or it does not match the filter. An alternative to exact matching is *approximate matching*. Liu and Jacobsen presented A-ToPSS [240], a publish/subscribe prototype with approximate matching. Yan and Garcia-Molina [403] discussed index structure for information filtering under the vector space model.

Matching Algorithms

Fabre et al. [132] and Pereira et al. [305] present matching algorithms which exploit similarities among predicates. In a first step the satisfied predicates are computed, and after that the number of predicates satisfied by a subscription are counted using an association table. Two variants of this algorithm are described that incorporate special treatment of equality tests and of constraints having only inequality tests.

A predicate matching algorithm for database rule systems is presented by Hanson et al. [187] that indexes the most selective predicate that is determined by the query optimizer. They use a special indexing data structure called interval binary search tree to support the efficient evaluation of interval tests.

Gough and Smith [181] present a matching algorithm that is based on automata theory. They show how a set of conjunctions of predicates, each dependent on exactly one attribute, can be transformed to a deterministic finite state automaton. In the paper different types of test predicates are considered and complexity results are obtained. Their algorithm is very efficient, but its worst-case space complexity is exponential. The proposed solution is also not suited for dynamic environments as the automaton has to be newly constructed from scratch if subscriptions change.

Pu et al. [241, 372] present indexing strategies for continual queries based on trigger patterns. In particular, a strategy which uses an index on the most selective predicate is described. More complex indexing strategies exploit similarities among trigger patterns to reduce the processing costs. They restrict optimizations to constraints which place a constraint on a single attribute involving at most one constant.

Gryphon uses the content-based matching algorithm presented by Aguilera et al. [6]. This algorithm traverses a parallel search tree, where nonleaf nodes correspond to simple tests and edges from nonleaf nodes represent results. Leafnodes are associated with matched subscriptions. Banavar et al. [26] present a multicast routing algorithm that executes the matching algorithm at each broker. The algorithm presented is limited to equality tests.

4

Distributed Notification Routing

In this chapter we describe how a simple event system can be implemented by distributed notification routing relying an overlay network of brokers. To emphasize that we focus on the communication in this chapter, we use the term publish/subscribe system instead of event system for the rest of this chapter. We first introduce our system model and a routing framework in Sect. 4.1 and Sect. 4.2, respectively. Then, we introduce the notion of valid and monotone valid routing algorithms that are sufficient for correct publish/subscribe systems in Sect. 4.3. Section 4.4 defines valid framework instantiations that implement monotone valid routing algorithms. A set of content-based routing algorithms is presented as instances of the routing framework, and their validity is shown in Sect. 4.5. Then, in Sect. 4.6 extensions of the basic framework are described informally. They deal with advertisements, hierarchical routing, rendezvous-based routing, topology changes, joining and leaving clients, routing in cyclic topologies, exploiting IP multicast, and topology maintenance.

4.1 System Model

In our model, the publish/subscribe system consists of a set of cooperating concurrent processes $B_1, \ldots, B_n$, called *brokers*,[1] that are arranged in a topology. If nothing else is said, we restrict ourselves to acyclic connected topologies. This restriction can be circumvented by running a spanning tree algorithm on the original (potentially cyclic) topology. Of course, routing algorithms that can deal more directly with cyclic topologies are desired (cf. Sect. 4.6.6). Since we focus more on the implementation in this chapter, we call the components (cf. Chap. 2) that connect to the event notification service *clients*. Each broker B manages a mutually exclusive set of *local clients* L_B that is a subset of all clients $\mathcal{C}$. Clients communicate with their broker using *local synchronous procedure calls*. Concurrent updates to local data structures are synchronized

---

[1] In this chapter, we do not distinguish among border, inner, and local brokers.

using a broker-specific monitor μ_B. Moreover, each broker is connected to a set of neighbor brokers N_B. Brokers communicate with their respective neighbors by *asynchronous message passing*. For this chapter, we refer with B_i to an arbitrary broker and by B_j to an arbitrary neighbor of B_i, i.e., $B_j \in N_{B_i}$.

Assumptions

The subsequent discussion is based on the following assumptions:

- Clients are stationary, i.e., they cannot disconnect from one broker and connect to another broker; client mobility is addressed in Sect 8.4.
- We first concentrate on a system without advertisements; their discussion is postponed to Sect. 4.6.1.
- The topology is static; topology changes are discussed in Sect. 4.6.4.
- The set of clients is static; clients that join and leave the system are discussed in Sect. 4.6.5.
- The system is not overloaded; congestion control is discussed in Sect. 8.3.
- The system is fault-free; fault tolerance is discussed in Sect. 8.2.
- The communication channels are reliable and respect FIFO message ordering; no messages are duplicated, lost, corrupted, or erroneously sent, and messages are received in the order in which they have been sent. These assumptions are not severe restrictions because they can easily be achieved by using transport layer functionality (e.g., TCP).
- The message delay is unbounded but finite. Since channels are reliable this implies that if a message is sent, it is eventually received.

Fairness Property

To allow us to prove liveness properties, the entire system has to satisfy a *fairness property*, namely that (a) no pending message (i.e., a message that was received but which has not yet been processed) and (b) no thread waiting to enter a monitor can be infinitely delayed by a process because of unfavorable scheduling choices of that process. The fairness property allows us (together with the reliable channel assumption and the finite message delay) to conclude that if a message was sent, it is eventually processed. In the implementation, the property is respected by using fair scheduling (e.g., round robin on all incoming channels). Furthermore, if no deadlocks can occur, it allows us to conclude that a thread waiting to enter a monitor will eventually enter this monitor.

Message Batching

In many cases when we discuss algorithms in the following, *message batching* could be used. In this case, the sending of a message is postponed until a timeout occurs. If more than one message is "sent" to the same destination before the timeout occurs, those message are combined into a single, larger message. While this may reduce the network overhead, it may introduce an additional delay to messages. We mainly refrain from using message batching and piggybacking in the following because it would overly complicate the discussion of the algorithms.

4.2 Routing Algorithm Framework

The pseudocode of the routing framework that runs on each broker is depicted in Figs. 4.1 and 4.2. The main program (lines 1–11) starts when the broker is created. It initializes the routing table of the broker, a monitor, and, for each local client, a delivery queue. Then, it enters an infinite loop (lines 4–10) that dispatches messages arriving from neighbor brokers to the *handleMessage* procedure. This is done in a fair way, e.g., by using round robin. The *handleMessage* procedure (lines 37–46) further dispatches a message based on its type. The framework uses two types of messages for its internal implementation that are exchanged among neighboring brokers using asynchronous message passing: (1) *forward*(n), which is used to disseminate a notification n in the broker network and (2) *admin*($\mathcal{S}, \mathcal{U}$), which is used to propagate routing table updates by interpreting the sets of filters $\mathcal{S}$ and $\mathcal{U}$ as subscriptions and unsubscriptions, respectively. The *handleMessage* procedure dispatches *forward* messages to the *handleNotification* procedure (lines 18–24), which notifies local clients and sends *forward* messages to neighbor broker. *admin* messages are dispatched to the administer procedure. The values returned by administer are used as input to the *handleAdminMessage* procedure that sends *admin* messages to neighbor brokers. The code of the administer procedure is not shown here because it is not part of the framework. It is implemented by a framework instantiation to realize a concrete routing algorithm. This allows a variety of routing algorithms (Sect. 4.5) to be implemented.

Besides the code that processes messages received from neighbor brokers, the framework comprises a set of interface procedures that correspond to the interface operations introduced in Sect. 2.5.2. The procedures *pub*, *sub*, and *unsub* (lines 47–62) are called by local clients to publish a notification and to subscribe and unsubscribe to a filter, respectively. The *notify* procedure (lines 13–16) is called by the broker itself to notify a local client about a notification. A notification is delivered to a client Y by appending the notification to the *delivery queue* Q_Y of the client.

4.2.1 Atomic Steps of the Implementation

At the implementation level, we distinguish the following six *atomic steps*: *pub*, *sub*, and *unsub* (corresponding to the interface operations called by local clients), *notify* (corresponding to the interface operation called by the broker),

```
 1  program ContentBasedRoutingFramework()
    begin
      initialize T_B and μ_B, and Q_C for all C ∈ L_B;
      loop
        wait until a message is available;
 6      sync(μ_B)
          m ← return next fairly selected message;
          handleMessage(m);
        endsync
      endloop
11  end


    procedure notify(Client Y, Notification n)
    begin
      Q_Y ← append(Q_Y, n);
16  end


    procedure handleNotification(Dest D, Notification n)
    begin
      send "forward(n)" to all neighbors in F_B(n) \ {D};
21    forall local clients C ∈ F_B(n) do
        notify(C, n);
      endforall
    end


26  procedure handleAdminMessage(Dest D, Set M_S, Set M_U, Bool b)
    begin
      forall H ∈ N_B \ {D}
        S ← {F | (F, H) ∈ M_S};
        U ← {F | (F, H) ∈ M_U};
31      if S ≠ ∅ ∨ U ≠ ∅ then
          send "admin(S, U)" to H;
        endif
      endforall
    end
36

    procedure handleMessage(Message m)
    begin
      if m is "forward(n)" message from neighbor U then
        handleNotification(U, n);
41    endif
      if m is "admin(S, U)" message from neighbor U then
        (F_S, F_U) ← administer(U, S, U);
        handleAdminMessage(U, F_S, F_U, 0);
      endif
46  end
```

Fig. 4.1. Content-based routing framework, part I

```
     sync(μ_B) procedure pub(Client X, Notification n)
     begin
        handleNotification(X, n);
50   end

     sync(μ_B) procedure sub(Client Y, Filter F)
     begin
        (F_S, F_U) ← administer(Y, {F}, ∅);
55      handleAdminMessage(Y, F_S, F_U, 0);
     end

     sync(μ_B) procedure unsub(Client Y, Filter F)
     begin
60      (F_S, F_U) ← administer(Y, ∅, {F});
        handleAdminMessage(Y, F_S, F_U, 0);
     end
```

Fig. 4.2. Content-based routing framework, part II

and *forward* and *admin* (corresponding to the two types of messages that can be sent and received by a broker).

The execution of *pub*, *sub*, and *unsub* steps starts when the calling thread of the respective client enters the body of the respective procedure and ends when the respective procedure returns. The execution of *notify* steps starts when the calling thread enters the *notify* procedure and ends when the *notify* procedure returns. Note that strictly speaking, *notify* steps are executed within a surrounding atomic *pub* or *forward* step. However, it is sufficient to model this by appending the *notify* steps directly to the corresponding surrounding step in the resulting trace.

The execution of the *forward* and *admin* steps starts when the thread of the broker enters the monitor (line 6) and ends when it leaves the monitor (line 9). Which of these two steps is executed depends on what type of message is received. We say that a *forward* (an *admin*) step is executed when a *forward* (an *admin*) message is received. The $forward(B_i, B_j, n)$ step takes three parameters: B_i is the broker at which the step is executed; B_j is the broker from which the $forward(n)$ message was received; n is the notification that was received as part of the $forward(n)$ message. The $admin(B_i, B_j, S, U)$ step takes four parameters: B_i is the broker at which the step is executed; B_j is the broker from which the $admin(S, U)$ message was received; S and U are the two sets of filters that were received as part of the *admin* message.

To ensure that the execution of a step is atomic, i.e., does not interleave with the execution of other steps at the *same broker*, the interface procedures that can be called by local clients (i.e., *pub*, *sub*, and *unsub*) and the code that receives and handles a messages from a neighbor broker (lines 6–9) are protected by a broker-specific monitor μ_B. Since we use only a single monitor, no deadlocks can occur. There is no need to protect the *notify* procedure because it is only called by threads that have already entered the monitor.

The trace for the whole system consists of the steps of all brokers and clients. It arises from interleaving the traces of the individual brokers and their clients.

4.2.2 Notification Forwarding and Delivery

In the following, we explain in more detail how routing tables are used in the *handleNotification* procedure to forward notifications to neighbor brokers and to deliver them to local clients. Each broker B manages a private *routing table* T_B that comprises a set of routing entries. Each *routing entry* is a pair (F, D) consisting of a filter F and a *destination* $D \in N_B \cup L_B$. The state of all routing tables determines the current *routing configuration* of a publish/subscribe system. Initially, each routing table is set to a predefined state that usually depends on the applied routing algorithm. This defines the *initial routing configuration*. The routing configuration of a single broker B consists of two disjoint parts: the *remote routing configuration* that comprises all routing entries whose destination is a neighbor and the *local routing configuration* consisting of all routing entries whose destination is a local client.

The routing configuration induces the set of notifications that a broker *potentially* forwards to a destination. In the following, we often need to refer to the filters that comprise the routing configuration of a broker regarding a single destination D and all but a single destination D:

$$T^{|D} \stackrel{\text{def}}{=} \{F \mid \exists (F, D) \in T\} \tag{4.1}$$

$$T^{\backslash D} \stackrel{\text{def}}{=} \{F \mid \exists (F, E) \in T \land E \neq D\} \tag{4.2}$$

The destinations to which a broker B forwards or delivers a given notification n is given by $F_B(n)$:

$$F_B(n) \stackrel{\text{def}}{=} \big\{ D \mid D \in N_B \cup L_B \land n \in N(T_B^{|D}) \big\}. \tag{4.3}$$

Now, we can describe how a broker forwards a notification to its neighbors and how a broker delivers a notification to its local clients:

- Calling $pub(X, n)$ leads to a call of *handleNotification*(X, n) (line 49).
- If a broker receives a *forward*(n) message from a neighbor U, it invokes *handleNotification*(U, n) (line 40).
- If *handleNotification*(D, n) is called at a broker B, a *forward*(n) message is sent to all of neighbors of B in $F_B(n) \setminus \{D\}$ (line 20) and all local clients of B in $F_B(n)$ are notified about n (lines 21–23).

For example, consider the situation depicted in Fig. 4.3. Here, B_1 delivers a notification received from X_1 to its local client X_2 due to the entry (F_1, X_2) and forwards n to its neighbor B_2 due to the entry (F_3, B_2).

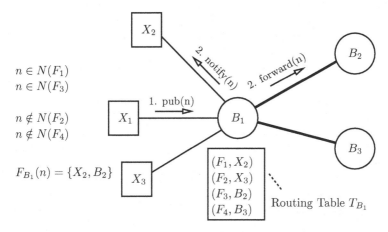

Fig. 4.3. Diagram explaining notification forwarding

4.2.3 Avoidance of Duplicate and Spurious Notifications

Now, we prove that duplicated and spurious notifications are avoided by the notification forwarding algorithm if the broker topology is acyclic. Duplication avoidance in cyclic topologies is discussed in Sect. 4.6.6.

Lemma 4.1. *If the topology is acyclic, notification forwarding satisfies*

$$\square\left[notify(Y, n) \;\Rightarrow\; \left[\bigcirc\square\neg notify(Y, n)\right] \wedge \left[n \in \cup_{X \in \mathcal{C}} P_X\right]\right] \qquad (4.4)$$

Proof. The algorithm never forwards a notification to a neighbor broker from which it received this notification. This fact, the reliable channel assumption, the fact that the topology is acyclic, and the fact that a notification cannot be published twice ensures that no duplicates are delivered to a client. To prove that no spurious notifications are delivered to a client, we argue backwards from the delivery of a notification to its publication. For every, $notify(n)$ a corresponding $pub(n)$ should exist. A broker only notifies a client if it either received $forward(n)$ message from a neighbor or if $pub(n)$ was called by a local client. In the former case, we have found the witness $pub(n)$. In the latter case, the same case distinction can be applied to the neighbor broker from which the $forward(n)$ message was received. As the topology is acyclic and has a finite diameter, this recursion must abort after a finite number of steps. Hence, a corresponding $pub(n)$ exists in any case. This concludes the proof. □

4.2.4 Routing Table Updates

In this section, we explain in more detail how the routing tables are updated by calling **administer** and by propagating *admin* messages using *handleAdminMessage*. Routing tables are exclusively updated by (calling) the `administer` procedure. The `administer` procedure is called at a broker B if an *admin* message from a neighbor is received (line 43) or, if *sub* (line 54) or *unsub* (line 60) is called by a local client. If its execution was triggered by an *admin* message, it is called with the broker S from which this message was received and the two filter sets S and U that were embedded in the message as parameters. If a local client Y calls $sub(Y, F)$ or $unsub(Y, F)$, then $\mathtt{administer}(Y, \{F\}, \emptyset)$ and $\mathtt{administer}(Y, \emptyset, \{F\})$ are called, respectively. `administer` can identify whether the call was triggered by a neighbor or by a local client by checking whether S is in N_B or in L_B. As result `administer` returns two sets which are both comprised of pairs of filters and neighbors. These sets are used as input to the *handleAdminMessage* procedure. To each neighbor apart from S, which is represented in either of both sets, exactly one $admin(S_H, U_H)$ message is sent. While S_H contains all filters F for which there is a pair (F, H) in the first returned set, U_H contains all filters F for which there is a pair (F, H) in the second returned set. Roughly speaking, S_H and S_U are the subscriptions and unsubscriptions which are forwarded to H, respectively.

4.3 Valid and Monotone Valid Routing Algorithms

A publish/subscribe system has to deal with new subscriptions and cancellation of existing subscriptions. A *routing algorithm* adapts, starting from an eligible initial routing configuration, the routing configuration to the changing set of active subscriptions. Intuitively, a routing algorithm is *valid* if it adapts the routing configuration such that the resulting system satisfies the safety and the liveness property of Def. 2.5. But can we express validity as property of the routing configuration?

4.3.1 Valid Routing Algorithms

A valid routing algorithm must lead together with the routing framework to a publish/subscribe system that satisfies the safety and the liveness properties of Def. 2.5. As duplicated and spurious notifications are already avoided by the notification forwarding algorithm, it suffices to require that the local routing configuration ensures that only matching notifications are delivered to imply safety. For a client Y, let $\theta(Y)$ be the broker that manages Y. Hence, $N(T_{\theta(Y)}^{|Y})$ should be a subset of $N(S_Y)$ for all clients.

To guarantee liveness we must show that when a client Y subscribes to a filter F and stays subscribed, then from some time (after the subscription was issued) on, every notification that is published at any broker B (i.e., by a local client of B) and that matches F should be delivered to Y. To achieve this such a notification must first be forwarded to the broker managing Y (i.e.,

$\theta(Y)$) and second be delivered to Y subsequently. The second requirement is easily ensured by keeping $N(T^{|Y}_{\theta(Y)})$ a superset of $N(S_Y)$. We achieve the first requirement by requiring that for each notification $n \in N(F)$ a simple[2] directed path exists connecting B with $\theta(Y)$ over which eventually always n is forwarded. By requiring the property to hold for all $n \in N(F)$ independently, the delivery of all notifications matching F can be split among multiple delivery paths in cyclic topologies. Thus, let $B_{i_1}, \ldots, B_{i_j}$ be a simple directed path in the broker network. Then,

$$\gamma(B_{i_1}, \ldots, B_{i_j}) \overset{\text{def}}{=} \cap_{1 \leq k \leq j} N(T^{|B_{i_{k-1}}}_{B_{i_k}}) \tag{4.5}$$

is the set of notifications that if a notification contained in this set is published at B_{i_j} and stays in this set, it reaches B_{i_1} over this path. Let $\nu(B_i, B_j)$ be the set of all simple directed paths connecting originating broker B_j to the receiving broker B_i.

Definition 4.1. *A routing algorithm is* valid *if the following conditions hold:*

- *(Local Subset Validity)*

$$\Box[N(T^{|Y}_{\theta(Y)}) \subseteq N(S_Y)] \tag{4.6}$$

- *(Eventual Superset Validity)*

$$\Box[\Box(F \in S_Y) \Rightarrow \Diamond\Box N(T^{|Y}_{\theta(Y)}) \supseteq N(F)] \tag{4.7}$$

$$\Box[\Box(F \in S_Y) \wedge B \neq \theta(Y) \wedge n \in N(F) \\ \Rightarrow \exists P \in \nu(\theta(Y), B). \Diamond\Box[n \in \gamma(P)] \tag{4.8}$$

In the following, we prove that a valid routing algorithm is sufficient for a correct publish/subscribe system.

Theorem 4.1. *A valid routing algorithm is sufficient for a correct publish/subscribe system.*

Proof. We have to show that if local subset validity and eventual superset validity hold, then safety and liveness according to Def. 2.5 are implied. As duplicates and spurious notifications are avoided by the notification forwarding algorithm in acyclic topologies (Lemma 4.1), it remains to be shown that only matching notifications are delivered to clients to imply safety. This follows directly from local subset validity. To prove liveness we assume that $\Box(F \in S_Y)$ and show that then $\Diamond\Box N(T^{|Y}_{\theta(Y)}) \supseteq N(F)$ and $\forall n \in N(F). \exists P \in \nu(\theta(Y), B).\Diamond\Box[n \in \gamma(P)]$ implies $\Diamond\Box[pub(X, n) \wedge n \in N(F) \Rightarrow \Diamond notify(Y, n)]$. Assume that X publishes a notification $n \in N(F)$. There are two cases: Case

---

[2] A *simple* path is a path in which no vertex occurs twice.

1: If $\theta(X) = \theta(Y)$ then n is delivered to Y because of Eq. (4.7) if it was published after $N(T_{\theta(Y)}^{|Y}) \supseteq N(F)$ began to hold. Case 2: If $\theta(X) \neq \theta(Y)$, then there exists at least one path connecting $\theta(X)$ to $\theta(Y)$ since the topology is connected. According to Eq. (4.8) for one of these paths $n \in \gamma(P)$ holds eventually always. Hence, if n is published after $n \in \gamma(P)$ began to hold, n is forwarded to $\theta(Y)$ due to the definition of γ. Then, n is forwarded to Y due to Eq. (4.7) if it was published after $N(T_{\theta(Y)}^{|Y}) \supseteq N(F)$ began to hold. Hence, n will be delivered to Y if n was published after $N(T_{\theta(Y)}^{|Y}) \supseteq N(F)$ and $n \in \gamma(P)$ began to hold. Both cases together prove liveness. Thus, liveness and safety hold and a correct publish/subscribe system is implied.   □

Note that while we restricted the discussion to acyclic topologies here, valid routing algorithms can also be used in cyclic topologies (without requiring a single spanning tree). In this case, notification forwarding must be changed such that duplicates are avoided (Sect. 4.6.6).

4.3.2 Monotone Valid Routing Algorithms

Theorem 4.1 reveals that valid routing algorithms are sufficient for a correct publish/subscribe system. The properties of validity, however, have the following disadvantages: Local subset validity does not require that the delivery of notifications that are published by local clients connected to the same broker as the subscribing client is guaranteed immediately after subscribing. This would, however, be feasible in our setting. Furthermore, eventual superset validity depends on individual subscriptions and is a property of the routing configuration of the entire topology. A property that only depends on the routing configurations of neighboring brokers would be much simpler to handle. This motivated us to look for stronger requirements which are nevertheless satisfied by most routing algorithms of practical relevance. This process leads to a stronger form of validity, called monotone valid routing:

Definition 4.2. *A routing algorithm is* monotone valid *if the following conditions hold:*

- *(Local Validity)*

$$\Box \left[N(T_{\theta(Y)}^{|Y}) = N(S_Y) \right] \tag{4.9}$$

- *(Eventual Monotone Remote Validity)*

$$\Box \left[\Box [n \in N(T_{B_i}^{\backslash B_j})] \Rightarrow \Diamond \Box [n \in N(T_{B_j}^{|B_i})] \right] \tag{4.10}$$

While monotone validity implies validity, the opposite is, in general, not true. First, validity allows the local delivery to be guaranteed eventually, while monotone validity requires immediate delivery. Second, validity only requires that those notifications are sent over a link between two brokers that are necessary to serve the respective subscription. Further assumptions are not made. Monotone validity, on the other hand, does not depend on individual subscriptions. Instead, it requires that at least those notifications that are sent over a link from B_{i+1} to B_i are sent over the link from B_{i+2} to B_{i+1}. Hence, the set of notifications forwarded is monotonically increasing for any path in the broker network. This led to the naming of monotone validity. Subsequently, we prove that monotone valid routing algorithms are a subclass of valid routing algorithms.

Lemma 4.2. *Every monotone valid routing algorithm is also valid.*

Proof. It is easy to see that local validity implies local subset validity. To show that also eventual superset validity is implied by monotone validity, assume that $\Box F \in S_Y$ for some client Y and consider an arbitrary notification $n \in N(F)$. Let $P = B_{i_1}, \ldots, B_{i_j}$ with $B_{i_1} = \theta(Y)$ and $B_{i_j} = \theta(X)$ be an arbitrary path that connects the broker $\theta(X)$ to the broker $\theta(Y)$. To prove that for all $n \in N(F)$. $\Diamond\Box n \in \gamma(P)$, we prove by an induction that $\Diamond\Box n \in T_{B_{i_k}}^{|B_{i_k-1}}$ for all directed edges $(B_{i_{k-1}}, B_{i_k}) \in P$. Due to local validity $n \in N(T_{B_{i_1}}^{|Y})$ holds. Due to eventual monotone remote validity $\Diamond\Box n \in N(T_{B_{i_2}}^{|B_{i_1}})$ is implied. This proves the base case. Now, assume that $\Diamond\Box n \in N(T_{B_{i_{k+1}}}^{|B_{i_k}})$ holds (induction assumption). This implies that $\Diamond\Box n \in N(T_{B_{i_{k+2}}}^{|B_{i_{k+1}}})$ due to eventual monotone remote validity. This proves the induction step. Since $\Diamond\Box n \in T_{B_{i_k}}^{|B_{i_k-1}}$ for all directed edges $(B_{i_{k-1}}, B_{i_k}) \in P$, $\forall n \in N(F)$. $\Diamond\Box n \in \gamma(P)$ is implied by the definition of γ. Hence, eventual monotone remote validity holds in addition to local subset validity. This concludes the proof. $\Box$

Corollary 4.1. *A monotone valid routing algorithm implies a correct publish/subscribe system.*

Proof. By Lemmas 4.1 and 4.2. $\Box$

The definition of monotone valid routing algorithms sets up the design space for valid framework instantiations, which are presented in the next section.

4.4 Valid Framework Instantiations

In this section, we derive general requirements for valid framework instantiations revealing new insights into the characteristics content-based routing algorithms have in common. All requirements are expressed as invariants of the framework which refer only to a single step of the system. This allows the correctness of concrete framework instantiations (Sect. 4.5) to be proved more easily. Two requirements are derived from those characterizing monotone valid routing algorithms, while the other requirements are framework specific.

First, we require that `administer` returns after a finite time. This guarantees that a broker is not blocked infinitely by processing a message. From the two requirements characterizing monotone valid routing algorithms, local validity (which is called *local invariant* here) is directly used as an invariant of the framework. Eventual monotone remote validity is mapped to an invariant of the routing framework called *remote invariant*. This is done by looking at the transformation that a sequence of *admin* messages received from a neighbor B_i causes on the routing table of a broker B_j. This transformation can be computed without considering messages that B_j receives from other destinations if we require that

1. an *admin*, *sub*, or *unsub* step regarding a destination D can only influence the part of the routing table dealing with destination D (i.e., $T_B^{|D}$) and leaves for all other destinations their respective part of the routing table unchanged and that
2. $(T'_{B_j})^{|D}$ only depends on $T_{B_j}^{|D}$ and the processed message and not on the rest of the routing table.

These requirements are called *restricted change* and *restricted impact*, respectively. They are satisfied by all routing algorithms that we will discuss later on. Now, we look in more detail on how the above-mentioned transformation can be computed to derive the desired invariant.

A call of `administer` is triggered by the receipt of an *admin* message or if *sub* or *unsub* is called by a local client. Each call of `administer` transforms the routing table of the respective broker from its current state T to its subsequent state T'. Now, assume that the routing table of B_j contains only routing entries regarding a destination D and that a given implementation of `administer` is called at B_j triggered by a destination D. In this case, we define δ as the transformation of T into T', i.e., the function such that $\delta(T_{B_j}^{|D}, m) = (T'_{B_j})^{|D}$. Multiple *admin* messages can be in transit simultaneously on the communication channel between a broker B_i and one of its neighbors B_j. To capture the change to the routing table of B_j triggered by this sequence of *admin* messages, let $\mathcal{K}_{B_i,B_j} = \langle m_1, \ldots, m_n \rangle$ be the sequence of *admin* messages that B_i sent to B_j that have not yet been processed by B_j, i.e.,that are still in transit. For sequences, we assume the existence of the functions *head*, *tail*, and *append*:

$$head(\langle m_1, m_2, \ldots, m_n \rangle) \overset{\text{def}}{=} m_1 \tag{4.11}$$

$$tail(\langle m_1, m_2, \ldots, m_n \rangle) \overset{\text{def}}{=} \langle m_2, \ldots, m_n \rangle \tag{4.12}$$

$$append(\langle m_1, m_2, \ldots, m_n \rangle, m_{n+1}) \overset{\text{def}}{=} \langle m_1, \ldots, m_{n+1} \rangle \tag{4.13}$$

The *admin* messages in transit will eventually trigger a call of `administer` at B_j, as we now explain. We define for a set of filters $\mathcal{A}$:

$$\Delta(\mathcal{A}, \langle \rangle) \stackrel{\text{def}}{=} \mathcal{A} \tag{4.14}$$

$$\Delta(\mathcal{A}, \langle m_1, \ldots, m_n \rangle) \stackrel{\text{def}}{=} \Delta(\delta(\mathcal{A}, m_1), \langle m_2, \ldots, m_n \rangle) \tag{4.15}$$

Hence, $\Delta(T_{B_j}^{|B_i}, \mathcal{K}_{B_i, B_j})$ contains all filters represented in the routing table of broker B_j regarding neighbor B_i that one would obtain if T_{B_j} contains only routing entries regarding neighbor B_i and the $admin$ messages in $\mathcal{K}_{B_i, B_j}$ are sequentially processed by the $\mathtt{administer}$ procedure at B_j. Note that $\Delta(T_{B_j}^{|B_i}, \mathcal{K}_{B_i, B_j})$ does not change if messages from $\mathcal{K}_{B_i, B_j}$ are processed by B_j; it only changes if a new message is appended. Now, assume that we require that $N(\Delta(T_{B_j}^{|B_i}, \mathcal{K}_{B_i, B_j}))$ is always a superset of $N(T_{B_i}^{\setminus B_j})$. This means that if $\Diamond \Box n \in N(T_{B_i}^{\setminus B_j})$, then $\Diamond \Box n \in N(T_{B_j}^{|B_i})$. Hence, monotone remote validity is implied. We have found the desired invariant. Now, we can define valid framework instantiations:

Definition 4.3. *An instance of the framework consisting of an implementation of* $\mathtt{administer}$ *and an initial routing configuration is valid if the following conditions hold:*

1. *(Progress) If called,* $\mathtt{administer}$ *eventually returns.*
2. *(Restricted Change)*

$$\Box \big[admin(B_i, B_j, \mathcal{S}, \mathcal{U}) \Rightarrow \forall D \neq B_j. \, (T'_{B_i})^{|D} = T_{B_i}^{|D} \big] \tag{4.16}$$

and

$$\Box \big[sub(Y, F) \vee unsub(Y, F) \Rightarrow \forall D \neq Y. \, (T'_{\theta(Y)})^{|D} = T_{\theta(Y)}^{|D} \big] \tag{4.17}$$

3. *(Restricted Impact)*

$$\Box \big[admin(B_i, B_j, \mathcal{S}, \mathcal{U}) \Rightarrow (T'_{B_i})^{|B_j} = \delta(T_{B_i}^{|B_j}) \big] \tag{4.18}$$

and

$$\Box \big[sub(Y, F) \vee unsub(Y, F) \Rightarrow (T'_{\theta(Y)})^{|Y} = \delta(T_{\theta(Y)}^{|Y}) \big] \tag{4.19}$$

4. *(Local Invariant)*
$$\Box \big[N(T_{\theta(Y)}^{|Y}) = N(S_Y) \big] \tag{4.20}$$

5. *(Remote Invariant)*

$$\Box \big[N(\Delta(T_{B_j}^{|B_i}, \mathcal{K}_{B_i, B_j})) \supseteq N(T_{B_i}^{\setminus B_j}) \big] \tag{4.21}$$

The individual properties of valid framework instantiations have the following informal meaning:

1. This property simply guarantees that $\mathtt{administer}$ terminates.

2. The restricted change property states that if the call of `administer` was triggered by a certain destination, only the part of the routing table regarding this destination can be affected.
3. The restricted impact property states that a change to a part of the routing table regarding a certain destination cannot be influenced by any part of the routing table dealing with other destinations.
4. The local invariant states that exactly those notifications should be delivered to a local client in which it is interested.
5. The remote invariant states that after B_j has processed all *admin* messages from B_i that are currently in transit, B_j will forward to B_i at least those notifications that B_i currently forwards to its other neighbors and local clients.

Next, we prove that a valid framework instantiation implies a monotone valid routing algorithm.

Lemma 4.3. *A valid framework instantiation implies a monotone valid routing algorithm.*

Proof. Local validity follows directly from the local invariant. It remains to be shown that also eventual monotone remote validity is implied. To prove this property assume that $\Diamond\square[n \in N(T_{B_i}^{\setminus B_j})]$. Then, the superset relation in the remote invariant implies that $\Diamond\square[n \in N(\Delta(T_{B_j}^{|B_i}, \mathcal{K}_{B_i, B_j}))]$. Termination, progress, restricted change and impact, and the reliable channel assumption then imply that $\Diamond\square[n \in N(T_{B_j}^{|B_i})]$, giving the desired property. $\square$

Theorem 4.2. *A valid framework instantiation implies a correct publish/subscribe system.*

Proof. By Lemmas 4.1 and 4.3. $\square$

4.5 Content-Based Routing Algorithms

In this section, a set of content-based routing algorithms is discussed. Each algorithm is given as an instance of the content-based routing framework presented in the previous section, i.e., as an instance of the `administer` procedure. The presentation follows a natural evolution in the development of routing algorithms from basic approaches to more advanced algorithms. We start with *flooding*. Then, we discuss in full detail *simple routing, identity-based routing, covering-based routing*, and *merging-based routing*. Using our framework theorems, the proof of correctness of these algorithms boils down to proving that the `administer` procedure is a valid framework instantiation.

```
procedure administer(Dest S, Set 𝒮, Set 𝒰)
begin
    T_B ← T_B ∪ {(F, S) | F ∈ 𝒮};
 4  T_B ← T_B \ {(F, S) | F ∈ 𝒰};
    return (∅, ∅);
end
```

Fig. 4.4. Flooding

4.5.1 Flooding

With flooding, the routing table of each broker B is initialized to the set $\{(F_T, U) \mid U \in N_B\}$ at system startup, where $F_T(n) = true$ for all $n \in \mathcal{N}$. Since $N(F_T) = \mathcal{N}$, this routing configuration implies that a broker forwards a notification received from a local client to all neighbors and a notification received from a neighbor to all other neighbors. Because the topology is acyclic and connected and since no messages are duplicated, flooding ensures that every notification is processed exactly once by every broker. Flooding is the only routing strategy that does not require the remote routing configuration to be updated. Therefore, the algorithm returns $(\emptyset, \emptyset)$ (Fig. 4.4, line 5) and no *admin* messages are exchanged. After the initialization, each broker solely adds and deletes routing entries regarding its local clients as they subscribe and unsubscribe:

- If a client Y subscribes to a filter F, the corresponding broker adds (F, Y) to its routing table (line 3).
- If a client Y unsubscribes to a filter F, the corresponding broker deletes (F, Y) from its routing table (line 4).

Correctness Proof

For flooding, we use the following initial state:

$$Init_F \stackrel{\text{def}}{=} T_B^0 = \{(F_T, H) \mid H \in N_B\} \wedge \mathcal{K}_{B_i, B_j}^0 = \langle\rangle$$
$$\wedge \ S_Y^0 = \emptyset \ \wedge \ P_X^0 = \emptyset \ \wedge \ D_Y^0 = \emptyset \tag{4.22}$$

Lemma 4.4 (Progress). *Each call of the flooding instantiation of* administer *returns.*

Proof. Obvious. □

Lemma 4.5. *Flooding satisfies the restricted change and the restricted impact property.*

Proof. An application of δ to T_B corresponds to a call of `administer`. In the `administer` procedure, the only code lines that manipulate routing entries are lines 3 and 4. These lines change only routing entries regarding the triggering destination S. Hence, the restricted change property holds. The above-mentioned lines do not take any routing entries regarding a destination distinct from the triggering destination into account. Hence, the restricted impact property holds, too. □

Lemma 4.6. $Init_F \Rightarrow \Box[T_{\theta(Y)}^{|Y} = S_Y]$

Proof. The property is shown by an induction. Initially, $(T_{\theta(Y)}^0)^{|Y} = S_Y^0$ holds due to $Init_F$, proving the base case. Now, assume that $T_{\theta(Y)}^{|Y} = S_Y$ holds and assume that the system executes a step. Only the $sub(Y, F)$ and $unsub(Y, F)$ steps change $T_{\theta(Y)}^{|Y}$ and S_Y because of Lemma 4.5. In the $sub(Y, F)$ case, $S_Y' = S_Y \cup \{F\}$ and $(T_{\theta(Y)}')^{|Y} = T_{\theta(Y)}^{|Y} \cup \{F\}$ (line 3) holds. In the $unsub(Y, F)$ case, $S_Y' = S_Y \setminus \{F\}$ and $(T_{\theta(Y)}')^{|Y} = T_{\theta(Y)}^{|Y} \setminus \{F\}$ holds (line 4). In both cases, $(T_{\theta(Y)}')^{|Y} = S_Y'$ is implied. This proves the induction step and concludes the proof. □

Lemma 4.7. $Init_F \Rightarrow \Box[T_{B_j}^{|B_i} = \{F_T\}]$

Proof. Due to $Init_F$, $(T_{B_j}^0)^{|B_i} = \{F_T\}$ holds initially. Due to $Init_F$ and because flooding always returns empty sets (line 5), $\mathcal{K}_{B_i,B_j} = \emptyset$ always holds. This implies that an $admin(B_j, B_i, S, \mathcal{U})$ step is never executed. Due to Lemma 4.5 this implies that $T_{B_j}^{|B_i}$ never changes. Hence, $T_{B_j}^{|B_i} = \{F_T\}$ always holds. □

Theorem 4.3. *Flooding is a valid routing algorithm.*

Proof. Lemma 4.4 and 4.5 imply the progress and the restricted change and impact property, respectively. Lemma 4.6 implies the local invariant because $T_{\theta(Y)}^{|Y} = S_Y$ implies $N(T_{\theta(Y)}^{|Y}) = N(S_Y)$. Lemma 4.7 implies the remote invariant because $T_{B_j}^{|B_i} = \{F_T\}$ implies that $N(T_{B_j}^{|B_i}) \supseteq N(T_{B_i}^{\setminus B_j})$. Hence, flooding is a valid framework instantiation which, by Theorem 4.2, yields a correct publish/subscribe system. □

4.5.2 Simple Routing

Simple routing uses filter forwarding to update the routing configuration in reaction to subscribing and unsubscribing clients: new and canceled subscriptions are flooded into the broker network such that they reach every broker. This allows the brokers to update their routing tables accordingly. Initially, the routing table T_B of each broker B is initialized to $\emptyset$. Simple routing assumes that each filter has a unique ID and that filters issued by different clients have disjoint sets of IDs. The filter ID is used to identify a filter when adding it to and deleting it from routing tables. The algorithm (Fig. 4.5) works as follows:

```
procedure administer(Dest S, Set S, Set U)
begin
    T_B ← T_B ∪ {(F,S) | F ∈ S};
 4  T_B ← T_B \ {(F,S) | F ∈ U};
    M_S ← {(F,H) | H ∈ N_B \ {S} ∧ F ∈ S};
    M_U ← {(F,H) | H ∈ N_B \ {S} ∧ F ∈ U};
    return (M_S, M_U);
end
```

Fig. 4.5. Simple routing

- The subscriptions in S are added to the routing table (line 3).
- The unsubscriptions in U are removed from the routing table (line 4).
- For each neighbor H except S, a tuple (F,H) is returned for each subscription F in S in the first returned set (line 5). Hence, each subscription is forwarded to all neighbors except S.
- For each neighbor H except S, a tuple (F,H) is returned for each unsubscription F in U in the second returned set (line 6). Hence, each unsubscription is forwarded to all neighbors except S.

This means that if $sub(Y,F)$ and $unsub(Y,F)$ are called, an $admin(\{F\}, \emptyset)$ and an $admin(\emptyset, \{F\})$ message are sent to all neighbors, respectively. The receipt of these messages causes the receiving broker to send the same message to its other neighbors. Hence, only these two types of $admin$ messages occur with simple routing, and administer is either called with $S = \{F\}$ and $U = \emptyset$ or with $S = \emptyset$ and $U = \{F\}$.

Example

Figure 4.6 shows an example using simple routing. X_1 subscribes to F. Then, B_1 inserts (F, X_1) into its routing table and sends messages $admin(\{F\}, \emptyset)$ to B_2 and B_3. On receipt of this message, B_2 and B_3 insert (F, B_1) into their routing table.

Correctness Proof

For better readability, we use the following abbreviations for all subsequent correctness proofs:

$$\alpha = T_{B_i}^{\backslash B_j} \tag{4.23}$$

$$\beta = \Delta(T_{B_j}^{|B_i}, \mathcal{K}_{B_i,B_j}) \tag{4.24}$$

For simple routing and all other subsequently discussed routing algorithms, we use the following initial state:

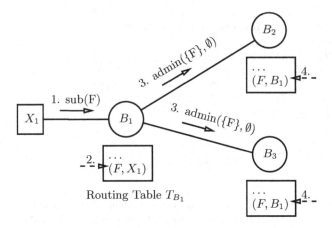

Fig. 4.6. Diagram explaining simple routing (new subscription)

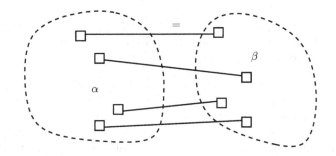

Fig. 4.7. Relation among α and β for simple routing

$$Init_E \overset{\text{def}}{=} T_B^0 = \emptyset \wedge \mathcal{K}_{B_i,B_j}^0 = \langle\rangle \wedge S_Y^0 = \emptyset \wedge P_X^0 = \emptyset \wedge D_Y^0 = \emptyset \quad (4.25)$$

The idea underlying the correctness proof is the following invariant: α is always equal to β (Fig. 4.7). This invariant is proved in the next lemma:

Lemma 4.8. *Simple routing satisfies the progress, restricted change, and the restricted impact property as well as the local invariant.*

Proof. Can be proved in the same way as in Lemmas 4.4, 4.5, and 4.6.

Lemma 4.9. *$Init_E \Rightarrow \square[\beta = \alpha]$*

Proof. This property is proved by an induction. Due to $Init_E$, initially $\beta = \emptyset$ and $\alpha = \emptyset$, proving the base case. Now, assume that $\beta = \alpha$. We have to show that $\beta' = \alpha'$ after the execution of an arbitrary step of the algorithm to prove the induction.

As a result of the restricted change and impact properties, we have to consider only four cases here: (1) $sub(Y, F)$ for a local client of B_i, (2) $unsub(Y, F)$ for a local client of B_i, (3) $admin(B_i, H, S, U)$ for a broker $H \in N_{B_i} \setminus \{B_j\}$, and (4) $admin(B_j, B_i, S, U)$. The steps 1–3 potentially change α and β, while step 4 could only but actually does not affect β.

Case (1) $sub(Y, F)$: According to simple routing, $\alpha' = \alpha \cup \{F\}$ (line 3) and $\mathcal{K}'_{B_i, B_j} = append(\mathcal{K}_{B_i, B_j}, (\{F\}, \emptyset))$ (line 5). The latter implies that $\beta' = \delta(\beta, (\{F\}, \emptyset))$. According to the induction assumption, this equals $\delta(\alpha, (\{F\}, \emptyset))$. According to simple routing this equals $\alpha \cup \{F\}$. Hence, β' equals α' giving the desired property.

Case (2) $unsub(Y, F)$: This case is analogous to case (1) except that $\alpha' = \alpha \setminus \{F\}$ (line 4) and $\mathcal{K}'_{B_i, B_j} = append(\mathcal{K}_{B_i, B_j}, (\emptyset, \{F\}))$ (line 6).

Case (3) $admin(B_i, H, S, U)$: Here, we must distinguish two cases: (3.1) $H = B_j$ and (3.2) $H \neq B_j$.
Case (3.1) $H = B_j$: The restricted change property implies that $\alpha' = \alpha$. Simple routing (lines 5+6) implies that $\mathcal{K}'_{B_i, B_j} = \mathcal{K}_{B_i, B_j}$ because no $admin$ message is passed back to the sender. Hence, the property holds.
Case (3.2) $H \neq B_j$: We must consider two cases: (3.2.1) $S = \{F\} \wedge U = \emptyset$ and (3.2.2) $S = \emptyset \wedge U = \{F\}$. In the former case, the same proof as in case 1 can be applied. In the latter case, the same proof as in case 2 can be applied. Hence, the desired property holds in both cases.

Case (4) $admin(B_j, B_i, S, U)$: This implies that $\mathcal{K}'_{B_i, B_j} = tail(\mathcal{K}_{B_i, B_j})$ and that $\beta' = \delta(\beta, head(\mathcal{K}_{B_i, B_j}))$. According to the definition of δ, this implies that $\beta' = \beta$. Hence, the desired property holds.

This finishes the case distinction and proves the induction step.   □

Theorem 4.4. *Simple routing is a valid routing algorithm.*

Proof. The progress, the restricted change, and the restricted impact property, as well as the local invariant hold due to Lemma 4.8. Since $\beta = \alpha$ implies that $N(\Delta(T_{B_j}^{|B_i}, \mathcal{K}_{B_i, B_j})) \supseteq N(T_{B_i}^{\setminus B_j})$, Lemma 4.9 implies the remote invariant. Hence, simple routing is a valid framework instantiation. By Theorem 4.2 this yields a correct publish/subscribe system.   □

4.5.3 Identity-Based Routing

We now begin to present routing algorithms that avoid global knowledge by taking *similarities* among the subscriptions into account. These algorithms are based on the following idea: The set of notifications that a broker B_j forwards to a broker B_i, i.e.,$N(T_{B_j}^{|B_i})$, is the set of all notifications that are matched by any routing entry (F, B_i) in T_{B_j}. In general, a subset of these routing entries might be sufficient to determine $N(T_{B_j}^{|B_i})$. For example, there can be two routing entries (F, B_i) and (G, B_i) with $N(F) = N(G)$. Clearly, one of these entries is sufficient as both have identical sets of matching notifications. This fact is used by the identity-based routing algorithm to avoid redundant routing entries and unnecessary forwarding of subscriptions and unsubscriptions. The basic idea of identity-based routing is the following:

- A subscription (unsubscription) is only forwarded to a neighbor U if there is no identical subscription in the routing table for a destination distinct from U. This test is evaluated before (after) the subscription (unsubscription) is added (removed) to (from) the routing table.

Formally, two filters F and G are *identical*, denoted by $F \equiv G$, if $N(F) = N(G)$. We define the set $C_B^I(F, D)$ (the superscript I stands for "identity") to be the set of all routing entries in T_B of which the filter is identical to a given filter F and of which the destination equals a given destination D. Moreover, we denote with $D_B^I(F)$ the set of all neighbors H for which there is no routing entry (G, D) in T_B, where G is identical to F and D is distinct from H:

$$C_B^I(F, D) \stackrel{\text{def}}{=} \{(G, D) \mid (G, D) \in T_B \ \wedge \ F \equiv G\}, \qquad (4.26)$$

$$D_B^I(F) \stackrel{\text{def}}{=} \{H \in N_B \mid \nexists G \in T_B^{\backslash H}. F \equiv G\}. \qquad (4.27)$$

We now describe identity-based routing (Fig. 4.8). If a broker B receives a subscription or unsubscription F from a neighbor or a local client S, it does the following:

1. B updates its routing table (lines 6–11):
   - If S is a neighbor, B removes all routing entries whose filters are identical to F and that refer to the destination S, i.e., $C_B^I(F, S)$ (line 8).
   - If S is a local client, B removes solely (F, S) (line 10).
2. B forwards F to all neighbors that are in $D_B^I(F)$ except S (lines 15/17 and 22).
3. If F is a subscription, B inserts a routing entry (F, S) into its routing table (line 18).

Examples

In Fig. 4.9, broker B_1 receives a new subscription F from a neighbor S. B_1 inserts (F, S) into its routing table and forwards F to its neighbors B_2 and B_3 because they are both in $D_B^I(F) \setminus \{S\}$.

In Fig. 4.10, broker B_1 receives a new subscription F from a local client S. Here, B_1 also inserts (F, S) into its routing table but forwards F only to its neighbor B_3, which is the only neighbor in $D_B^I(F) \setminus \{S\}$. B_2 is not in that set due to the routing entry (F', B_3), where $F' \equiv F$.

```
  procedure administer(Dest S, Set S, Set U)
2 begin
     M_S ← ∅;
     M_U ← ∅;

     forall F ∈ S ∪ U do
7      if S ∈ N_B then
           T_B ← T_B \ C_B^I(F, S);
       else
           T_B ← T_B \ {(F, S)};
       endif
12
       A ← {(F, H) | H ∈ D_B^I(F) \ {S}};
       if F ∈ U then
           M_U ← M_U ∪ A;
       else
17          M_S ← M_S ∪ A;
           T_B ← T_B ∪ {(F, S)};
       endif
     endforall

22   return (M_S, M_U);
  end
```

Fig. 4.8. Identity-based routing

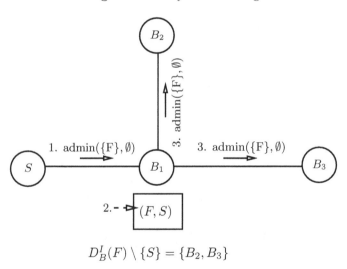

$$D_B^I(F) \setminus \{S\} = \{B_2, B_3\}$$

Fig. 4.9. Identity-based routing: Processing a new subscription from a neighbor

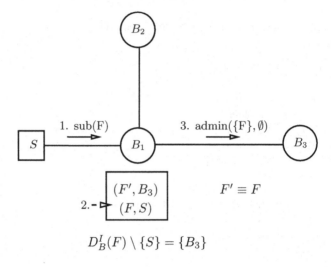

$$D^I_B(F) \setminus \{S\} = \{B_3\}$$

Fig. 4.10. Identity-based routing: Processing a new subscription from a client

Correctness Proof

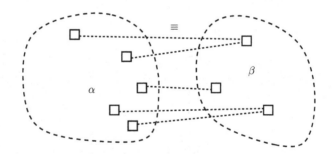

Fig. 4.11. Relation among α and β for identity-based routing

In the following, the correctness of identity-based routing is proved. The idea underlying the correctness proof is the following invariant: For every filter in α there is a filter in β that is identical to the former (Fig. 4.11). This invariant is stated in the next lemma. The proof is a lengthy case distinction in the spirit of and similar to the proof of Lemma 4.9.

Lemma 4.10. $Init_E \Rightarrow \Box[\forall F \in \alpha. \exists G \in \beta. G \equiv F]$

Proof. This property is proved by an induction. Due to $Init_E$, initially α and β are empty. Hence, the property is satisfied. This proves the base case. To prove the induction step assume that the property holds for α and β. We have to show that the property also holds after an arbitrary step was executed, i.e., for α' and β'. Again, we have to consider four cases here:

Case (1) $sub(Y, F)$: According to identity-based routing, $\alpha' = \alpha \cup \{F\}$ (line 18). Now, we must distinguish two cases (line 13): (1.1) $B_j \in D_{B_i}^I(F)$ and (1.2) $B_j \notin D_{B_i}^I(F)$.

Case (1.1) $B_j \in D_{B_i}^I(F)$: This implies $\mathcal{K}'_{B_i,B_j} = append(\mathcal{K}_{B_i,B_j}, (F, \emptyset))$. Hence, $\beta' = \delta(\beta, (\{F\}, \emptyset)) = \beta \cup \{F\}$. Therefore, the desired property holds.

Case (1.2) $B_j \notin D_{B_i}^I(F)$: This implies $\mathcal{K}'_{B_i,B_j} = \mathcal{K}_{B_i,B_j}$. Hence, $\beta' = \beta$. But this also implies that there is a filter $G \in \alpha$ with $G \equiv F$ (lines 10+13). According to the induction assumption this implies that there is also a filter $H \in \beta$, where $H \equiv F$. Hence, the desired property holds.

Case (2) $unsub(Y, F)$: According to identity-based routing, $\alpha' = \alpha \setminus \{F\}$ (line 10). Now, we must distinguish two cases (line 13): (2.1) $B_j \in D_{B_i}^I(F)$ and (2.2) $B_j \notin D_{B_i}^I(F)$.

Case (2.1) $B_j \in D_{B_i}^I(F)$: This implies $\mathcal{K}'_{B_i,B_j} = append(\mathcal{K}_{B_i,B_j}, (\emptyset, F))$. Hence, $\beta' = \delta(\beta, (\emptyset, \{F\})) = \beta \setminus \{F\}$. The case assumption also implies that there is no filter $G \in \alpha'$ with $G \equiv F$. Hence, the desired property holds.

Case (2.2) $B_j \notin D_{B_i}^I(F)$: This implies $\mathcal{K}'_{B_i,B_j} = \mathcal{K}_{B_i,B_j}$. Hence, $\beta' = \beta$. Hence, the desired property holds.

Case (3) $admin(B_i, H, \mathcal{S}, \mathcal{U})$: Here, we must distinguish two cases: (3.1) $H = B_j$ and (3.2) $H \neq B_j$.

Case (3.1) $H = B_j$: The restricted change property implies that $\alpha' = \alpha$. Identity-based routing implies that $\mathcal{K}'_{B_i,B_j} = \mathcal{K}_{B_i,B_j}$ because no $admin$ message is passed back to the sender. Hence, the property holds.

Case (3.2) $H \neq B_j$: We must consider two cases: (3.2.1) $\mathcal{S} = \{F\} \wedge \mathcal{U} = \emptyset$ and (3.2.2) $\mathcal{S} = \emptyset \wedge \mathcal{U} = \{F\}$. In the former case, the same proof as in case 1 can be applied. In the latter case, the same proof as in case 2 can be applied. Hence, the desired property holds in both cases.

Case (4) $admin(B_j, B_i, \mathcal{S}, \mathcal{U})$: This implies that $\mathcal{K}'_{B_i,B_j} = tail(\mathcal{K}_{B_i,B_j})$ and that $\beta' = \delta(\beta, head(\mathcal{K}_{B_i,B_j}))$. According to the definition of δ, this implies that $\beta' = \beta$. Hence, the desired property holds.

This finishes the case distinction and proves the induction step. Hence, the validity of the induction is implied concluding the proof. $\square$

Theorem 4.5. *Identity-based routing is a valid routing algorithm.*

Proof. The progress, the restricted change, and the restricted impact property, as well as the local invariant can be proved in the same way as in Lemmas 4.4, 4.5, and 4.6. Lemma 4.10 implies that the remote variant holds because $\forall F \in \alpha. \exists G \in \beta. G \equiv F$ implies that $N(\Delta(T_{B_j}^{|B_i}, \mathcal{K}_{B_i,B_j})) \supseteq N(T_{B_i}^{\setminus B_j})$. Hence, identity-based routing is a valid framework instantiation that (following Theorem 4.2) yields a correct publish/subscribe system. $\square$

```
   procedure administer(Dest S, Set S, Set U)
 2 begin
     M_S ← ∅;
     M_U ← ∅;
     P ← ∅;

 7 if U = ∅
     // handle subscriptions
     forall F ∈ S do
       if S ∈ N_B then
         T_B ← T_B \ C_B^L(F, S);
12     else
         T_B ← T_B \ {(F, S)};
       endif
       M_S ← M_S ∪ {(F, H) | H ∈ D_B^U(F) \ {S}};
       T_B ← T_B ∪ {(F, S)};
17     endforall
   else
     // handle unsubscriptions
     forall F ∈ U do
       if S ∈ N_B then
22       T_B ← T_B \ C_B^L(F, S);
       else
         T_B ← T_B \ {(F, S)};
       endif
       M_U ← M_U ∪ {(F, H) | H ∈ D_B^U(F) \ {S}};
27     P ← P ∪ (C_B^L(F) \ C_B^I(F));
     endforall

     // handle uncovered subscriptions
     T_B ← T_B ∪ {(F, S) | F ∈ S};
32   P ← P ∪ {(F, S) | F ∈ S};
     forall (F, U) ∈ P do
       k ← |{H | (G, H) ∈ P  ∧  G ≡ F}|;
       P ← P \ {(G, H) | (G, H) ∈ P  ∧  G ≡ F};

37     A ← D_B^{PU}(F) \ {S};
       if k = 1 then
         A ← A \ {U};
       endif
       M_S ← M_S ∪ {(F, H) | H ∈ A};
42     endforall
   endif
   return (M_S, M_U);
 end
```

Fig. 4.12. Covering-based routing

4.5.4 Covering-Based Routing

After discussing identity-based routing, an obvious idea is to exploit more complex similarities among subscriptions. The next step is to take advantage of covering among filters, a concept that was first mentioned in the area of notification services by Carzaniga [65]. A filter *covers* another filter if the former matches all notifications the latter matches. Therefore, a routing entry (F, U) is obsolete if there exists a routing entry (G, U), where G covers F. This fact is used by the covering-based routing algorithm to further reduce redundant routing entries and unnecessary forwarding of subscriptions and unsubscriptions. The basic idea of covering-based routing is the following:

- A subscription (unsubscription) is only forwarded to a neighbor U if there is no covering subscription in the routing table for a destination distinct from U. This test is evaluated before (after) the subscription (unsubscription) is added (removed) to (from) the routing table.
- A broker receiving a subscription deletes all routing entries whose filters are covered by the new subscription that refer to the same destination. This is done to get rid of the obsolete routing entries.
- If an unsubscription is forwarded to a neighbor, the sending broker also forwards a possibly empty subset of subscriptions in the same *admin* message to ensure the delivery of all needed notifications.

Formally, a filter F *covers* a filter G, denoted by $F \sqsupseteq G$ iff $N(F) \supseteq N(G)$. F is a *proper cover* of G, denoted by $F \sqsupset G$, iff $N(F) \supset N(G)$. We define the set $C_B^L(F)$ (the L stands for "lower") to comprise the set of all routing entries in the routing table of a broker B that are covered by a given filter F. We also define $C_B^L(F, D)$ as the restriction of $C_B^L(F)$ to a given destination D. Additionally, we denote with $D_B^U(F)$ (the U stands for "upper") as the set of all neighbors H for which no routing entry (G, D) in the routing table of B exists, where G covers F and D is distinct from H. With $D_B^{PU}(F)$ (the PU stands for "proper upper") the set of all neighbors H for which no routing entry (G, D) in the routing table of B exists, where G is a proper cover of F and D is distinct from H:

$$C_B^L(F) \stackrel{\text{def}}{=} \{(G, U) \in T_B \mid F \sqsupseteq G\} \tag{4.28}$$

$$C_B^L(F, D) \stackrel{\text{def}}{=} \{(G, D) \in C_B^L(F)\} \tag{4.29}$$

$$D_B^U(F) \stackrel{\text{def}}{=} \{H \in N_B \mid \nexists G \in T_B^{\backslash H} . G \sqsupseteq F\} \tag{4.30}$$

$$D_B^{PU}(F) \stackrel{\text{def}}{=} \{H \in N_B \mid \nexists G \in T_B^{\backslash H} . G \sqsupset F\} \tag{4.31}$$

Covering-based routing either processes (1) a single subscription or (2) a single unsubscription that comes along with a set of uncovered subscriptions (Fig. 4.12). These cases are described in the following.

Processing of a Subscription

If a broker B receives a new subscription F from a neighbor or a local client S, B first updates its routing table: If S is a neighbor, B removes all entries whose filters are covered by F that refer to S, i.e., $C_B^L(F, S)$, to get rid of the obsolete routing entries (line 11). If S is a local client, B removes solely (F, S) (line 13). Next, B forwards F to all neighbors which are in $D_B^U(F)$ except S (line 15). Finally, B inserts (F, S) into its routing table (line 16).

Processing of an Unsubscription

The fact that complicates covering-based routing is that to forward an unsubscription F to some neighbors is not sufficient. Instead, to each neighbor to which F is forwarded, also a possibly empty subset of filters which are properly covered by F has to be forwarded. These subscriptions are called *uncovered* subscriptions. Without forwarding these subscriptions, it is not ensured that the receiving broker forwards all notifications matching these subscriptions. This is because the receiving broker has either not ever received these subscriptions or they have been dropped when F arrived. It is important that the unsubscription and the corresponding uncovered subscriptions are forwarded in a single message in order to guarantee that the change to the routing table of the receiving broker is atomic. Otherwise, in the intermediate time between the cancellation of the unsubscription and the time at which the uncovered subscriptions become effective, notifications may be lost.

The basic processing of an unsubscription (lines 20–28) is similar to the handling of a subscription. First, the routing table is updated (lines 21–25) and the destinations to which the unsubscription is forwarded are determined (line 26), as described above. Finally, all routing entries in $C_B^L(F) \setminus C_B^I(F)$ are added to a temporary storage $\mathcal{P}$ (line 27). These routing entries are potentially newly uncovered subscriptions because their filter is properly covered by F.

Processing of Uncovered Subscriptions

First, all old uncovered subscriptions in $\mathcal{S}$ received from S are added to the routing table (line 31), and the routing entries representing these subscriptions are added to $\mathcal{P}$ (line 32). Now, it is determined which subscriptions represented in $\mathcal{P}$ have to be forwarded to which destinations (lines 33–42): For each entry $(F, U) \in \mathcal{P}$, F is forwarded to neighbor H if H is in $D^{PU}(F) \setminus \{S\}$ (line 37). However, F is only forwarded to U if additionally there is a second routing entry (G, I) in $\mathcal{P}$, where $G \equiv F$ and $I \neq U$ (lines 34+38–40). This is the case if $k \neq 1$. To ensure that identical subscriptions are only forwarded once, all entries whose filters are identical to F are removed from $\mathcal{P}$ (line 35).

This approach ensures that (a) all subscriptions in the set of uncovered subscriptions are covered by the handled unsubscription and that (b) in this set there are no two subscriptions, where one covers the other. Hence, an unsubscription that is received from a neighbor comes along with a possibly empty set of uncovered subscriptions and may generate new uncovered subscriptions. To every neighbor to which the handled unsubscription is forwarded, a possibly empty subset of these two sets is forwarded that also satisfies the requirements stated above.

Examples

In Fig. 4.13, B_1 receives a new subscription F from a local client S. Therefore, B_1 adds (F, S) to its routing table. Moreover, B_1 forwards F only to its neighbor B_3 because B_3 is the only neighbor in $D_B^U(F) \setminus \{S\}$. B_2 is not in this set because of the routing entry (G, B_3), where $G \sqsupseteq F$.

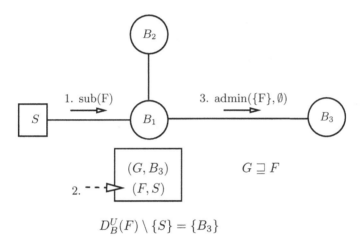

$$D_B^U(F) \setminus \{S\} = \{B_3\}$$

Fig. 4.13. Covering-based routing: Processing of a new subscription from a client

In the next example (Fig. 4.14), B_1 receives a subscription F from a neighbor S. B_1 removes the entry (G, S) from its routing table because the entry is in $C_B^L(F, S)$. Moreover, B_1 inserts (F, S) into its routing table. Finally, B_1 forwards F to its neighbors B_2 and B_3 because they are both in $D_B^U(F) \setminus \{S\}$.

In Fig. 4.15, broker B_1 receives an unsubscription F from a neighbor S. Hence, B_1 removes (F, S). Furthermore, B_1 forwards the unsubscription to its neighbors B_2 and B_3 as both are in $D_B^U(F) \setminus \{S\}$.

In the next example (Fig. 4.16) B_1 receives an unsubscription F from a local client S. Hence, B_1 removes (F, S). In this case, B_1 forwards the unsubscription only to B_3 because it is the only broker in $D_B^U(F) \setminus \{S\}$. B_2 is not in this set because of the routing entry (G, B_3), where $G \sqsupseteq F$.

In Fig. 4.17, broker B_1 receives an unsubscription F from a local client S. Hence, it removes (F, S) from its routing table. In this example the unsubscription F uncovers a subscription G. While the subscription F is forwarded

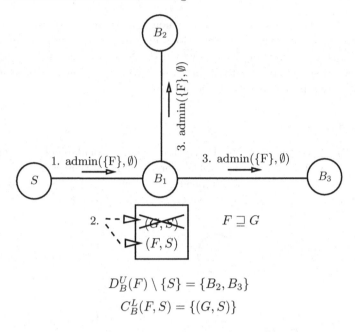

Fig. 4.14. Covering-based routing: Processing of a new subscription from a neighbor

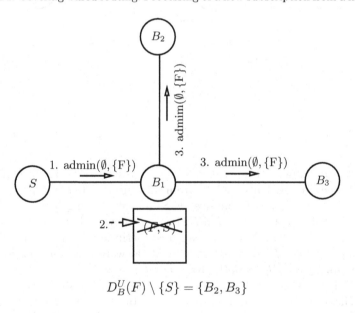

Fig. 4.15. Covering-based routing: Processing of an unsubscription from a neighbor

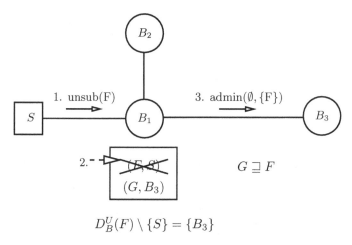

$$D_B^U(F) \setminus \{S\} = \{B_3\}$$

Fig. 4.16. Covering-based routing: Processing of an unsubscription from a client

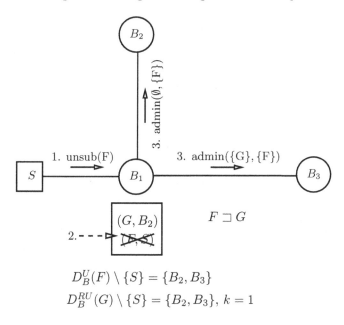

$$D_B^U(F) \setminus \{S\} = \{B_2, B_3\}$$
$$D_B^{RU}(G) \setminus \{S\} = \{B_2, B_3\}, \, k = 1$$

Fig. 4.17. Covering-based routing: Processing of an unsubscription from a client

to B_2 and B_3, the uncovered subscription G is solely forwarded to B_3. G is not forwarded to B_2, although it is in $D_B^{RU}(G) \setminus \{S\}$ because $k = 1$.

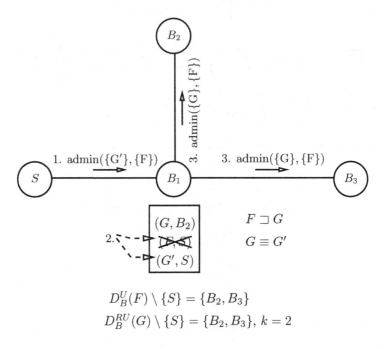

$$D_B^U(F) \setminus \{S\} = \{B_2, B_3\}$$
$$D_B^{RU}(G) \setminus \{S\} = \{B_2, B_3\}, \, k = 2$$

Fig. 4.18. Covering-based routing: Processing of an unsubscription from a neighbor, example 2

In the last example (Fig. 4.18), broker B_1 receives an unsubscription F that comes along with an uncovered subscription G'. Moreover, in the routing table of B_1 there is an entry (G, S), where $G \equiv G'$. Here, B_1 removes (F, S) from and inserts (G', S) into its routing table. The unsubscription F and the uncovered subscription G are sent to B_2 *and* B_3. G is forwarded to B_2 and B_3 because they are both in $D_B^{RU}(G) \setminus \{S\}$ and additionally $k = 2$ holds.

Correctness Proof

The idea underlying the correctness proof is the following invariant: For every filter in α there is a filter in β that covers the former (Fig. 4.19). This invariant is stated in the next lemma. The proof is a lengthy case distinction in the spirit of and similar to the proof of Lemma 4.9.

Lemma 4.11. $\mathit{Init}_E \Rightarrow \square[\forall F \in \alpha. \exists G \in \beta. G \sqsupseteq F]$

Proof. Proof: This property is proved by an induction. Due to Init_E, initially α and β are empty. Hence, the property is satisfied. This proves the base

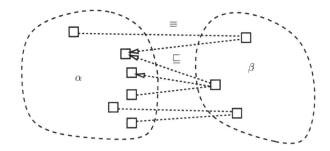

Fig. 4.19. Relation among α and β for covering-based routing

case. To prove the induction step assume that the property holds for α and β. We have to show that the property also holds after an arbitrary step was executed, i.e., for α' and β'. Again, we have to consider four cases here:

Case (1) $sub(Y, F)$: According to covering-based routing, $\alpha' = \alpha \cup \{F\}$ (line 16). Now, we must distinguish two cases (line 15): (1.1) $B_j \in D^U_{B_i}(F)$ and (1.2) $B_j \notin D^U_{B_i}(F)$.

Case (1.1) $B_j \in D^U_{B_i}(F)$: This implies $\mathcal{K}'_{B_i,B_j} = append(\mathcal{K}_{B_i,B_j}, (\{F\}, \emptyset))$. Hence, $\beta' = \delta(\beta, (\{F\}, \emptyset)) = \beta \cup \{F\}$. Therefore, the desired property holds.

Case (1.2) $B_j \notin D^U_{B_i}(F)$: This implies that $\mathcal{K}'_{B_i,B_j} = \mathcal{K}_{B_i,B_j}$. Hence, $\beta' = \beta$. But this also implies that there is a filter $G \in \alpha$ with $G \sqsupseteq F$. According to the induction assumption this implies that there is also a filter $H \in \beta$, where $H \sqsupseteq F$. Hence, the desired property holds.

Case (2) $unsub(Y, F)$: According to covering-based routing, $\alpha' = \alpha \setminus \{F\}$ (line 24). Now, we must distinguish two cases (line 26): (2.1) $B_j \in D^U_{B_i}(F)$ and (2.2) $B_j \notin D^U_{B_i}(F)$.

Case (2.1) $B_j \in D^U_{B_i}(F)$:
Here, $\mathcal{K}'_{B_i,B_j} = append(\mathcal{K}_{B_i,B_j}, (\{F_1, \ldots, F_n\}, \{F\}))$, where $\forall F_i. F \sqsupseteq F_i$. Hence, $\beta' = \delta(\beta, ((\{F_1, \ldots, F_n\}, F), \{F\})) = \beta \setminus \{F\} \cup \{F_1, \ldots, F_n\}$. The case assumption also implies that there is no filter in α' which is identical to F. But there may be filters in α' for which F was the only properly covering filter. These filters are included in those routing entries that are stored in P (line 27). Line 31 does not change P because $\mathcal{S} = \emptyset$. A filter G represented in P (as P is at line 33) is forwarded to B_j if $B_j \in D^{PU}_{B_i}(F)$ and if G originates from a destination different from B_j. This implies that there is no filter in α' that covers G and that F was the only filter in α that covers G. The filters satisfying the same conditions as G are exactly those filters $\{F_1, \ldots, F_n\}$ introduced above. Note that identical filters are only forwarded once to a destination due to line 35. Hence, the desired property holds.

Case (2.2) $B_j \notin D^U_{B_i}(F)$: This implies that $\mathcal{K}'_{B_i,B_j} = \mathcal{K}_{B_i,B_j}$. Hence, $\beta' = \beta$. So the desired property holds.

Case (3) $admin(B_i, H, \mathcal{S}, \mathcal{U})$: Here, we must distinguish two cases: (3.1) $H = B_j$ and (3.2) $H \neq B_j$.

Case (3.1) $H = B_j$: The restricted change property implies that $\alpha' = \alpha$. Covering-based routing implies that $\mathcal{K}'_{B_i,B_j} = \mathcal{K}_{B_i,B_j}$ because no *admin* message is passed back to the sender. Hence, the property holds.

Case (3.2) $H \neq B_j$: We must consider two cases: (3.2.1) $\mathcal{S} = \{F\} \wedge \mathcal{U} = \emptyset$ and (3.2.2) $\mathcal{S} = \{F_1, \ldots, F_n\} \wedge \mathcal{U} = \{F\}$, where $\forall F_i.\ F \sqsupseteq F_i$.

Case (3.2.1) $\mathcal{S} = \{F\} \wedge \mathcal{U} = \emptyset$: In this case, the same proof as in case 1 can be applied.

Case (3.2.2) $\mathcal{S} = \{F_1, \ldots, F_n\} \wedge \mathcal{U} = \{F\}$, where $\forall F_i.\ F \sqsupseteq F_i$: The proof here is similar to those of case 2 except that $\alpha' = \alpha \setminus \{F\} \cup \{F_1, \ldots, F_n\}$ (lines 16 + 22).

Case (4) $admin(B_j, B_i, \mathcal{S}, \mathcal{U})$: This implies that $\mathcal{K}'_{B_i,B_j} = tail(\mathcal{K}_{B_i,B_j})$ and that $\beta' = \delta(\beta, head(\mathcal{K}_{B_i,B_j}))$. According to the definition of δ, this implies that $\beta' = \beta$. So the desired property holds.

This finishes the case distinction and proves the induction step. Hence, the validity of the induction is implied, concluding the proof. $\square$

Theorem 4.6. *Covering-based routing is a valid routing algorithm.*

Proof. The progress, the restricted change, and the restricted impact property, as well as the local invariant can be proved in the same way as in Lemmas 4.4, 4.5, and 4.6. Lemma 4.11 implies that the remote variant holds because $\forall F \in \alpha.\, \exists G \in \beta.\, G \sqsupseteq F$ implies that $N(\Delta(T_{B_j}^{|B_i}, \mathcal{K}_{B_i,B_j})) \supseteq N(T_{B_i}^{\setminus B_j})$. Hence, covering-based routing is a valid framework instantiation and, by Theorem 4.2, yields a correct publish/subscribe system. $\square$

4.5.5 Merging-Based Routing

Merging-based routing is a whole class of routing algorithms rather than a single routing algorithm. It is based on creating new, broader filters, called mergers, from existing filters. These mergers are then forwarded instead of the original filters. In the following, a concrete merging-based routing algorithm is presented. It is implemented on top of covering-based routing and allows every broker solely to merge routing entries that refer to the *same* destination. This keeps the algorithm simple enough to be applied in a dynamic publish/subscribe system. The algorithm presented by Handurukande et al. [186] can also be seen as a variant of merging-based routing.

Formally, a filter F is a *merger* (or covers) a set of filters $\mathcal{F} = \{F_1, \ldots, F_n\}$, iff $N(F) \supseteq N(\mathcal{F})$. F is a *perfect merger* if the equality holds and an *imperfect merger*, otherwise. In order to enable filter merging as sketched above, a broker can replace a set of routing entries $\{(F_1, D), \ldots, (F_n, D)\}$ with the *same* destination D by a single merged entry (F, D) if F is a merger of $\{F_1, \ldots, F_n\}$. The merged routing entries are removed from the routing table, and (F, D) is added to the routing table instead. If F is a perfect merger this does not affect the set of notifications that B is forwarding to D, i.e., $N(T_B^{|D})$. Otherwise,

$N(T_B^{|D})$ might increase. This might violate the safety condition if D is a local client. If D is a neighbor broker, imperfect merging can be applied trading routing tables sizes against network bandwidth. We assume perfect merging for the sake of simplicity in the following. Imperfect merging algorithms, especially those that are adaptive, are subject to future research.

```
    procedure administer(Dest S, Set S, Set U)
    begin
      if U = ∅ then
         S ← handlesubs(S, S);
 5    else
         (S, U) ← handleunsubs(S, S, U);
      endif

      S ← prune_co(S);
10    U ← prune_co(U);

      return administer_co(S, S, U);
    end
```

Fig. 4.20. Merging-based routing

Now an exemplary routing algorithm based on merging (Fig. 4.20) is described in full detail. The algorithm stores what filters a merger is constituted of in case the merger has to be canceled. The set of filters that constitute a merger M is given by $c(M)$. Note that whether or not a filter is a merger can only be detected at the broker that generated the merger. The set of all mergers of a broker regarding a destination D is denoted by $\mathcal{M}_B^{|D}$.

```
    procedure prune_co(Set A)
 2  begin
      forall F ∈ A do
         A ← A \ {G ∈ A | G ≠ F ∧ F ⊒ G};
      endforall
      return A;
 7  end
```

Fig. 4.21. Merging: deletion of covering filters

The merging-based algorithm works on top of covering-based routing (cf. line 12). Therefore, the calls of **administer** triggered by the former algorithms have to be compatible with the latter one. Our algorithm either sends a single subscription or a set of unsubscriptions accompanied by a set of covered subscriptions. The algorithms determines which message type is processed by checking whether $\mathcal{U} = \emptyset$ (line 3). Depending on the result, either the procedure *handlesubs* (line 4) or the procedure *handleunsubs* (line 6) is called. These procedures are described in the next two subsections. After, the called procedure returned, the updated sets $\mathcal{S}$ and $\mathcal{U}$ are pruned (lines 9–10) by calling the procedure $prune_{co}$ (Fig. 4.21). This procedure removes from both sets those filters which are covered by another filter of the respective set. Finally, the pruned sets $\mathcal{S}$ and $\mathcal{U}$ are used as input to the covering-based routing algorithm (line 12).

```
    procedure getcoveringmerger(Filter F, Dest D)
    begin
3       forall M ∈ M|D_B do
            if M ⊒ F then
                return M;
            endif
        endforall
8       return ∅;
    end
```

Fig. 4.22. Merging: searching for a covering merger

Processing of a Subscription

Every time, a new subscription is received, the following (Fig. 4.23) is done by the *handlesubs* procedure:

- If $S \in N_B$, those filters and mergers regarding this neighbor that are covered by the new subscription are removed from the routing table (lines 3–7).
- After that, it is checked whether the new subscription is covered by any existing merger regarding the same destination. This is done by calling the *getcoveringmerger* (Fig. 4.22). If a covering merger is found, the new subscription is added to one of these mergers and is removed from $\mathcal{S}$ (lines 10–14).
- If the new subscription is not covered by any existing merger, it is checked whether an existing merger regarding the same destination can be extended to include the new subscription. If this succeeds, the merger is updated and added to $\mathcal{S}$, and the new subscription is removed from $\mathcal{S}$ (lines 15–18).
- If the new subscription could also not be used to extend an existing merger, it is tried to generate a new merger from the new subscription and existing filters (which are not mergers) regarding the same destination. If a new

```
 1 procedure handlesubs(Dest S, Set S)
   begin
     if  S ∈ N_B  then
       forall  F ∈ S  do
         T_B ← T_B \ C_B^L(F, S);
 6     endforall
     endif

     forall  F ∈ S  do
       M ← getcoveringmerger(F, S);
11     if  M ≠ ∅  then
         c(M) ← c(M) ∪ {F};
         S ← S \ {F};
       else
         M ← tryadd(F, S);
16       if  M ≠ ∅  then
           S ← S \ {F} ∪ {M};
         else
           M ← trynew(F, S);
           if  M ≠ ∅  then
21           S ← S \ {F} ∪ {M};
             T_B ← T_B \ {(G, S) | G ∈ c(M)};
           endif
         endif
         if  M ≠ ∅  then
26         forall  G ∈ {H | (H, S) ∈ C_B^L(M, S)}  do
             if  G ∈ M_B^{|S}  then
               c(M) ← c(M) ∪ c(G);
             else
               c(M) ← c(M) ∪ {G};
31           endif
             T_B ← T_B \ (G, S);
           endforall
           if  S ∈ N_B  then
             forall  G ∈ c(M)  do
36             c(M) ← c(M) \ {H ∈ c(M) | G ⊒ H};
             endforall
           endif
         endif
       endif
     endif
41   endforall

     return S;
   end
```

Fig. 4.23. Merging: handling of subscriptions

merger can be generated, the other constituting filters are removed from the routing table. Furthermore, the new subscription is removed from S and the new merger is added instead (lines 19–23).

- If an extended or a new merger was generated, it is checked whether any filters (or other mergers) regarding the same destination are covered by this merger. The covered filters (mergers) are removed from the routing table and (their constituting filters) are added to the new merger (lines 26–32). If $S \in N_B$, from a new or extended merger those constituting filters are removed that are covered by another constituting filter (lines 34–38).
- The updated set S is returned to the **administer** procedure (line 43).

The code for the procedures *tryadd* and *trynew* is not given here because they largely depend on the details of the underlying filter model.

Processing of a Set of Unsubscriptions Accompanied with a Set of Covered Subscriptions

Every time a set of unsubscriptions accompanied with a set of covered subscriptions is received, the following (Fig. 4.24) is done by the *handleunsubs* procedure:

- If $S \in N_B$, those filters and mergers regarding the same destination that are covered by one of the subscriptions or unsubscriptions are removed from the routing table (lines 3–7).
- Now the set of constituting filters of those mergers which are affected by one of the unsubscriptions is updated (lines 11–18).
  - If $S \in N_B$, those filters are removed from the set of constituting filters of a merger that are covered by an unsubscription but not covered by any subscription (line 12).
  - If $S \notin N_B$, only the unsubscriptions are removed from an affected merger (line 15).
- If a merger from which some constituting filters were removed is afterwards no longer a perfect merger of its remaining constituting filter, the merger is removed from the routing table and added to the set of unsubscriptions. This is determined by the **disintegrated** procedure. Its remaining constituting filters are added to a set B (lines 19–23).
- After all mergers have been processed, all filters in B are added to the set of subscriptions (line 26).
- The updated sets S and U are returned to the **administer** procedure (line 30).

The code for the procedure *disintegrated* is not given here because it largely depends on the details of the underlying filter model.

```
 1  procedure handleunsubs(Dest S, Set S, Set U)
    begin
      if  S ∈ N_B  then
        forall  F ∈ S ∪ U  do
          T_B ← T_B \ C_B^L(F, S);
 6      endforall
      endif

      B ← ∅;
      forall  M ∈ M_B^{|S}  do
11      if  S ∈ N_B  then
          A ← {F ∈ c(M) | ∃G ∈ U. G ⊒ F
                        ∧∄H ∈ S. H ⊒ F};
        else
          A ← c(M) ∩ U;
16      endif
        if  A ≠ ∅  then
          c(M) ← c(M) \ A;
          if  disintegrated(M)  then
            T_B ← T_B \ {M, S};
21          B ← B ∪ {G | G ∈ c(M)};
            U ← U ∪ {M};
          endif
        endif
      endforall
26    S ← S ∪ B;

      return (S, U);
    end
```

Fig. 4.24. Merging: handling of unsubscriptions

Correctness

Since the correctness of the merging-based algorithm is based largely on the correctness of the covering-based routing scheme, we only give the main ideas for the correctness of merging-based routing here. Our algorithm solely merges routing entries regarding the destination S that triggered the call of administer at B. The main arguments for the correctness of our algorithm are the following:

- The routing entry of a new, updated, or covering merger causes B to forward exactly those notifications to S that match any of its constituting filters (including F).
- Forwarding a new or updated merger M (that covers F) instead of F ensures that the neighbors forward all notifications that match F or any other filter in $c(M)$ to B.

- If F is added to a covering merger M, neither F nor M need to be forwarded because M was already forwarded, ensuring that all neighbors of B except S forward to B all notifications that match F.
- If some mergers are canceled, those of their remaining constituting filters that are not covered by another filter of this set are inserted into the routing table and forwarded as subscriptions. This ensures that all notifications matching any of those filters are (a) forwarded by B to S and (b) forwarded to B by any neighbor of B except S.

4.5.6 Discussion

Overview and Use Cases of Algorithms

We now briefly recall the individual routing algorithms and describe their advantages and disadvantages. This helps engineers to choose a particular algorithm for different practical scenarios. One helpful indication are metrics for the efficiency of routing algorithms. Two main metrics have emerged in the past: the routing table sizes and the filter forwarding overhead [267]. The filter forwarding overhead is the number of *admin* messages needed for changing the routing tables in accordance with the used routing algorithm if a new subscription is issued or an existing subscription is revoked. A summary of the discussions is shown in Table 4.1.

Table 4.1. Portfolio of content-based routing algorithms

name	use case
flooding	Easy to implement, subscriptions become effective immediately, but has worst-case notification forwarding overhead
simple	Significantly reduces notification forwarding overhead if subscriptions and clients are sparsely distributed. Routing table sizes grow linearly with the number of subscriptions. Every routing table is affected by a new or canceled subscription
identity-based	Reduces routing table sizes and filter forwarding overhead if set of subscriptions contains a lot of identical entries; may degenerate to simple routing otherwise. Identity test must be efficiently computable
covering-based	Efficient for intervallike subscriptions. May degenerate to identity-based routing if subscriptions do not cover each other. Covering test must be efficiently computable
perfect merging	Reduce routing table sizes if subscriptions can often be merged perfectly; may degenerate to covering-based routing if not. May increase the filter forwarding overhead
imperfect merging	Allows users to trade accuracy against efficiency. Degenerates to flooding if too much imperfection is tolerated

With flooding, the routing tables have only local entries and no *admin* messages must be handled, so in terms of the efficiency metrics it can be regarded as a lower bound to the other algorithms. Flooding is, however, a degenerated case of the other algorithms and can be used to determine the worst-case notification forwarding complexity (all other algorithms try to decrease the overall number of forwarded notifications). This is the main disadvantage of flooding. Flooding is advantageous because of its simplicity which makes it is easy to implement correctly. Moreover, new subscriptions become effective immediately.

Because simple routing enforces that every broker has knowledge about all active subscriptions, the size of each routing table grows linearly with the number of active subscriptions. Moreover, all routing tables are affected if a subscription is issued or revoked. In our framework this means that the number of *admin* messages necessary to carry out such a change is independent of the number of active subscriptions and equals the number of links in the broker topology. Simple routing is preferable to flooding if the set of subscribing clients is very sparse and if subscriptions do not change very often.

Identity-based routing degenerates to simple routing if distinct filters are never identical. Hence in this worst-cast, routing table sizes and the filter forwarding overhead of identity-based routing are the same as for simple routing. However, our experimental findings [267] suggest that both numbers can be much smaller in practice. For example, if the number of different filters is bounded, the remote part of the routing tables grows only sublinearly in the number of active subscriptions and converges to a limit for large numbers of active subscriptions. This is because identity-based routing maintains the following invariant, which can be proved by a simple induction: In a routing table, there are never two distinct entries (F, H) and (G, H) for a neighbor H for which $F \equiv G$. This limits the size of the remote part of the routing table regarding a certain neighbor to at most the number of different filters.

Compared to simple routing, which forwards filters unselectively to all neighbors, identity-based routing forwards a filter selectively only to those neighbors that are in $D_B^I(F)$. This accounts for the observation that the filter forwarding overhead is lower than for simple routing. Because the probability that for a filter there is an identical filter increases with the number of active subscriptions, the forwarding overhead monotonically decreases. However, to use identity-based routing, a method to efficiently compute the identity relation is necessary. If such a method is not available, we must revert to simple routing (this observation also holds for the more refined algorithms that follow). A use-case for identity-based routing is a stock exchange quote service, where only individual stocks can be subscribed to.

Covering-based routing degenerates to identity-based routing if no filter properly covers another filter. In this case, $C_B^L(F, S) = C_B^I(F, S)$ and $D_B^U(F) = D_B^I(F)$. Hence, $C_B^L(F) \setminus C_B^I(F) = \emptyset$, implying that $\mathcal{P} = \emptyset$, resulting in identity-based routing. However, if filters properly cover each other, covering-based routing does better than identity-based routing. Both the routing table sizes and the filter forwarding overhead are reduced [267]. The reason for this is that covering-based routing also maintains an invariant: In a routing table there are no two distinct entries (F, H) and (G, H) for a neighbor H, where $F \sqsupseteq G$. This invariant can also be proved by a simple induction. Compared to identity-based routing this stronger invariant leads to a better behavior. Covering-based routing is advantageous over simple or identity-based routing in case of intervals, for example, if stocks can be subscribed to for special intervals (e.g., show me stock x if its value is between y and z).

Merging-based routing degenerates to covering-based routing if filters are never merged. If filters are merged, the routing table size can be reduced substantially. The reduction ratio, however, depends on the degree of imperfection that is tolerated (if any) and the filter predicates that are issued [262]. Our experimental findings suggest that a reduction in the routing table sizes can be achieved but that the filter forwarding overhead might increase in turn.

The Design Evolution of Content-Based Routing

As mentioned above, identity-based routing can be regarded as a special case of covering-based routing: whenever an identity-based routing algorithm processes a new filter that is identical to an existing filter, a covering-based routing algorithm would also process that new filter with the same effect on the routing table. From an implementation perspective, on the one hand, covering-based routing can be regarded as an "add-on" to identity-based routing (it handles all the cases of identity-based routing, but also more). On the other hand, identity-based routing can be achieved by "restricting" the power of covering-based routing. Interestingly, we can extend this relation to *all* other presented routing schemes, which results in a circular evolution hierarchy that we now explain (Fig. 4.25).

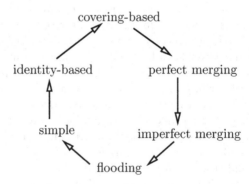

Fig. 4.25. Circular evolution of CBR algorithms

Simple-routing results from restricting the power of identity-based routing, i.e., by removing its potential to process identical filters. Therefore, simple routing can be regarded as a special case of identity-based routing. Similarly, covering-based routing results from a perfect merging algorithm by restricting the types of filters which are merged: if only those filters are merged that cover an existing filter we have covering-based routing. Obviously, imperfect merging algorithms can be similarly regarded as a generalization of perfect merging.

Interestingly, flooding can be regarded as both the starting and the ending point of this design evolution (Fig. 4.25). Historically, it is the starting point since the first schemes were based on flooding and more refined algorithms were developed to prevent the deficiencies of this scheme. More formally, flooding seems to be incomparable to, e.g., simple routing, because flooding does not have remote routing table entries. However, flooding can be regarded as a generalization of imperfect merging: consider an imperfect merging algorithm in which there exists a special initial action that spontaneously adds the filter that matches the set of all notifications to the routing table (i.e., performs an imperfect merge) and never removes any filters. Clearly, this results in flooding. In this sense flooding can be regarded as a generalization of imperfect merging.

4.6 Extensions of the Basic Routing Framework

We present — on a less formal level — three important extensions of the basic routing framework. It is shown how advertisements can be integrated into the framework, how hierarchical versions of the routing algorithms versions can be obtained, and how changes to the topology can be dealt with. The use of advertisements can enhance the efficiency of the systems by limiting the propagation of subscriptions into those subnets, where matching notifications are potentially produced. Hierarchical routing algorithms reduce the filter forwarding overhead and the routing table sizes but require the root broker to handle every notification. This can be superior in some environments. Coping with topology changes is important dealing with a changing system.

4.6.1 Routing With Advertisements

With the subscription-based routing algorithms presented in Sect. 4.5, subscriptions are forwarded regardless of whether or not matching notification are potentially produced in the respective subnet. Advertisements allow the propagation of subscriptions to be limited to those subnets, where matching notifications are potentially produced. The only assumption for the use of advertisements is that it can be detected whether or not a subscription and an advertisement overlap, i.e., whether there is a notification matching both filters. Formally, two filters F_1 and F_2 *overlap* iff $N(F_1) \cap N(F_2) \neq \emptyset$. We also say that a subscription *can be served* by an advertisement if both overlap.

```
 1  sync(μ_B) procedure  pub (Client X, Notification n)
    begin
        if  ∃(F, X) ∈ T_B^A . n ∈ N(F)  then
            handleNotification(X, n);
        endif
 6  end


    sync(μ_B) procedure  adv(Client X, Filter F)
    begin
        (ℱ_S, ℱ_U) ← administer_A(X, {F}, ∅);
11      handleAdminMessage(X, ℱ_S, ℱ_U, 1);
    end


    sync(μ_B) procedure  unadv(Client X, Filter F)
    begin
16      (ℱ_S, ℱ_U) ← administer_A(X, ∅, {F});
        handleAdminMessage(X, ℱ_S, ℱ_U, 1);
    end


    procedure  handleMessage(Message m)
21  begin
        switch
            case  m is "forward(n)" message from neighbor U :
                handleNotification(U, n);
            break
26          case  m is "admin_S(𝒮, 𝒰)"  message from neighbor U :
                (ℱ_S, ℱ_U) ← administer(U, 𝒮, 𝒰);
                handleAdminMessage(U, ℱ_S, ℱ_U, 0);
            break
            case  m is "admin_A(𝒮, 𝒰)" message from neighbor U :
31              𝒜 ← prune(P_B(𝒮, U));
                forall F ∈ 𝒜 do
                    send "admin_S({F}, ∅)" to U ;
                endforall
                (ℱ_S, ℱ_U) ← administer(S, 𝒮, 𝒰);
36              handleAdminMessage(S, ℱ_S, ℱ_U, 1);
                T_B^S ← T_B^S \ O_B ;
            break
        endswitch
    end
```

Fig. 4.26. Routing using advertisements, part I

```
      procedure handleAdminMessage(Dest D, Set M_S, Set M_U, Bool b)
      begin
        forall H ∈ N_B \ {D}
45         A ← {F | (F, H) ∈ M_S};
           B ← {F | (F, H) ∈ M_U};
           if A ≠ ∅ ∨ B ≠ ∅ then
             if b = 0 then
               A ← Q_B(A, H);
50             B ← Q_B(B, H);
               if A ≠ ∅ ∨ B ≠ ∅ then
                 send "admin_S(A, B)" to H;
               endif
             else
55             send "admin_A(A, B)" to H;
             endif
           endif
         endforall
      end
```

Fig. 4.27. Routing using advertisements, part II

If advertisements are used, each broker manages two routing tables, the known *subscription routing table* T_B^S (formerly T_B) and an additional *advertisement routing table* T_B^A. While the former is used (as described before) to route notifications from producers to interested consumers, the latter is used to route subscriptions and unsubscriptions from interested consumers to producers. Both routing tables have to be updated as clients issue or revoke subscriptions and advertisements, respectively. This now takes places on two cooperating levels. The first level is responsible for updating the subscription table, while the second level keeps the advertisement table up to date. For each of both levels one of the routing algorithms presented in Sect. 4.5 (except flooding) can be chosen. For example, simple routing can be used to update the subscription table, while at the same time covering-based routing is applied to the advertisement table. The basic idea of advertisements is that:

- A (un)subscription is only forwarded to a neighbor if it overlaps with an advertisement from this neighbor.
- If a new advertisement is received from a neighbor H, subscriptions from other neighbors that previously could not be served by any advertisement from H but that now can be served are forwarded to H.
- If an advertisement is canceled by neighbor H, those subscriptions that can no longer be served by any other but the originating neighbor are removed from the routing table.

A disadvantage of advertisements is that notifications which only match an advertisement that has been recently issued by a producer may not be delivered to all interested consumers. This is because the propagation of the respective advertisement triggers the forwarding of newly overlapping subscriptions. In the meantime, before this process has terminated, notifications may be dropped or may not be forwarded to all neighbors which have consumers with matching subscriptions in their subnet. This was also the main reason to apply a weakened liveness condition (Sect. 2.5.4) if advertisements are used. Indeed, with the proposed solution, delivery is only guaranteed after the new advertisement has been propagated *and* the subscriptions that are forwarded in turn have also been propagated. Both processes are guaranteed to terminate after a finite time. Hence, the proposed solution satisfies Def. 2.9, which defined simple event system with advertisements.

Integration into the Framework

Advertisements can easily be integrated into our framework. Two categories of *admin* messages are used to distinguish among *admin* messages related to subscriptions and those related to advertisements: $admin_S$ and $admin_A$. The existence of two instances of `administer` is now assumed: $administer_S$ and $administer_A$. The former defines the applied subscription routing algorithm. It is called if a *sub*, an *unsub*, or an $admin_S$ message is received; it only works on the subscription table. The latter defines the used advertisement routing algorithm. It is called if *adv* or *unadv* is called by a local client, or if an $admin_A$ message is received from a neighbor; it only works on the advertisement table.

In Figs. 4.26 and 4.27, the advertisement-enabled instantiation of those parts of the framework are shown which replace the ones shown in Figs. 4.1 and 4.2. Most of the code has already been discussed. The more interesting parts are (a) the forwarding of newly servable subscriptions (lines 31–34), (b) the dropping of unservable subscriptions (line 37), and (c) the postprocessing of subscriptions before the respective $admin_S$ messages are sent out (lines 49–50). These are described in the following.

```
   procedure prune_si(Set A)
 2 begin
     return A;
   end
```

Fig. 4.28. prune for simple routing

Forwarding of Newly Servable Subscriptions

An $admin_A$ message containing new advertisements might make some subscriptions newly servable. If the $admin_A$ message is received from a neighbor

```
1 procedure prune_id(Set A)
    begin
      forall F ∈ A do
        A ← A \ {G ∈ A | G ≠ F ∧ F ≡ G};
      endforall
6     return A;
    end
```

Fig. 4.29. prune for identity-based routing

H, this concerns all subscriptions of other neighbors which previously were not served by any advertisement from H but which are served by one of the new advertisement in S. We denote this set of newly serviceable subscriptions with P_B:

$$P_B(S, H) = \{F \mid (F, I) \in T_B^S \land H \neq I$$
$$\land \nexists (G, H) \in T_B^A. \, N(F) \cap N(G) \neq \emptyset \qquad (4.32)$$
$$\land \exists G \in S. \, N(F) \cap N(G) \neq \emptyset\}.$$

The subscriptions in P_B are pruned by calling the **prune** procedure that is tuned to the used subscription routing algorithm. For simple routing, it simply returns the unchanged set (Fig. 4.28). For identity-based routing (Fig. 4.29), it removes for each filter all identical filters. For covering and merging-based routing (Fig. 4.21), it removes all filters that are covered by any other filter.

Dropping of Unservable Subscriptions

After an $admin_A$ message was processed, all routing entries corresponding to subscriptions of neighbor brokers which cannot be served anymore are removed from the routing table. Subscriptions of local clients are not dropped. A subscription routing entry $(F, H) \in T_B^S$ cannot be served if there is no $(G, I) \in T_B^A$ such that $H \neq I$ and $N(F) \cap N(G) \neq \emptyset$. The set of all such routing entries regarding neighbors of B is given by:

$$O_B = \{(F, H) \in T_B^S \mid H \in N_B \land$$
$$\nexists (G, I) \in T_B^A. \, H \neq I \land N(F) \cap N(G) \neq \emptyset\}. \qquad (4.33)$$

Postprocessing of Subscriptions

With advertisements, a (un)subscription is only forwarded to a neighbor if it overlaps with an advertisement from this neighbor. This is achieved by removing from the set of subscriptions and the set of unsubscriptions being forwarded to a neighbor H those filters that do not overlap with an advertisement of H. This is done by evaluating:

$$Q_B(\mathcal{A}, H) = \{(F, H) \mid F \in \mathcal{A} \;\wedge\; \exists (G, H) \in T_B^A.$$
$$H \in N_B \wedge N(F) \cap N(G) \neq \emptyset\}. \tag{4.34}$$

4.6.2 Hierarchical Routing Algorithms

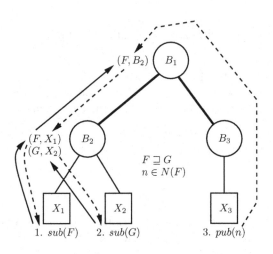

Fig. 4.30. Hierarchical covering-based routing

The routing algorithms discussed so far are called *peer-to-peer routing* algorithms because no brokers are distinguished and filters are exchanged between neighbors in both directions. With *hierarchical routing*, one broker is distinguished as *root* of the broker topology and every notification that is published is always forwarded stepwise to this root node. Hence, it is sufficient to forward subscriptions and unsubscriptions only in the direction pointing to the root broker (Fig. 4.30). Carzaniga has presented a hierarchical version of covering-based routing [65] which is also used by JEDI [92]. With hierarchical routing, every broker has to process every notification that is published in its respective subtree, but its routing table only contains filters originated in its subtree, too. Compared to peer-to-peer routing, hierarchical routing reduces the sizes of the routing tables substantially. For a topology being a balanced tree[3] with n brokers, a subscription is only present in about $O(\log n)$ instead of n routing tables in the worst-case, i.e., in a system with no other subscriptions. When the number of subscriptions increases, the advantage of hierarchical routing over peer-to-peer routing decreases. However, in a saturated system with many subscriptions, hierarchical routing only saves 50% of the routing

---

[3] Note that in this scenario the number of brokers grows exponentially in the number of hierarchy levels.

table sizes. For an individual broker, the reduction of its routing table size
corresponds to its level in the broker hierarchy. Its routing table only contains
filters that originated in its subtree. Hence, smaller routing tables are traded
for higher notification loads. For the root node, the size of its routing tables is
therefore not reduced, although it has to handle all notifications published in
the system. Hence, this node might possibly be overloaded. A possible solution
to this could be to replicate the root node and some of its child nodes exposed
to a higher load.

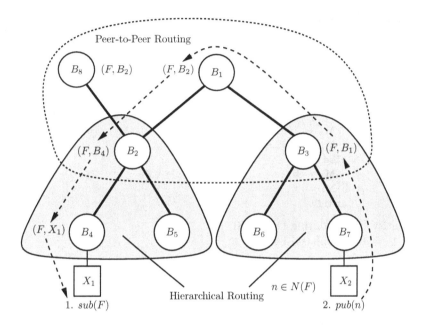

Fig. 4.31. Hybrid routing

Another potential solution is *hybrid routing* [65], which combines hierarchical and peer-to-peer routing. In this case, for certain subtrees hierarchical
routing is used as described above, while in the other parts of the topology
peer-to-peer routing is used (Fig. 4.31). In the part of the topology, where
peer-to-peer routing is used, advertisements can be used as described previously. In a subtree, where hierarchical routing is used, advertisements are only
propagated to the parent node, and the advertisement routing table is ignored
for subscription routing by every broker except the root node of this subtree.
The root node uses the advertisement table for deciding which subscriptions it
forwards to its peer nodes to which it is connected. Directly combining hierarchical routing with advertisements is not sensible because hierarchical routing
is essentially the same as if peer-to-peer routing is used and the respective root
node issues an advertisement that overlaps with all subscriptions.

Changes for Hierarchical Routing

Now let us look at how we can obtain the hierarchical variants of the routing algorithms presented in the previous sections. We have only to slightly modify notification forwarding such that a notification is always forwarded to the parent broker and the individual routing algorithms such that a filter is only propagated to the parent broker.

Let R be the root broker. For a broker B, let $P(B)$ be the parent broker of B if $B \neq R$ and B, otherwise.

- Framework algorithm (Figs. 4.1 and 4.2):
  - Line 20 is replaced by:

    send *"forward(n)" to all neighbors in* $F_B(n) \cup \{P(B)\} \setminus \{D\}$;

- Simple routing (Fig. 4.5):
  - Lines 5 and 6 are replaced by:

    ```
    if  B ≠ R then
        𝓜_S ← {(F, P(B)) | F ∈ 𝓢};
        𝓜_U ← {(F, P(B)) | F ∈ 𝓤};
    4 else
        𝓜_S ← ∅;
        𝓜_U ← ∅;
    endif
    ```

- Identity-based routing (Fig. 4.8):
  - Line 13 is replaced by:

    ```
    if  B ≠ R then
        if  P(B) ∈ D_B^I(F)  then
    3       𝓐 ← {(F, P(B))};
        else
            𝓐 ← ∅;
        endif
    endif
    ```

- Covering-based routing (Fig. 4.12):
  - Line 15 is replaced by:

    ```
    if  B ≠ R then
        if  P(B) ∈ D_B^U(F)  then
    3       𝓜_S ← 𝓜_S ∪ {(F, P(B))};
        endif
    endif
    ```

  - Line 26 is replaced by:

    ```
    if  B ≠ R then
        if  P(B) ∈ D_B^U(F)  then
            𝓜_U ← 𝓜_U ∪ {(F, P(B))};
        endif
    5 endif
    ```

– Lines 31–43 are replaced by:

```
   if  B ≠ R then
       P ← P ∪ {(F, S) | F ∈ S};
   endif
   T_B ← T_B ∪ {(F, S) | F ∈ S};
 5 endif
   if  B ≠ R then
      forall  (F, U) ∈ P do
          P ← P \ {(G, H) | (G, H) ∈ P  ∧  G ≡ F};
          if  P(B) ∈ D_B^{PU}(F) then
10            M_S ← M_S ∪ {(F, P(B))};
          endif
      endforall
   endif
```

- For merging-based routing, no changes are necessary.

4.6.3 Rendezvous-Based Routing

A complementary class of routing algorithms that can be combined with the previous approaches follows a rendezvous-based routing strategy. Rendezvous-based routing schemes derive from the observation that any content-based routing algorithm has to set up routing paths from publishers to subscribers. In the previous routing framework, this was achieved by propagating state about subscriptions to all nodes in the system (subject to covering among subscriptions). An alternative approach is to designate explicit nodes in the network that act as "meeting points" for notifications and matching subscriptions.

In rendezvous-based routing, a *rendezvous node* ensures that all interested brokers agree on the same dissemination tree for events. This means that a notification message that is sent to the rendezvous node is guaranteed to encounter all relevant subscription states in the network. In the worst-case, a notification will only a find matching subscription state once it reaches the rendezvous node. When constructing a dissemination tree, subscriptions and notifications are routed to the rendezvous node using the overlay network. The rendezvous node must exist at a globally known location in the network.

Any broker in the system must have a way to send a message to the broker acting as the rendezvous node for a given event type. A scalable implementation of such a scheme can be based on the routing substrate that is provided by a *distributed hash table (DHT)* [316]. For example, the rendezvous node can be chosen by using a unique event type name as a key for a lookup in a DHT. The broker in the DHT responsible for this key then becomes the rendezvous node. Due to the properties of the DHT, the chosen event broker will be globally agreed upon by all brokers so that every broker can use the peer-to-peer routing substrate to send messages to this rendezvous node. We will describe this technique for building a content-based publish/subscribe system on top of a peer-to-peer routing substrate in more detail in Sect. 4.6.8.

The idea of rendezvous nodes was introduced [24] in the context of *core-based trees* for building multicast trees. However, core-based trees require all messages to be routed via the rendezvous node, potentially creating a bottleneck at the node. In contrast, rendezvous-based routing can take advantage of subscription state to reduce the load on the rendezvous node. Notifications can be delivered directly to subscribers when matching subscriptions are encountered on the routing path to the rendezvous node.

A publish/subscribe system usually maintains multiple rendezvous nodes, for example, one per event type or class of events. This enables the set-up of multiple dissemination trees to balance the routing effort. Any broker in the system can assume the role of a rendezvous node for one or more event types. A rendezvous node is automatically created when a new event type is added. Once a broker has become a rendezvous node, it is responsible for managing that particular event type.

A rendezvous node can also be used to store metadata about a class of events and manage the authoritative version of the event type schema used for type-checking. Note that rendezvous nodes do not contain any state about the event dissemination trees itself, which makes them simple to replace in case of failure. When a rendezvous node fails, a new rendezvous node can take over if the peer-to-peer routing substrate is capable of adaptation. To prevent the event type metadata from being lost, they can be replicated across multiple nodes. There exist several strategies for managing redundant rendezvous nodes to achieve fault tolerance with rendezvous-based routing [310].

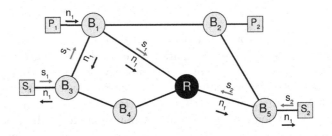

Fig. 4.32. Rendezvous-based routing

Figure 4.32 shows an example of an overlay network of brokers $B_{1...5}$ with one rendezvous node R. It illustrates how subscription ($s_{1,2}$) and notification messages (n_1) are routed toward the rendezvous node R and how the subscription messages $s_{1,2}$ establish routing state at brokers along the path. At first, a subscription message is routed toward the rendezvous node. The subscription is stored at every broker along the path ($B_{1,3,5}$ and R). After the subscription message has reached the rendezvous node, it is discarded.

Notification messages are also routed to the rendezvous node. Whenever they encounter a broker with matching subscriptions, they follow the reverse path of the subscription. Note that no state is created at brokers that process notification messages, and messages are never forwarded to a broker that was the previous hop on the path.

In this rendezvous-based routing scheme, a notification must reach the rendezvous node but may be discarded there because it has already encountered all matching subscriptions on its path to the rendezvous node. This has the drawback that the rendezvous node may become a bottleneck in the system when a large number of notifications are flowing through it. To address this issue, advertisements (as introduced in Sect. 4.6.1) can be used to establish more complete routing states in the system. This then enables notifications to follow the reverse path of subscriptions without necessarily traversing the rendezvous node [311].

Rendezvous-based routing usually has a lower message forwarding overhead than the other schemes because only brokers that are part of an event dissemination tree need to maintain routing state [312]. In other words, an inner broker that is not on the routing path from publishers and subscribers to their rendezvous node does not need to store any state and can be oblivious to the ongoing routing of notifications. This property is especially beneficial in large-scale networks, in which much of the event dissemination is geographically localized and the creation of globally consistent state at all brokers is an expensive operation. The price to be paid for this reduction in state is the complexity of managing one (or more) rendezvous nodes and the global dissemination of their identities. Often, this is achieved by a peer-to-peer routing substrate.

Even though rendezvous-based routing is not based on the flooding of subscriptions or notifications, it can be extended in a similar fashion, as shown in Fig. 4.25. Simple rendezvous-based routing can be combined with techniques from identity-, covering-, and merging-based routing to exploit commonality among subscriptions and reduce the amount of subscription states in the network. In contrast to the flooding-based schemes, rendezvous-based routing only installs filtering state along the routing paths from publishers and subscribers to rendezvous nodes. This means that even with imperfect merging, a rendezvous-based routing scheme will never have the high message forwarding overhead of flooding.

4.6.4 Topology Changes

So far we have assumed a static broker topology. The topology changes if a new broker connects to or a connected broker disconnects from the broker network. Note that we do not deal with transient disconnections due to system faults here but with desired connects and disconnects. In the following, we do not elaborate how connect and disconnect decisions are made. Instead, we assume that the system administrator makes these decisions. The administrator must take care to avoid cycles in the topology and undesired partitioning. For brevity, we focus our discussion on peer-to-peer routing. Topology changes in hierarchical routing can be handled similarly. We do not discuss how to enforce ordering requirements which is an additional challenge if the topology changes dynamically [303].

When a connection is established between two brokers, they exchange their active subscriptions to establish the desired delivery paths. Note that, similar to the case of advertisements, notification delivery can only be guaranteed eventually after a connection has been established. When an existing connection among two brokers is removed, both brokers cancel the subscriptions of the other broker at their remaining neighbors and delete the affected subscriptions from their routing tables.

In Fig. 4.33 the code of the procedures needed for connection management is shown. It assumes that the used routing algorithm is able to process individual (un)subscriptions from neighbor brokers (lines 11+25). A new connection is established by calling the *connect* procedure at a broker. This sends a *connect* message to the desired neighbor broker and then also forwards the active subscriptions by calling the *forwardFilters* procedure. The set of subscriptions forwarded is reduced by applying the **prune** procedure corresponding to the used routing algorithm. An existing connection is canceled by calling the *disconnect* procedure at a broker. This procedure sends a *disconnect* message to the desired neighbor broker and then calls the *dropFilters* procedure. This cancels all subscriptions of the neighbor broker as if an unsubscription was received. The *forwardFilters* and the *dropFilters* procedures also have to be executed by the neighbor broker when it receives a *connect* and *disconnect* message, respectively. Therefore, the following code is inserted into the framework (Fig. 4.1) after line 38:

```
1   case m is "connect" message from neighbor U :
        if U ∉ N_B then
            N_B ← N_B ∪ {U};
            forwardFilters(U);
        endif
6   break


    case m is "disconnect" message from neighbor U :
        if U ∈ N_B then
            N_B ← N_B \ {U};
11          dropFilters(S);
        endif
    break
```

Note that if advertisements are used, they are processed similarly, as described above. In this case, advertisements instead of subscriptions are exchanged among the brokers when a new connection is established. This subsequently leads to the exchange of servable subscriptions. When a connection is canceled, both affected brokers cancel their respective advertisements at their

```
   procedure  connect(Broker I)
 2 begin
     send  "connect" to I;
     forwardFilters(I);
   end


 7 procedure  forwardFilters(Broker I)
   begin
     A ← prune({F | (F, D) ∈ T_B ∧ D ≠ I});
     forall  F ∈ A do
       send  "admin({F}, ∅)" to I;
12   endforall
   end


   procedure  disconnect(Broker I)
   begin
17   send  "disconnect" to I;
     dropFilters(I);
   end


   procedure  dropFilters(Broker I)
22 begin
     A ← {F | (F, I) ∈ T_B};
     forall  F ∈ A do
       (M_S, M_U) ← administer(∅, {F});
       forall  H ∈ N_B \ {I}
27       S ← {F | (F, H) ∈ M_S};
         U ← {F | (F, H) ∈ M_U};
         if  S ≠ ∅ ∨ U ≠ ∅ then
           send  "admin(S, U)" to H;
         endif
32     endforall
     endforall
   end
```

Fig. 4.33. Managing connects and disconnects

remaining neighbors and delete the affected advertisements from their routing tables. This subsequently leads to the deletion of unservable subscriptions.

4.6.5 Joining and Leaving Clients

In a dynamic system, clients can join and leave the system. To support joining and leaving clients, we change the routing framework in the following way:

- Each broker B manages a set C_B containing B's current set of local clients. Initially, C_B is the empty set.

- If a client X calls one of the interface operations (e.g., *pub*, *sub*, *unsub*) and X is not in $\mathcal{C}_B$, then X is added to $\mathcal{C}_B$ and X's delivery queue is initialized.
- If a client wants to leave the system, it calls the new *leave* interface operation. This operation removes X from $\mathcal{C}_B$ and cancels all active subscriptions (and advertisements, if advertisements are used). Furthermore, X's delivery queue is freed.

4.6.6 Routing in Cyclic Topologies

Up to now, we have restricted the discussion to acyclic topologies. However, the definition of valid routing algorithms also makes sense in cyclic topologies (cf. Sect. 4.3.1). In this case, duplicates may be delivered to brokers if we would apply notification forwarding without changes. To ensure safety, it must be guaranteed that these duplicates do not reach a client. Next, we describe how duplicates can be eliminated by using notification ID histories. Then, we discuss routing algorithms for cyclic topologies. We do not discuss how to enforce ordering requirements which is an additional challenge in cyclic topologies.

Avoidance of Duplicates

Duplicates can be avoided in the following way: To detect duplicates each broker stores the ID of every notification it processes. If a broker receives a notification more than once, the broker ignores this notification.

Storing notification IDs for the whole lifetime of the system would sooner or later consume all the memory of a broker. In order to avoid the case that a broker has to store notification IDs forever, a broker must be able to detect that a duplicate corresponding to a stored ID can no longer reach this broker. In this case, the broker can delete this ID from its history. To make notification ID history cleanup possible, notifications carry a timestamp that is filled in at the time the notification is published by the broker hosting the publishing client. We assume that the clocks of the brokers are approximately synchronized and that notifications that are consecutively published at a broker get distinct and increasing timestamps.

Each broker stores for each neighbor and itself the maximum timestamp it has received from this neighbor in *timestamp message* and which corresponds to the last notification published by a local client, respectively. Initially, the maximum timestamp vector is initialized with sufficiently small timestamps. Each broker computes from the maximum timestamp vector a minimum maximum timestamp for each of its neighbors by taking the minimum of all but this neighbor's maximum timestamps. If a broker receives a timestamp message, it updates the respective neighbor's vector component. Similarly, it updates its own component in the vector if a local client publishes a notification. If for a too long a period of time no local client has produced a notification, the broker also updates its own component. If the minimum maximum timestamp of some neighbors has increased due to an update, the respective new minimum maximum is sent to these neighbors. This way, every broker has a current minimum maximum timestamp of all components of its vector that monotonically increases. For a broker, it is safe to discard those notification IDs whose timestamp is smaller than its current minimum maximum timestamp. Of course, timestamp messages can be piggybacked to *forward* and *admin* messages. Message batching can be applied, too.

Routing Algorithms

Similar to acyclic topologies, the simplest routing algorithm for cyclic topologies is flooding. Flooding (Fig 4.4) can be reused for cyclic topologies without changes. Applying flooding has the advantage of a maximum of fault tolerance. As long as the topology is connected, every notification will reach every broker.

```
    procedure administer(Dest S, Set S, Set U)
  2 begin
      S ← S \ {F | ∃(F, D) ∈ T_B  ∧ D ≠ S};

      T_B ← T_B ∪ {(F, S) | F ∈ S};
      T_B ← T_B \ {(G, S) | G ∈ U};
  7   M_S ← {(F, H) | H ∈ N_B \ {S}  ∧  F ∈ S};
      M_U ← {(F, H) | H ∈ N_B \ {S}  ∧  F ∈ U};
      return (M_S, M_U);
    end
```

Fig. 4.34. Simple routing in cyclic topologies: algorithm

Figure 4.34 shows simple routing adapted to cyclic topologies. There are only slight changes necessary to use simple routing in acyclic topologies (Fig. 4.5). In line 3, those filters are removed from S for which a routing entry from another destination already exists. The rest of the code is not changed. This algorithm actually creates a separate spanning tree for every subscription (Fig. 4.35). The advantage of this approach is that the load caused by notification is more balanced among the network connections than if a single spanning tree was used. This is especially true if a network has many more links than nodes. Note that simple routing in cyclic topologies has much in common with *directed diffusion* [204]. Simple routing can be used for advertisements and subscriptions. In this case, for each advertisement a separate spanning tree is built and this spanning tree is used to propagate the sub-

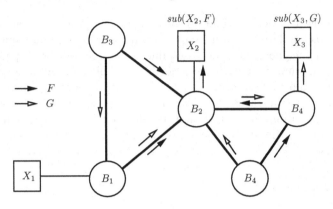

Fig. 4.35. Example of simple routing in cyclic topologies

scriptions to the consumers. How advanced routing algorithms can be used in cyclic topologies without relying on spanning trees is still an open issue.

4.6.7 Exploiting IP Multicast

Opyrchal et al. [291] described how IP multicast can be used in a publish/subscribe system and in which cases it reduces the consumed network bandwidth. They compare flooding to four multicast-enabled routing algorithms:

- *ideal multicast* assumes that for each set of brokers (having clients with a matching subscription) a (perfect) multicast group exists. In this case, a notification can be forwarded to all these brokers with a single send operation. This strategy is only realistic for a small number of brokers because for N brokers 2^N groups are needed.
- *clustered group multicast (CGM)* divides the set of all brokers into several mutually exclusive subsets called clusters. Then, for each cluster ideal multicast is used. For C equally large clusters, this strategy needs C sends, decreasing the efficiency of the multicast. The number of groups necessary is reduced by a factor of $2^C/C$, i.e., $c \cdot 2^{N/C}$ groups are needed.
- *Threshold Clustered Group Multicast (TCGM)* sends a notification to *all* members of all clusters if the number of receiving brokers exceeds a threshold T. This approach reduces the number of groups to

$$C \cdot \sum_{1 \leq i \leq T} \binom{N/C}{i}, \tag{4.35}$$

but further reduces the efficiency of the multicast because now brokers may receive notifications for which they do not have a local client with a matching subscription.

- *neighbor matching multicast* forwards a notification in multiple steps from the broker to which the publishing client is connected to the brokers which have clients with a matching subscription. In each step, the sending broker determines those neighbor brokers to which it should forward this notification. This strategy has the disadvantage that those links that connect brokers with (multicast routers) are traversed several times.

From the investigated multicast routing algorithms, the neighbor matching algorithm can be directly integrated into our routing framework. Instead of sending a notification to individual neighbors, now a corresponding multicast group is used that contains all neighbors to which the notification should be forwarded. If it is not possible to reserve a multicast group for each subset of neighbors, threshold clustering can be used. The authors state that neighbor matching is superior to flooding under conditions of high selectivity and high locality of subscriptions. It can be expected that their results are too pessimistic because their work depends on simple routing, i.e., the routing algorithm does not exploit covering and merging. They also assume that event brokers are not placed nearby to the multicast routers and therefore the use of multicast may even introduce a bandwidth penalty. They also did not investigate the use of advertisements.

The other two multicast-enabled routing algorithms (i.e., CGM, TCGM) assume that each broker has global knowledge about all active subscriptions because notifications are forwarded in only one step from the producer's broker to the consumers' brokers. These algorithms can easily be integrated with simple routing to fit into our routing framework.

4.6.8 Topology Maintenance

From the previous discussion it becomes clear that the maintenance of an overlay topology in the light of network and node failures and nodes joining and leaving the system (also known as *churn*) can be complex. Therefore, a publish/subscribe system benefits from a routing abstraction that handles the maintenance of the overlay network of broker transparently to the higher content-based routing layers. Recently, DHTs [316, 323, 331, 351] were introduced as scalable data structures for building large distributed applications. The multihop routing abstraction implemented by a DHT integrates naturally with the need for globally unique rendezvous nodes in rendezvous-based routing approaches (Sect. 4.6.3). In this section, we briefly introduce DHTs and explain how they can be used to implement rendezvous-based event dissemination.

Distributed Hash Tables (DHTs)

A DHT maps a key to a value that is stored at a particular node in the network. Rather than having global knowledge, nodes only need to know about a small subset of all existing nodes when performing key lookups. Lookup requests are routed via the overlay network to the destination node that is responsible for the key, even when nodes are constantly joining and leaving the DHT. The load of storing data in the hash table is therefore spread across all nodes in the system. The routing algorithm for the DHT builds a *small-world network* [391], which has a small diameter but is highly clustered, so that every node can be reached in a logarithmic number of hops.

Pastry [331], developed at Microsoft Research Cambridge, is an example of a DHT with locality properties that forms a self-organizing, resilient overlay network, which can potentially scale to millions of nodes. Its main operation is a route(message, key) function that reliably routes a message to the Pastry node that is responsible for storing the key. Messages take $O(\log N)$ hops on average, where N is the number of nodes in the Pastry network. The overlay network of nodes is organized so that routes with a lower proximity metric, such as latency or bandwidth, are preferred.

The routing algorithm of Pastry relies on the fact that each Pastry node has a unique node identifier, called a *nodeID*. NodeIDs populate a 128-bit namespace that is uniformly distributed; they are grouped into digits with base 2^b for a given value of b. DHT keys can be transformed into nodeIDs by using a hash function. The functionality of a DHT is implemented by routing a message to a live node with a nodeID that is numerically closest to the hashed key. The routing of messages relies on two data structures, a *routing table* and a *leaf set*, maintained by each node.

Routing Table. The routing table has $\log_{2^b} N$ rows with $2^b - 1$ columns. The rows contain entries for nodes whose nodeID matches the local node's nodeID in the first d digits but then differs afterwards. Among several candidate nodeIDs for an entry in the routing table, the one with the minimum proximity metric is chosen. Secondary entries are kept as backup in case the primary node fails.

Leaf Set. The leaf set has l nodeIDs as entries, which are the $l/2$ closest, numerically larger and smaller nodeIDs with respect to the local nodeID. This invariant must be maintained at all times, and routing will fail if more than $l/2$ nodes with consecutive nodeIDs fail. The leaf set can be used for data replication.

Routing in Pastry is a generalization of prefix routing: A message is forwarded to a node that shares a longer prefix with the destination nodeID than the current node. If such a node does not exist in the routing table, the message is sent to a node with a nodeID that is numerically closer to the destination. If the destination nodeID falls within the range of the leaf set, the message is sent directly to the numerically closest nodeID. The process of routing a message from node 123 to the key 333 with $b = 2$ is illustrated in Fig. 4.36. The message is first forwarded to node 311, which is obtained from the routing table at node 123. Each hop moves the message closer to the destination node.

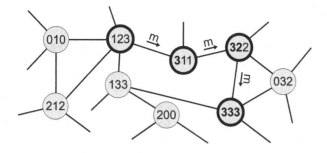

Fig. 4.36. Routing a message in a Pastry network

Rendezvous Nodes in a DHT

Rendezvous-based routing relies on globally known rendezvous nodes that ensure that publishers and subscribers agree on the same dissemination tree. If the overlay network of brokers forms a DHT, then its routing properties can be exploited to create rendezvous nodes, as follows: DHT routing has the property that a lookup of a nonexisting key will deterministically return the numerically closest existing key in the system. To create a unique rendezvous node for a given event type (or class of events), the event type name is used as the lookup key in the DHT. The broker that is responsible for this key is then designated as the rendezvous node for the event type. The load balancing properties of a DHT encourage a uniform distribution of rendezvous nodes in the system. Advertisement and subscription messages are routed using the DHT and create filtering state along the path as explained in Sect. 4.6.3. Notification messages then follow the reverse path and are filtered according to subscriptions.

4.7 Further Reading

Epidemic Multicast

The idea of epidemic multicast algorithms was introduced by Demers et al. [108] in 1987. The basic idea is very simple: The source of a notification sends it to some randomly chosen brokers. A broker that receives a notification for the first time also sends it to a number of randomly chosen brokers. This way, all brokers receive the notification with a certain probability. The algorithm can be tuned to make the probability that a broker misses a notification as low as desired. Many recent approaches use some sort of epidemic algorithm to distribute information [41, 86, 122, 123, 125, 128].

Evaluation of Routing Algorithms

Carzaniga, Rosenblum, and Wolf [65, 71] presented performance results which are based upon a simulation framework. Their work investigated two variants of covering-based routing, a peer-based and a hierarchical version. The simulated algorithms are also incorporated into their publish/subscribe prototype called *Siena*. Other routing algorithms are not considered.

The simulations investigated the total cost induced by the notification service, the cost induced on individual brokers (and its variance), the average cost per subscription (and its worst-case), and the per-notification cost. Unfortunately, it is not easy to interpret their results because the setup of the main parameters influencing the results are not described. This includes the metric underlying their cost analysis, the structure of the notifications, subscriptions, and advertisements, and the rates of subscribing/unsubscribing and advertising/unadvertising.

The current implementation of JEDI exploits a hierarchy of event brokers in conjunction with the hierarchical version of covering-based routing [65]. The algorithm implies that a notification is always propagated to the root broker regardless of the interests of the consumers. Moreover, an improved version is suggested that extends the hierarchical algorithm by using advertisements, and simulations have been carried out to compare the original with the improved version [51, 52]. Bricconi, Di Nitto, and Tracanella [52] also presented the analytical model that underlies their simulations and which allows the average number of notifications that is processed by an event broker to be estimated.

Performance results related to the prototype of the Gryphon notification service are presented by Banavar et al. [26] and Opyrchal et al. [291]. The routing algorithm exploited by Gryphon is similar to simple routing without advertisements. Their work concentrates on the use of multicast and efficient matching of events to subscriptions [6]. The matching algorithm clearly outperforms the simple sequential algorithm, but it depends on and supports only a few types of attribute filters, limiting its usability. Moreover, updating the matching data structure if clients subscribe and unsubscribe is costly.

The load caused at the individual brokers was investigated in the first article mentioned above [26]. The results presented show that flooding overloads at the same publishing rate regardless of the percentage of matches or the number of active subscriptions. Filtering-based routing, on the other hand, can handle much higher publication rates if subscriptions are highly selective or highly local, which can be expected in large-scale publish/subscribe systems.

The second article [291] concentrates on bandwidth utilization. It compares flooding to four multicast-enabled routing algorithms and ideal multicast, which assumes that for each event a perfect multicast group exists. The authors state that filtering-based routing is superior to flooding under conditions of high selectivity and high locality of subscriptions. This opinion supports the findings of this work. Nevertheless, it can be expected that their results are still too pessimistic because their work depends on simple routing, i.e., the routing algorithm does not exploit covering and merging. They also assume that event brokers are not placed nearby to the multicast routers and therefore the use of multicast may even introduce a bandwidth penalty. Moreover, they did not investigate the use of advertisements. Mühl et al. [263, 267] have investigated a set of routing algorithms and their effect on the routing table sizes and the filter forwarding overhead.

5

Engineering of Event-Based Systems

In the previous chapters we have learned what the infrastructure of a distributed notification service looks like. This chapter starts to look at the engineering issues in event-based systems.

The first part of the chapter presents main engineering problems, which are partly derived from experience in request/reply-based systems. Looking at example scenarios we see that current functionality is well suited for simply structured systems, but essential software engineering paradigms are hardly supported, which makes the engineering of complex systems very hard. Chapter 6 will detail these higher-level engineering issues.

In Sect. 5.2 we describe different forms of application programming interfaces (APIs) and Sect. 5.3 concentrates on how applications use the API. Besides directly accessing an API, code instrumentation and aspect-oriented programming are candidates for adding publishing functionality to existing application code. Some programming languages even provide intrinsic event handling mechanisms, like C# or some extensions of the Java language. Furthermore, we discuss what data items contribute to an event and does every change lead to a publication?

5.1 Engineering Requirements

This section analyzes engineering issues and points out shortcomings of many current services that make them difficult to maintain, let alone control, and that impede their use in complex application scenarios. The deficiencies are illustrated with the help of example scenarios, and a set of engineering requirements are inferred that should be addressed by event systems. Two main problems are identified. The first is that event-based systems basically do not imply other requirements for designing and engineering than those already known from engineering request/reply systems. The second observation is that while supporting abstractions are available for the latter, they are missing for event-based systems.

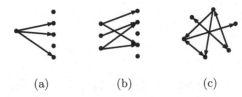

(a)   (b)   (c)

Fig. 5.1. Data flow graphs of applications: bipartite single (a) and mult source (b), and a general group (c)

5.1.1 Application Examples

A taxonomy of application scenarios is created according to the complexity of interaction between application components. A data flow graph describes who is sending notifications to whom: one-to-many, many-to-many, and repeated, "stateful" communication.

Information Dissemination

The simplest and most obvious application scenario of event-based communication is information dissemination and push services. It is typically characterized by a single, well-defined information source publishing notifications toward consumers (one-to-many communication). Applications are oblivious to the actual set of receivers and typically require high scalability. The call graph is bipartite, cf. Fig. 5.1a, which means it consists of two distinct sets of components and messages are sent only between, not within the sets. Example applications are:

- *monitoring* of stock prices, sensor data, real-time control systems, process execution, etc. [177, 224, 255]
- *push services* in electronic commerce, news feeds like weather forecasts and sports [73, 109]
- *content delivery networks* [8, 333]

This is the classic application domain of event-based systems, and also of network-level multicast [321]. However, even in this simple scenario issues arise that are not covered by typical event services. The weather information may contain temperatures in Fahrenheit, whereas consumers expect degrees centigrade. Stock quotations may be published using an established financial markup language like FIXML [273] to facilitate interoperability with external system, whereas internal communication stick to more efficient binary representations. The heterogeneity of data models and the limited support thereof often demands manual adaptations before connecting components to information buses.

Furthermore, security in event-based systems is a critical open issue. Who is allowed to view sensor data that monitors a person's presence or health?

Access to real-time stock quotations may be restricted, requiring subscriptions with additional fees.

Groups of Producers

In Fig. 5.1b a slightly more complex scenario is depicted that includes multiple producers publishing similar notifications. This raises new problems if it is necessary to distinguish the sources, especially when systems evolve from the type shown in Fig. 5.1a to that shown in Fig. 5.1b. Consider

- *multiple* stock markets or auction platforms publishing similar information [47, 138]
- *multiple* application-specific beacons or sensors that are deployed somewhere in the infrastructure [18]

When a system implementing one stock market is connected to another market, measures must be taken to prevent unintended effects on existing consumers. It must be possible to restrict communication to one market so that components do not react incorrectly to external events. The necessary distinction of markets is often achieved by simply having producers annotate notifications with a name or an ID (of the market, for example). Here, producers encode the context of an event in the notifications, e.g., the market from which it originated. Consumers operate in a specific context if they test for this information in their subscriptions.

This is a straightforward approach, of course, but it draws context knowledge into application components that pertains to the interaction and not to the component's implementation. Moreover, this context specification not only counteracts the characteristics of event-based systems, but it is unnecessary *within* the respective context. Consider the second example where presence awareness sensors inform about people/objects moving within a building. The notifications include an ID of the object tracked and a room number. If events from multiple buildings are integrated in a facility management application, an identifier of the "source building" must be included in the notifications. This approach would increase the coupling as it influences the internal configuration of components when applications are integrated.

Therefore, application components should not be forced to deal with their execution context. They would have to consider all possible contexts, which inhibits runtime evolution and is neither desirable nor needed.

Complex Interaction

The third class considered comprises complex applications that have arbitrary call graphs and include bidirectional communication (Fig. 5.1c). Examples are:

- chat groups, multiplayer games, or computer-supported cooperative work (CSCW) tend to cluster interacting groups of components [117, 159].

- *virtual marketplaces* exhibit complex interactions where sequences of published notifications are interrelated, e.g., auctions [47, 138].
- wireless sensor networks [9, 205] convey data from sources to sinks and process and filter data within the network.

Apart from the last example, such scenarios are seldom considered in the context of event-based systems. They are typically based on request/reply, although their interaction is often event-based in essence: the initiator is the producer of data and destinations are chosen indirectly, e.g., based on roles or interests. Producers may get some information back from their consumers, but not necessarily by replies. Such feedback is due to events triggering other events and notifications following loops in the data flow graph. This should not preclude such applications from exploiting the flexibility of the event-based architectural style.

The requirements posed in these scenarios, however, exceed pure scalability considerations. The examples show that the principle of locality is important in event-based systems, too. Clusters of interacting participants can be identified as part of larger applications; the data flow graphs are more dense within these clusters than toward the outside. And within such groups often more stringent requirements are placed on communication quality. For instance, a chat application exchanging user input via notifications will certainly gain from ordering guarantees for notification delivery, e.g., atomic broadcast providing each participant with the same perceived order of inputs. In general, intracluster communication may require dedicated services, whereas interaction with the remaining system gets by with the basic functionality of notification dissemination.

Virtual marketplaces illustrate the need to group notifications. Producers and consumers do not know each other but must establish a conversation[1] by relating notifications that belong to the same auction. Again, a simple workaround is directly found by inserting identifiers in notifications, and the same counterarguments as above still apply. Identifiers may be viable in this simple case, but in more general terms the context of notifications must be distinguished to relate bids to auctions, reactions to actions, and events to transactions.

5.1.2 Requirements

The above discussion exemplifies the problems raised by the loose coupling of the event-based style: effects and side effects, design, implementation, and engineering, management, and security issues. From these problem domains four requirements for the engineering of event-based systems are inferred: bundling of components, support for heterogeneity, flexible customization, and support for activities.

---

[1] Repeated, possibly bidirectional communication.

Illustrative Example

A stock trading application will be used as an illustrative example. It shall not, of course, describe a perfect architecture for stock trading. The example illustrates most of the aforementioned problems and helps underline the requirements of engineering event-based systems.

The following components of a stock market can be identified (Fig. 5.2):

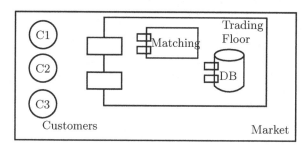

Fig. 5.2. An example stock trading application

- Customers monitor quotations and issue orders to buy or sell shares.
- A central matching engine implements the matching algorithm and generates quotations.
- A database logs the generated data to ensure consistency and persistence, and to audit the operation.

Nearly all parts of a stock trading application are inherently event based. The dissemination of stock quotations from the central trading floor (or its computerized equivalent) to the market participants is an accepted and plausible example of applying event notification services. The database and matching engine are composed into the virtual trading floor, a component which consumes orders and publishes notifications carrying share prices of successfully executed trades.

Bundling of Related Components

Locality, encapsulation, and the composition of existing components into higher-level units are well-known concepts for mastering complexity and for supporting evolution [300]. These concepts are used in request/reply systems, but they are equally important here. The grouping of components that share some commonality or achieve a common goal is a prerequisite for reasoning about effects and side effects, and it is the basis for addressing both engineering and management issues.

Bundling is both a syntactic and a semantic abstraction. From the syntactic point of view such a bundle limits the distribution of notifications produced within; it identifies notification delivery localities. The bundling mechanism should be orthogonal to any subscription mechanism so that grouping is independent from component implementation and it should not influence the subscriptions issued by them. This is important to draw locality not only based on the described interests of consumers but also on other criteria, such as organizational and geographical constraints of a company or some other application-specific semantics.

From the semantic point of view, bundles of components must be components themselves with their own semantics. The bundles should not only limit distribution, but should also publish notifications themselves as the result of notifications produced within the bundle, indicating important state changes of the bundle as a whole. Similarly, they should consume notifications from the outside by further propagating them to their internal participants. This opens the possibility to recursively create higher-level components and to hierarchically structure an event-based system.

Consider the running example. The virtual trading floor in the stock trading application is the first candidate of a component bundle. One can imagine a "verbose" matching engine producing detailed notifications about the progress of the matching algorithm, of which the majority is only relevant for logging purposes (e.g., for auditing system operations) and only a few are relevant for customers. Hence, it makes sense to constrain the visibility of most of the notifications to the database component and to allow only a few of them to pass the boundary of the trading floor bundle.

The next reasonable structuring step would be to bundle the trading floor and a set of customers (i.e., the participants in the market described in Fig. 5.2) into a higher-level syntactic and semantic market component. In this way multiple trading floors could be supported without having customers receive duplicate and inconsistent notifications. Such duplication cannot be avoided in a flat design space, where all components in the system are visible to each other. The absence of market bundles would require users to encode knowledge about the market structure into the subscriptions of individual components, which impedes reuse and system evolution (cf. Sect. 2.1.3).

Supporting Sessions and Activities

The engineering of complex systems benefits not only from bundling related components according to application structure but also from grouping notifications into sessions. Be it because notifications originate from the same source or because they belong to a set of cooperating components, sometimes it is necessary to distinguish sessions of dependent interactions to identify conversational state. This is especially important in event-based systems, where the identity of peers is unknown. That is, without any additional information consecutive notifications cannot be related to each other. The publish/subscribe paradigm does not offer any intrinsic means to identify conversational state other than introducing IDs manually.

An example for sessions is a stockbroker who listens to a specific share traded on two stock markets. Obviously, notifications distributed in one market must, generally, be invisible in the other. However, the stockbroker should be able to observe and distinguish both. In general terms, individual components should identify and participate in multiple sessions, delimiting them from each other to support session state. However, taking up the discussion about IDs in Sect. 5.1.1, it is generally undesired to have components do session handling on their own. From an engineering point of view, it complicates their implementation.[2] More importantly, it reduces the loose coupling of publish/subscribe by explicitly tangling notifications and interaction control. Using IDs is an ad hoc approach to distinguish groups of producers, but it hides the fact that the underlying problem of session handling is not yet addressed directly.

Furthermore, activities comprising bundles of notifications can be modeled as well-defined structures as described for bundles of application components above. Activities structure the interaction in the system and in themselves are components with well-defined semantics. Drawing on localities of distribution, they can determine when "internal" notifications are to be made visible to the outside. This will help to prevent side effects, to build structured, hierarchical sessions, and to customize and orchestrate them. Activities thus correspond to a simplified version of the notion of transactions known from the world of request/reply-based systems [102, 182].

Mastering Heterogeneity

A single uniform event notification service with uniform syntax and semantics is hardly able to cope with the diverging requirements of large distributed systems, which typically operate in heterogenous environments [80]. As pointed out in the examples of Sect. 5.1.1, an event service that, e.g., relies on a global naming scheme is not scalable and complicates system integration. Furthermore, syntax and semantics of notifications are likely to vary and there are inevitably different data models in use, which can be induced by hardware-dependent issues (like bounded message size) or by middleware or application-layer differences. While heterogeneity is a well-known problem in other areas of computer science, it only recently started gaining attention in the context of notification services [80, 146, 185].

From the observations above an apparent conclusion is that bundling of related components should not only encapsulate functionality but also delimit common syntax and semantics. This requires mechanisms to support adapting data that cross boundaries of component bundles by mapping content and representation. To motivate the requirement consider again the running example. For efficiency reasons it is reasonable to distinguish between low-volume external representations in XML versus more optimized internal representations.

---

[2] Enterprise JavaBeans introduce session beans as a remedy to this problem in the request/reply approach.

The matching and database components may use a binary representation, while stock quotations are published using an established financial markup language like FIXML [273] to facilitate interoperability. Hence, transformation between the external XML representation and the internal binary representation would be needed for notifications crossing the border of a trading floor composite.

Flexible Configuration and Customization

Similar to the heterogeneity discussion, a static definition of notification transmission semantics is not adequate either. The service must be adaptable, and it must be configured to meet applications needs. As pointed out in Sect. 5.1.1, subsets of closely interacting participants often rely on communication guarantees that differ from those of basic notification dissemination. This includes ordering or real-time guarantees that refine the specification of the simple event-based system given in Sect. 2.5. But application-specific needs may also demand deviation from this basic specification. For example, instead of the default "broadcast" of notifications to all eligible consumers with matching subscriptions, only a specific subset of them may be selected due to an application-specific policy. An 1-of-n policy realizes load balancing within a bundle of components, and outside of the components themselves.

In the stock trading application, the matching engine might be replicated to distribute processing load over multiple instances using a delivery policy that routes orders to instances dedicated to the respective share. Furthermore, if the structure of the bundles is not static, security policies must control who is allowed to join. The trading floor component could be compromised if everyone is allowed to join and issue notifications influencing the matching engine. On a lower level of adaptation the implementation of the trading floor will use broadcast mechanisms of a local area network, whereas the dissemination of price information on the Internet has to use other techniques.

In general, the ability to adapt and program bundles of components tackles the design, implementation, and engineering problems stated above. The whole event service is subject to customization with respect to these bundles: API, syntax, and semantics of subscriptions and notifications, security policies, and implementation techniques of notification dissemination must be tailored to fit the needs of evolving complex systems.

5.1.3 Existing Support

The bundling of components is the basic requirement presented in the previous paragraphs, and it complies with the fact that information hiding and abstraction have long been identified as a fundamental principle in software engineering [300]. In request/reply-based distributed systems, like the CORBA platform [283], solutions exist for all of the outlined requirements. Object-oriented programming and decomposition, heterogeneity by standardized interconnection protocols (e.g., CORBA-IIOP, SOAP [400] based on XML), bundling of activities with the help of transactions [37, 281], and security services, e.g., Kerberos [270], provide the appropriate support.

However, comparable hierarchical structuring mechanisms are missing in event-based systems. The missing knowledge about communicating peers leads to the desired separation of communication from computations. But control of component interaction is drawn out of the application components themselves, and any adequate support for the mentioned requirements must respect and facilitate the external control of interaction. Unfortunately, existing services recognize and address these issues only partially.

A first approach to achieving these goals would be to build on existing features of notification services. For example, one could make use of content-based filtering mechanisms [71, 262] to decompose and delimit sets of components and notifications from each other. Subscriptions can be adapted to encode additional constraints on the decomposed structure. This approach of modifying application components counteracts the stated separation. Knowledge about the application structure is put into the components, contradicting the idea of components being loosely coupled and self-focused. Furthermore, the structure is not explicitly enforced by the system so that components can deliberately modify their subscriptions to evade security measures. Subject-based addressing is too limited to implement any sensible structuring in addition to existing subscriptions, because different points of view are not supported. Event channels like in the CORBA Notification Service support structuring in addition to notification selection to some extent. However, individual components still have to select channels manually.

The above points showed shortcomings of the plain API, which makes publish/subscribe communication difficult to maintain, let alone control, and that impedes its use in complex application scenarios. The deficiencies are analyzed with the help of example scenarios, and a set of engineering requirements are inferred that should be supported by event systems. The next chapter introduces a scoping concept that addresses the underlying problem of controlling notification visibility and serves as a tool of both application design and event system implementation.

5.2 Accessing Publish/Subscribe Functionality

5.2.1 Generic APIs

In Sect. 2.1 we described the constituents of a publish/subscribe system and sketched its minimum functionality. Figure 5.3 depicts the interface operations a generic implementation should provide. A number of standards are available that define publish/subscribe APIs. Most notably, there are the Java Message Service (JMS) [364] and the CORBA Notification Service [287], which we will detail later in Chap. 9. They both include this core set of functions, but differ in the filter models they support and other higher level functionality. The characteristic differences of publish/subscribe services can be summarized in the following points:

- `publish` and `subscribe` are mandatory API operations. All implementations must provide these operations.
- Advertisements are optional because they are not necessary for the main functionality. Hence, `advertise` and `unadvertise` are optional.
- Data and filter models are an important factor distinguishing notification services. In most cases they are predetermined and only one model is available.
- Notification services differ in the quality of service (QoS) they offer. Under a similar API, reliability, performance, and dissemination semantics vary largely.
- Black box or open implementation decide about adaptability. The layout of the underlying infrastructure may be completely hidden from the application, or it may be open for adaptation to better cope with aspects of heterogeneity and customization.

The availability of `publish` and `subscribe` API calls is an obvious necessity. Only their signature varies because of differences in QoS as well as in data and filter models as described below. If the `unsubscribe` operation is not provided, the subscription will usually be valid only for a limited time. Advertisements are not mandatory for publish/subscribe communication, and many systems do not offer them. However, advertisements help to improve resource usage (e.g., by enabling routing optimization [267]) and offer means for implementing other higher level features like security [34]. They also allow clients to determine the potential notifications that might be published by producers.

The data and filter models offered by a notification service are critical for its usefulness regarding a given application. Nearly all major standards and nonstandard implementations differ in this respect. However, from an API point of view these models can be transformed into each other with the help of additional wrappers. Of course, this affects performance, but such wrapping is often necessary anyway. Complex applications hardly fit into only one data and filter model.

`publish(n)`	publishes a notification n
`subscribe(F)`	subscribes to a filter F
`unsubscribe(F)`	revokes a subscription
`advertise(F)`	all publications will conform to filter F
`unadvertise(F)`	revokes an advertisement

Fig. 5.3. Generic publish/subscribe interface

The more important distinguishing factor is the level of QoS offered by the publish/subscribe API. The definition of communication semantics in Sect. 2.5 is only an outline that any concrete implementation will refine in one way or another.[3] If the API offers more than one default behavior, it should not be the `publish` and `subscribe` methods that take the QoS parameters. First, it is unlikely that each publication is issued with different QoS parameters. And second, in event-based systems producers do not know how important their notifications are for the consumers, and thus they should not determine the QoS. For example, if the notification service uses channels, we can determine the QoS characteristics at channel creation time. And we can provide the channels to the producers from the outside, which is also current software engineering best practice and is known as dependency injection [153].

Open implementations [209, 220] are a more generic alternative to QoS-rich APIs. An open implementation enables system engineers to not only wrap existing functionality but also to extend the service from within. This approach is also known as reflective middleware [96, 223]. Interceptors, hooks, aspects, dependency injection, etc., are the vehicles to insert code into the implementation of an existing notification service. These techniques change the internal behavior and help to adapt the middleware below the API. Currently, only some research prototypes of notification services use an open implementation. On a lower layer, active networks also exploit this idea to construct network infrastructures that are open for customization even after deployment [373].

5.2.2 Domain-Specific APIs

When focusing only on one specific application domain, it is often convenient to offer publish/subscribe functionality through domain-specific communication APIs. We can distinguish two different approaches: the APIs either act as wrappers, using other terms for generic communication facilities, or they really provide an implementation tailored to domain characteristics. The first alternative is often used for implementing typed on top of untyped eventing.

If different terminology is used to offer an otherwise generic publish/ subscribe service, the classic wrapper or adaptor pattern is used [57, 161]. Internet NewsNet, newsgroups, and bulletin boards are examples.[4] News is posted without destination and is classified in newsgroups, and readers must select the posts in which they are interested. Technically similar approaches are Linda Tuple Spaces with their `in` and `out` commands [7, 64].

Other examples include building technology and control systems in general. In nearly all of these systems, state changes are signaled: elevators move, lights are turned on, temperature is monitored. Signaling is usually done using APIs and messages that correspond to the domain entities, but those applications are essentially event-based. In order to easily set up and configure a

---

[3] Note that performance metrics are not considered as part of this API discussion.
[4] In fact, they also add some persistence mechanisms that we will not consider here.

specific configuration of the control system, system engineers exploit the indirection of publish/subscribe. In business applications, databases often act as central information hubs. Newly entered tuples trigger other operations, and so the database can be seen as the notification service conveying data and invoking reactions in consumers, see [172, 385].

Since events are used in many areas of computer science, APIs were created with different terminology but the same notion of communication. Implicit invocation is an early example that views publish/subscribe from a software engineering perspective [164].

The other approach to domain-specific APIs is to utilize specialized implementations that are not generic anymore but are instead tailored to the specific domain. Signaling in telecom networks is like publish/subscribe, but typically relies on specific assumptions about the hierarchical structure of the network. Alarms in network management and SNMP traps similarly employ an event-based style, yet alarm processing often is handled along a chain of management stations. Each station may escalate event handling by forwarding the notification to the next station.[5] A pragmatic implementation can exploit this domain knowledge to send alarm notifications just to the next management station.

Domain specific APIs and implementations are good for optimizing the specific application. Nevertheless, system engineers should not forget that, in principle, they still follow an event-based style. Publishing an alarm notification is inherently event-based; even if we know that it is delivered to only one consumer, the producer should not rely on this structural information. As pointed out in Chaps. 1 and 2, the benefits of loose coupling can only be fully exploited if the event-based architectural style is clearly identified in the participating components; mixing styles complicates changes. And one way to clearly identify the style is to use the plain publish/subscribe API somewhere in the application stack.

5.3 Using the API

Whatever the API looks like and whatever QoS it offers, the application programmer has to decide how to use the API. We look at patterns for using a publish/subscribe API and instrumentation techniques that "automatically" invoke the API, and we raise the questions whether all changes shall be published and how long notifications are *in* the system. These issues go deep into the software engineering aspects of event-driven systems, and we can only touch upon them in this book.

---

[5] We deliberately disregard the annotations commonly added in each step.

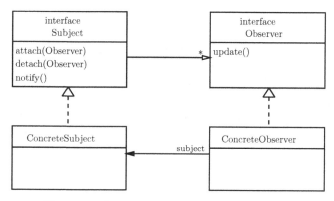

Fig. 5.4. The structure of the observer pattern

5.3.1 Patterns and Idioms

Best practices in designing software are compiled in software patterns. The observer pattern [161] or publish/subscribe pattern [57] are two examples that summarize the idea of having a producer that sends notifications to registered consumers (Fig. 5.4). Observers are consumers and they register their interest in receiving notifications at specific subjects, which are the producers of data. Each subject maintains its own list of registered observers. Whenever the subject changes, it calls the observers stored in its list.

In both patterns, however, the producer itself stores the list of consumers, and the consumers have to register at individual producers. The patterns correspond to the callback interaction model described in Sect. 2.2. The characteristic decoupling of event-based communication is very restricted in this way. Nevertheless, this pattern is used extensively in contemporary systems, ranging from graphical user interfaces to the listener and delegate concepts in Java and C# (see below).

The *Event Channel* variant offers complete decoupling [57]. It essentially employs the idea of the mediator and broker patterns to decouple producers and consumers, and it corresponds to the channel filtering described in Sect. 2.1. The *Event Notification* pattern [327] extends the observer pattern. Producers and consumers define different types of events they are going to produce and consume as part of their interface—but it also contains the direct reference from consumer to producer.

On the subscriber side, the *Reactor* pattern [337] dispatches incoming notifications to one or more handlers. One reactor thread waits on operating system handles for indications of incoming notifications and calls appropriate handlers to fetch and further process the data. The single-threaded dispatching limits concurrency, but other patterns (e.g., *Leader/Followers* [337]) mitigate this problem, leading up to the implementation of scalable Internet services, which is detailed elsewhere [389, 392]. The *Proactor* pattern [337] describes how to handle replies of asynchronous requests, but does not focus on event-based systems.

The term *implicit invocation* [164] describes loose coupling in the context of classic procedure calls. Unlike the above patterns, the invoked procedures (i.e., the consumers) do not determine the calling procedures (i.e., the producers) a priori. The invocation of a procedure is divided into three parts: (i) a call on the caller's side is bound at runtime to a set of procedures, introducing a one-to-many indirection; (ii) the bound procedures are invoked concurrently; and (iii) multiple replies are handled.

An alternative approach is to offer publish/subscribe functionality through data structures, for which distributed asynchronous collections are one example [126]. A publish/subscribe API is not explicitly visible, but it is exploited to realize a distributed shared memory with a generic collection API. Distributed hash tables (DHT) follow essentially the same idea, but typically have a completely different implementation [316, 351]. They store data in a distributed data structure and we can disseminate content in this way, cf. [85, 374].

Next, we will take a look at how current programming languages incorporate asynchronous communication and invocation. A patternlike implementation specific to a programming language is called indexidiom*idiom* [57]. In Java, listeners implement the observer pattern. The Swing GUI library uses listeners, and the `java.nio` networking code offers selectors as needed in the Reactor pattern. Both may serve as basis for a subscriber implementation. However, standard Java does not provide any dedicated support for notification delivery and handler methods. Eugster et al. [127] and Damm et al. [99] added `publish` and `subscribe` keywords to the Java language and generated standard Java code with a precompiler.

In C++, function pointers are a low-level primitive, e.g., for referring to handler functions. The Qt library also relies on a precompiler to extend the language with signals and slots [383]. These indicate notification sources and destinations, respectively. Slots are explicitly connected to signal sources, following the observer pattern. The Boost library uses templates to build signals and slots without a precompiler [4].

C# introduces the delegates concept for event handling [395]. Delegates are a type-safe way to treat methods as objects, which can be passed and stored like any other object. The following snippet defines a delegate and calls `Method1` by invoking `d1`:

```
public delegate String SomeDelegate (int x, float y);
SomeDelegate d1 = new SomeDelegate(Method1);
String result = d1(42, 3.14);
```

Additionally, delegates support list operations of C#. We concatenate two delegates like this:

```
SomeDelegate d2 = new SomeDelegate(Method2);
SomeDelegate d3 = d1 + d2;
```

```
result = d3(42, 3.14);
```

which calls `Method1` and then `Method2`, returning the result of the latter invocation. The **event** keyword defines an instance variable that stores delegates and can only be invoked from within the defining class.

Functional constructs like Lambda expressions or closures [152], which are available in programming languages like Smalltalk, Python, Ruby, or Perl, can be exploited to achieve similar solutions [272].

5.3.2 Emitting Notifications

From the preceding discussion we know various mechanisms for accessing the publish/subscribe API. The important questions from an engineering point of view are now:

- What are the appropriate data sources for detecting events?
- What additional information is put into a notification?
- Publish all changes of the identified data sources or only a subset?
- Publish notifications immediately or defer publication?
- Publish if there are no consumers?

Event Sources and Notification Content

In short, appropriate data sources for events are application specific and there is no generic rule for selecting them. Event "detection" includes the case where the producer modifies data and then publishes a change notification.

In most cases, however, events come from central parts of the software architecture. That is, application engineers can identify relevant events within the domain model and other high-level diagrams of the application. For example, it should be easy to identify classes as possible event sources in UML class diagrams [288].

There is currently no established way for identifying events and notifications in UML diagrams, but stereotypes and tagged values can be used informally to annotate classes as event sources. A stereotype is usually defined as part of a UML profile, which extends standard UML, and it can be added to classes, associations, and attributes. An «event-source» stereotype added to a `thermometer` class identifies a source of temperature events (Fig. 5.5). And a «notification» stereotype put on an association to a `TempNotification` class defines the corresponding notification.

Code Instrumentation

Fortunately, old ideas from code instrumentation for system monitoring (e.g., [231]) and new ideas from aspect-oriented programming (AOP, [119]) can be combined to generate code for publishing notifications. AOP separates code

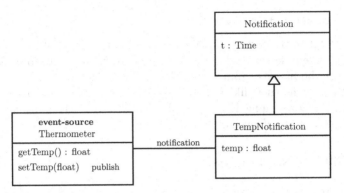

Fig. 5.5. Event and notification in a UML class diagram

that deals with different aspects of a program without destroying an underlying object-oriented design. Common examples are business functionality versus logging or persistence—or eventing. At deployment the separated code is woven together to accomplish the combined functions.

Once event sources and notifications are identified as shown above, we are able to publish a `TempNotification` whenever the temperature changes on a thermometer, for example, with the following AspectJ advice [226]:

```
public aspect TempNotificationAspect {
  after(Temp t) : call (Thermometer.setTemp(float))
                  && args(t) {
    psService.publish (new TempNotification(t));
  }
}
```

Of course, if publishing must be announce with advertisements, aspects can be used for instrumenting constructors accordingly. With dynamic AOP, we would be able to do such modifications even at runtime [45].

Change Encoding

The application developer has to decide how the observed changes are specified in notifications. A notification can carry

- a copy of the data item, e.g., a Thermometer object
- the delta of the change, i.e., the difference to the previous publication
- a copy of both the old and the new value

The first approach simply publishes the new data and consumers have to store and compare old values if they are interested in the difference. Filters cannot test relative changes as they are usually stateless. Sending only the differences in the second approach minimizes message size, but makes it harder for consumers to maintain the absolute value. They must reliably receive and process the complete stream of notification to not run out of synch. *And* they must initialize themselves through another path to the data. The last alternative certainly contains all possible information, but increases the message size considerably. In object-oriented settings, the developer additionally has to choose whether to send the complete object or just the parts that changed. In this case, we might end up with a combination of the first two points.

Adding context information to notifications

A notification should contain as much additional information as necessary to enable consumers to correctly interpret the context of the event. Therefore notifications typically contain more data than only the changed value. This could be:

- time of event occurrence
- source identifier
- sequence number
- geographical context, e.g., room location of a thermometer data source
- organizational context, e.g., security domain of detected intrusion

Unfortunately, the downside of loose coupling is that producers do not know what additional information their possible consumers need. So far, we cannot solve this conflict and notifications have to carry all information that *might* be necessary—but we will discuss one solution in the next chapter.

Omit Some Changes

So far, a properly set up consumer gets every notification that matches its subscription. This may lead to situations in which too many notifications are delivered. We discuss stateful filters and rate control as countermeasure.

Consumers do not always need every single update. Consider, for instance, a producer publishing temperature notifications with a precision of $0.1°C$ while consumers are only interested in changes of $1°C$. Or the consumers are interested in changes of a certain percentage only. Otherwise, one oscillating event source might continuously send notifications triggering other parts of an event-based application unnecessarily. Generally, if producers work with higher data precision than their consumers, this not only wastes network and processing resources, but may lead to instable systems.

This example already indicates that stateful filters are necessary to smooth the data stream. Filters without state evaluate each notification independently and they are inadequate for the mentioned examples. If a filter keeps the last n values, it is able to compute average and deviation. Shah et al. [342] uses coherency constraints to limit the maximal difference between observed and actual values. The stateful filters are either deployed within the publish/ subscribe service and help reduce the network load, or they reside on the clients. Discarding messages on the client side (client-side filtering) makes sense in cases where filtering capacity of the publish/subscribe middleware is the limiting factor.

The other obvious problem of too many notifications is that consumers get overloaded and are unable to process all incoming notifications. If available, we can use flow control in the underlying communication layer to throttle incoming load. For example, if consumers are connected to the broker network via TCP connections, delivery is delayed as long as the client TCP buffers are full. This handles short load peaks, but when the in-network queues get filled messages must be discarded. Several approaches exist to do this intelligently. As usual, filters and consumers can be prioritized to keep important messages. In this way, all notifications, e.g., of type `CriticalAlert`, and all temperature notifications with values above a certain threshold are not discarded. We get a completely different approach if notifications can be summarized, e.g., to deliver only the average of a sequence of temperature notifications.[6] The specification in Chap. 2 currently forbids this behavior, because of the performance impact on brokers and since interference with different subscriptions is hard to predict. Yet, it might be a helpful extension for some application scenarios.

In cases where the size of the message is the main load problem, notification summaries are used to forward only a small portion of the data to the consumer. The remaining part is stored in the network (e.g., in a database) and the notification carries an identifier or uniform resource locator (URL) to access this data instead.

Deferring Publication

It is not always appropriate to publish notifications directly when observing the event. Consider a component that changes multiple data items, publishes each of the changes, and then detects a failure that invalidates the computations. If the modifications include the addition of an item to a list, for instance, it should not be on the list after the failure. In short, we need a transaction [182] to make several **publish** calls succeed or fail in combination.

The Java Message Service (JMS) defines local transactions that support exactly this behavior. The producer associates itself with a transaction and all subsequently published notifications are held back until the producer confirms the publication in a second step, i.e., commits the transaction. The transaction is local because it cannot incorporate resources from other nodes. The data distribution service (DDS) has such a function, too, cf. Sect. 9.1.4.

This simple support for deferring notifications is a building block for other, more sophisticated services. If the local event broker commits the transaction after successfully transmitting the messages to the border broker of the network, the producer reliably handed over the data to the publish/subscribe service. Such acknowledged handshakes are important for building reliable distributed systems [239].

---

[6] In signal processing such transformations are called downsampling.

Publication on Demand

Event-based systems often run at maximum load. Components operate on all incoming data and produce new notifications without caring whether any consumers are willing to accept them. Furthermore, complex components possibly create many different kinds of notifications on different levels of detail. And we rarely need all state changes on any level of detail. This is situation dependent and changes during the execution of the system. Consequently, producers probably send many notifications for which no consumers exist.

The Elvin system has introduced *quenching* to reduce the unnecessary load [341]. If the last consumer unsubscribes, the producer's notifications are discarded. It continues processing its own incoming notifications, but a `publish` call has no effect. The producer may even be notified about the last consumer leaving so that it can inactivate itself. If it unsubscribes in turn, complete sequences of producers/consumers stop.

Once a new subscription enters the system, the inactive producers are reactivated (recursively) to continue normal operation. This is essentially an open field of future work.

5.4 Further Reading

The discussion of engineering requirements can only be preliminary, showing an initial approach to understanding the problems inherent to the design of loosely coupled systems. In general, the engineering of event-based systems is possibly the field that can do most for the broad adoption of event-based systems, but which is, at the same time, the one understood least. Essentially, we hardly know the event-based analogs of object-oriented design [378], transactions [237], and security [290], to name just a few well-known concepts from the "classic" world of computing.

The discussion of publish/subscribe APIs can be supplemented with a review of existing standards, like CORBA Notification Service, JMS, DDS, etc., which are detailed in Sect. 9.1. Chapter 9 also includes reviews of selected notifications services. Alternatives to the plain publish/subscribe API can be found in the area of domain-specific languages (DSLs, [129, 379]). They are an interesting starting point for elaborating new APIs.

Best practices for using publish/subscribe can be found in the pattern community, e.g., [56, 57, 161, 337] and the respective conferences and journals, e.g., EuroPLoP, OOPSLA, etc.

Central engineering questions were touched when we discussed selecting appropriate data sources, notification content, and when to publish the data. These are issues of data and control flow modeling, i.e., software design. There is little experience on methodologies for designing events or to what extent object-oriented methods are suited in the context of event-based systems. A good starting point to follow up on this topic is the book of Luckham [242], which focuses on complex event processing. Another good complement are more formal treatments like [396].

We shortly presented the idea of quenching producers when no consumers are active. This idea can be generalized to sequences or networks of producers/ consumers that are turned on and off. They have their own requirements on shutdown timeouts and restart times.

Furthermore, if we do not restrict ourselves to a predefined filter model, we can generalize filters as remotely executed code. Eager handlers [411] ship event handling code toward the sources to reduce network usage and responsiveness. The idea of exploiting mobile code in distributed "incident" handling was also considered in tuple spaces [61, 160], agent systems [62, 175, 393], and active networks [373].

6

Scoping

So far, the presented simple event systems merely provide the functionality to distribute notifications, but still fails to offer any support for coping with the complexities of designing and engineering distributed systems. The main deficiency is the missing control of the interaction in the system, which is only given implicitly. The resulting problems were recognized in different contexts, and the means to address the missing control are centered around encapsulation and information hiding, principal engineering techniques that are relevant here, too.

This chapter investigates visibility as central abstraction to cope with engineering complexity and introduces a scoping concept for event-based systems. As an design and engineering tool, scopes offer a module construct to structure applications and compose new functionality. Second, scopes reify aspects of event communication and thus make them adaptable within the composed modules, e.g., access to underlying communication technologies, delivery to module members, forwarding of events out of the module scope, transforming heterogeneous data sources, etc.

The first section analyzes the notion of visibility in event-based systems and relates it to the requirements defined in Sect. 5.1. The scoping concept is defined in Sect. 6.2, including a formal specification of scoped event-based systems that refines the specification of simple systems given in the previous chapter. Scopes reintroduce control on communication, which was drawn out of the components in event-based interaction, without impairing the benefits of loose coupling. The concept is extended in Sects. 6.3 and 6.4 to include interfaces and mappings; the former further refine visibility control, the latter generalize interfaces to transform notifications at scope boundaries, coping with heterogeneous data models. While communication within scopes is by default like in traditional publish/subscribe systems, the transmission policies presented in Sect. 6.5 adapt the semantics of notification dissemination within scopes. In Sect. 6.6 we sketch a development process for scopes and present a declarative scope language for defining and manipulating scope graphs. Finally, we investigate implementation strategies for scopes in Sect. 6.7 and discuss combining these They open the publish/subscribe service implementation and allow for the integration of a wide variety of communication techniques.

6.1 Controlling Cooperation

The visibility of transmitted data is of little concern in request/reply systems where destinations are explicitly addressed. In event-based systems, however, the visibility of notifications complements subscription techniques, for it determines which subscriptions have to be evaluated at all. Surprisingly, visibility was rarely considered so far.

6.1.1 Implicit Coordination and Visibility

The problems of current event-based systems, which are described in the previous chapter, stem from the loss of control of interaction. This control has been relinquished deliberately in favor of the loose coupling. It is withdrawn from the components, replacing explicit addressing with the matching of notifications to subscriptions. The explicit control of interaction given in request/reply approaches is replaced by the implicit interaction in event-based systems.

The implicit interaction is characterized by an indirection of communication. Producers make notifications available and consumers select with the help of subscriptions. This indirection gives room for a concept complementary to the notification selection done by consumers. The *visibility* of a notification limits the set of consumers that may pick this notification. If a notification is not visible to a consumer, its subscriptions need not be tested at all. Notifications and subscriptions are unaltered, and matching takes place as before but under the constraints of visibility limitations. Clearly, visibility influences the interaction of components; it can even be seen as a means to govern implicit coordination.

The implicit coordination[1] of the components offers the desired loose coupling but makes the overall functionality an *implicit* result of *all* the participating components. However, extracting control from application components must not necessarily mean to have it nowhere. In fact, the requirements posed in Sect. 5.1 demand some form of control on event-based communication. Visibility may offer such a control of notification dissemination.

The implications are twofold. First, visibility is an important factor of implicit coordination, and second, it promises to be an important abstraction in event-based systems. While subscriptions are related to the function of individual consumers, visibility governs the interaction in the system. Hence, the visibility of notifications is essential for the overall function of an event-based system.

---

[1] Explicit and implicit coordination are also termed objective and subjective coordination in coordination theory [326].

6.1.2 Explicit Control of Visibility

The key to exploiting visibility is to regard it as a first-class citizen. While existing work has addressed some facets of visibility, it was never taken as a fundamental concept in event-based systems. Nevertheless, it will prove to be the basis for both controlling and extending dissemination functionality.

Explicit visibility control constrains the areas where loose coupling and implicit coordination are applied. It makes bundles of implicitly interacting components explicit, and these bundles reify the structure of applications. They serve as a tool for designing and programming event-based systems, because once the interaction is localized at well-defined points, additional mechanisms can be applied to control the interaction within and between definite parts of the system.[2]

But how is visibility actually represented in an event-based system? Where is it exposed? Any form of reintegrating control into the components counteracts the event-based paradigm. Whenever notifications are annotated to reach a specific set of consumers, external dependencies are encoded in application components, which defeats the benefits of the event paradigm. Visibility of notifications is not a matter of producers because it concerns interaction and communication, but not the computation within the component. Thus, the necessary control must be exerted outside of the components themselves.

6.1.3 The Role of Administrators

When designing and engineering event-based systems, only the *roles* of producers and of consumers were considered so far. They represent the tasks of designing and programming individual application components. The self-focus of event-based components is mirrored in these roles. They concentrate on internal computation alone and disregard interaction. Due to the implicit coordination, responsibility for the overall functionality is not assigned to any specific role. It is delegated to producers and consumers, but with no adequate support. The preceding discussion corroborates that an additional role in the system to handle visibility is needed.

The obvious implication is to introduce the role of an *administrator* which is responsible for orchestrating components in an event-based system. An administrator may be human, but it can also be comprised of programs and rules that maintain some system properties (cf. autonomic computing).

The main objective of this role is to support component assembling and the management of their interrelationships. This role is employed to associate visibility control with a distinguished role different from producers and consumers. It is similar to those identified in component-based development or in reference architectures of open systems [206]. In terms of coordination theory, administrators are a means of objective coordination providing an exogenous

---

[2] Technically, this is the essence of the scope concept presented in the following.

extension of event-based interaction [36], which separates the shaping of interaction from, and generally makes it invisible to, the computation in the base entities.

Effective means to control visibility in event-based systems are necessary to support the administrator's role, and with respect to the requirements given in Sect. 5.1, such a control is a prerequisite to solving the underlying problems of current event systems. The demanded bundling of related components is directly addressed by the visibility of notifications. Heterogeneity issues can only be solved if communication is intercepted and converted, which requires a limited visibility in the first place. The same holds for the customization and configuration of the event service itself. With limited visibility the interaction within certain system parts may receive a dedicated service tailored to its needs, whereas interaction with the outside is handled differently, like the case of heterogeneous data models.

Unfortunately, current work disregards this important role and does not provide any appropriate support. The scoping concept presented in the next section, however, describes visibility in event-based systems and offers the explicit control needed by administrators.

6.2 Event-Based Systems With Scopes

This section formally introduces the notion of scoping in event-based systems.[3] It extends the specification of the simple event system presented in Sect. 2.5.2 and is the basis for further extensions and reasoning about scoping functionality.

6.2.1 Visibility and Scopes

The notion of scoping in event-based systems is introduced to realize the visibility of notifications. A *scope* bundles a set of producers and consumers and limits the visibility of notifications to the enclosed components. The event-based style of matching notifications and subscriptions is still used within the scope, whereas the interaction of this bundle with the outside is no longer implicit; it is prohibited at first. The notion of scopes serves two purposes. The term is used to describe the visibility of notifications and to name the entity that defines visibility.[4]

Scopes have interfaces to regulate the exchange of notification with the remaining system. Scopes forward external notifications to their members and republish internal ones to the outside if they match the output and input interfaces of the scope. In addition, scopes can recursively be members of

---

[3] see also [135, 146].

[4] In fact, in most cases we refer to the entity, which implies the scope of notifications in the former meaning.

higher level scopes and in this way offer a powerful structuring mechanism. Scopes thus act as components in an event-based system. They publish and consume notifications and can be deemed equivalent to the simple base components considered so far. So, the system consists of simple components and of complex components that bundle other simple or complex components.

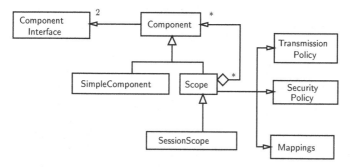

Fig. 6.1. A metamodel of scopes

The concept of scopes as illustrated in Fig. 6.1 includes further features that will be described in the course of this chapter. *Transmission policies* can be applied between scopes and within a scope to adapt notification forwarding, allowing for tailoring notification delivery semantics to application needs in a restricted part of the system. Furthermore, *event mappings* at scope boundaries generalize scope interfaces and are capable of transforming between different data models of notifications. Security policies are a straightforward way to control the access to the scoping structure.

6.2.2 Specification

The notion of components is extended to distinguish simple and complex components. The set of all simple components $\mathcal{C}$ includes any possible software entity that accesses the notification service API. The set of all complex components $\mathcal{S}$ describes all possible scopes. The set of all components $\mathcal{K}$ is defined to be the union of the disjoint sets of simple components $\mathcal{C}$ and complex components $\mathcal{S}$, $\mathcal{K} = \mathcal{C} \cup \mathcal{S}$.

A scope bundles a set of components, and a component can be a member of multiple scopes. To denote the relationship between components and scopes, a graph of scopes is defined.

Definition 6.1 (scope graph). *Let* $\mathcal{K} = \mathcal{C} \cup \mathcal{S}$ *be the set of all simple and complex components. A* scope graph *is an acyclic directed graph* $G = (C, E)$. *The graph consist of a set of components* $C \subseteq \mathcal{K}$ *as nodes and a relation* $E \subset \mathcal{K} \times \mathcal{K}$ *as edges between the nodes so that* $(C_1, C_2) \in E \Rightarrow C_2 \in \mathcal{S}$.

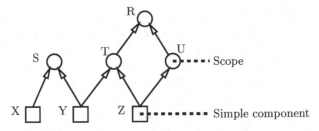

Fig. 6.2. An exemplary scope graph

A scope graph denotes the scope-component relationship. An edge (C, S) from node C to node S indicates that C is a component of *scope S*.[5] The stated property $(C_1, C_2) \in \mathsf{E} \Rightarrow C_2 \in \mathsf{S}$ ensures that a simple component cannot be a superscope of any node in G. C is a subscope if $C \in \mathsf{S}$. Conversely, the scope of a component C is any S such that $(C, S) \in \mathsf{E}$. S is also called superscope of C to emphasize the relationship between S and C, e.g., in cases where C is a scope itself. In Fig. 6.2, X is a component of S, Y is a component of both S and T, and T is a component/subscope of R and superscope of Y and Z.

The edges of the scope graph describe a partial order $\leq$ on C, where $C_1 \leq C_2$ iff $(C_1, C_2) \in \mathsf{E} \vee C_1 = C_2$. Avoiding the reflexivity of $\leq$, the scope-component relation is described by $\lhd$, where $C_1 \lhd C_2 \Leftrightarrow (C_1, C_2) \in \mathsf{E}$. The transitive closure of $\lhd$ is denoted by $\overset{*}{\lhd}$; $\rhd$ and $\overset{*}{\rhd}$ are defined accordingly. In the example of Fig. 6.2, $Y \lhd T$ and $Y \overset{*}{\lhd} R$ hold. According to the partial order, the simple components are the minimal elements and those scopes having no superscopes are the maximal elements of C. Additionally, the following terms are borrowed from graph theory. T is a *parent* of Y, and Y is a child of T. Y is a sibling of Z, and vice versa, i.e., they have the same parent.

Based on these definitions, visibility can be defined formally. In the first instance, the visibility of components is defined, which implies a visibility of notifications.[6] Informally, component X is visible to Y iff X and Y "share" a common superscope.

Definition 6.2 (visibility of components). *The* visibility of components *is a reflexive, symmetric relation v over $\mathcal{K}$, also written as $v(X, Y)$, and is recursively defined as:*

---

[5] Edges could have been defined in the inverse direction to emphasize that components do not need to know their scopes and how they are aggregated. However, the presented notation follows the one originally published in Fiege et al. [140].

[6] The more general visibility of individual notifications is discussed in Sect. 6.3.1.

$$v(X,Y) \Leftrightarrow X = Y$$
$$\vee\, v(Y,X)$$
$$\vee\, v(X',Y) \ \text{with} \ X' \triangleright X$$
$$\Leftrightarrow \exists Z.\ X \stackrel{*}{\triangleleft} Z \wedge Y \stackrel{*}{\triangleleft} Z$$

In the graph of Fig. 6.2, for example, $v(X,Y)$ and $v(Y,U)$ hold, but not $v(X,Z)$.

Using this visibility, the specification of simple event-based systems given in Def. 2.5 of Sect. 2.5 can be refined. For presentation purposes, the specification is at first restricted to static scopes, i.e., the scope hierarchy and membership cannot change once the first notification has been published. This restriction is relaxed later.

Definition 6.3 (scoped event system). *A scoped event system ES^S is a system that exhibits only traces satisfying the following requirements:*

- *(Safety)*

$$\Box\Big[notify(Y,n) \ \Rightarrow \ \big[\bigcirc\Box\neg notify(Y,n)\big]$$
$$\wedge \big[\exists X.\, n \in P_X \ \wedge \ v(X,Y)\big]$$
$$\wedge \big[\exists F \in S_Y.\, n \in N(F)\big]\Big]$$

- *(Liveness)*

$$\Box\Big[sub(Y,F) \Rightarrow$$
$$\Big(\Diamond\big[\Box v(X,Y) \ \Rightarrow \ \Box\big(pub(X,n)\wedge n \in N(F) \ \Rightarrow \ \Diamond notify(Y,n)\big)\big]\Big)$$
$$\vee \Big(\Diamond unsub(Y,F)\Big)\Big]$$

Definition 6.3 differs only slightly from Def. 2.5 in Sect. 2.5. The safety requirement contains an additional conjunct $v(X,Y)$. This means that in addition to the previous conditions, the producer and the subscriber must also be visible to each other when a notification is delivered. The liveness requirement has an additional precondition $\Box v(X,Y)$ that can be understood in the following way: If component Y subscribes to F, then there is a future point in the trace such that if X remains visible to Y every publishing of a matching notification will lead to its delivery at Y. The *always* operator requires the scope graph to be static.

Note that Def. 6.3 is a generalization of Def. 2.5. A simple event system can be viewed as a system in which all components belong to the same "global" scope. This implies a "global visibility," i.e., $v(X,Y)$ holds for all pairs of components (X,Y) and can be replaced by the logical value *true* in the formulas of Def. 6.3, resulting in Def. 2.5.

6.2.3 Notification Dissemination

According to the previous definition, a published notification is delivered to all visible consumers that have a matching subscription. In order to clarify the impact of the scoping structure and the dissemination of notifications through the scope graph, the visibility of notifications is analyzed in the following.

The visibility of a notification n to a component C determines C's ability to deliver this notification at all, and is denoted by $\overset{n}{\leadsto} C$. Visibility is a test that precedes any subscription matching. Subscriptions decide in a second step whether to deliver a visible notification or not. The visibility of notifications in the scope graph is directly related to the visibility of components, of course. The visibility of a notification n, which is published by X, to a specific component Y is denoted by $X \overset{n}{\leadsto} Y$, where

$$pub(X, n) \wedge v(X, Y) \Rightarrow X \overset{n}{\leadsto} Y.$$

A published notification is made visible in the scopes the producer belongs to. $Y \overset{n_1}{\leadsto} S$ in Fig. 6.3a, or simply $\overset{n_1}{\leadsto} S$ to denote the visibility alone if the specific producer is not important. This rule is applied recursively to make notifications visible in all further superscopes; $Y \overset{n_1}{\leadsto} T$ and $Y \overset{n_1}{\leadsto} T'$. On the other hand, if a notification is visible within a scope S, $\overset{n}{\leadsto} S$, it is visible to all its children. Recursively applying this rule yields in Fig. 6.3b $X \overset{n}{\leadsto} T \Rightarrow X \overset{n}{\leadsto} S \Rightarrow X \overset{n}{\leadsto} Y$. Note that edge direction indicates scope membership but notifications can travel in both directions. In summary, notification dissemination is governed by two rules, a publishing policy PP and a delivery policy DP:

$$\textbf{PP}: \quad X \overset{n}{\leadsto} S \wedge X \lhd S \lhd T \Rightarrow X \overset{n}{\leadsto} T \tag{6.1}$$

$$\textbf{DP}: \quad \overset{n}{\leadsto} T \wedge S \lhd T \Rightarrow \overset{n}{\leadsto} S \tag{6.2}$$

Consider Fig. 6.3. A notification n_1 published by Y is forwarded to S and to all children of S, and from S to T and T' and to all of their children, i.e., to all siblings of S. n_1 is an internal notification of S, T, and T', which means it is visible to their children. $X \overset{n_2}{\leadsto} S$ is at first an external notification to S and is made internal by the delivery policy of Eq. (6.2). A notification forwarded in the direction of an edge, e.g., $(S, T) \in E$, is an *outgoing notification* with respect to S; it leaves the scope of S. Conversely, a notification that travels against an edge is an *incoming notification*, e.g., from T to X in Fig. 6.3a or from T to S in Fig. 6.3b; in the latter case n_2 is external to S.

The semantics of notification dissemination is that incoming notifications are forwarded to all children of a scope, and outgoing notifications are forwarded to superscopes and to all siblings. Note that incoming notifications are not forwarded to superscopes; n_2 is not visible to T' in Fig. 6.3 as X is not visible to T'. This default transmission of notification dissemination is the consistent extension of the semantics of simple event systems. The intuitive meaning of scope membership corresponds to this definition. That is, (i)

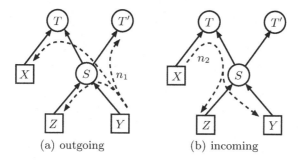

(a) outgoing   (b) incoming

Fig. 6.3. Outgoing and incoming notifications

siblings are eligible consumers as they are in the same scope, (ii) being a sub-scope also denotes a part-of relationship, which makes it obvious that internal notifications are also forwarded to superscopes, and (iii) external notifications are made visible to members of complex components.

Visibility is a set inclusion test so far, which disregards the way a notification becomes visible. In practice, however, the paths of dissemination in the scope graph are of great importance for any analysis of system behavior.

Definition 6.4. *A* delivery path *p between two components X and Y is a sequence of components $p = (C_i) = (X, C_2, \ldots, C_{n-1}, Y)$ for which holds:*

1. p is an undirected path in the graph of scopes.
2. p obeys the visibility v in that $v(C_i, C_j)$ holds for all $1 \leq i < j \leq n$.

Delivery paths are not directed, which means that either $(C_i, C_{i+1}) \in \mathsf{E}$ or $(C_{i+1}, C_i) \in \mathsf{E}$. The dissemination in the scope graph is described by the following

Lemma 6.1. *Every delivery path $p = (C_1, \ldots, C_n)$ can be subdivided into two, possibly empty, parts: an upward path $(C_1, \ldots, C_j)$ where $(C_i, C_{i+1})_{i<j} \in E$, i.e., $C_i \lhd C_{i+1}$, and a downward path $(C_j, \ldots, C_n)$ where $(C_{i+1}, C_i)_{i \geq j} \in E$.*

Proof. Show that p turns at most once. A delivery path $p = (C_1, \ldots, C_n)$ connects two components C_1 and C_n that are visible, $v(C_1, C_n)$. If $C_1 \lhd C_n$, the downward path is empty and C_n is reached by forwarding notifications to superscopes according to Eq. (6.1). If $C_1 \stackrel{*}{\rhd} C_n$, the upward path is empty and C_n is reached by propagating visible notifications to children according to Eq. (6.2). Otherwise, the path turns at least once and two cases can be distinguished: p starts with an upward or a downward edge.

Assume p starts with a downward edge, $C_1 \rhd C_2$. Select d such that $1 \leq d \leq n$ and $C_i \rhd C_{i+1}$ for all $i \leq d$. If $d \neq n$, the downward path is $(C_1, \ldots, C_d)$ and $C_d \lhd C_{d+1}$. However, Eq. (6.1) allows this upward delivery only if the notifications originated in C_d. This is not the case and by contradiction the downward path ends at $C_d = C_n$.

Assume p starts with an upward edge, $C_1 \lhd C_2$. In the same way p starts with an upward path of length $u \leq n$ such that $C_i \lhd C_{i+1}$ for all $i \leq u$. If $u \neq n$, $C_u \rhd C_{u+1}$. However, the path $p' = (C_u, \ldots, C_n)$ starts with a downward edge and from the preceding arguments follow that p' consists only of downward edges.

If p starts downwards, $C_1 \overset{*}{\rhd} C_n$. If p starts upwards, either $C_1 \overset{*}{\lhd} C_n$ or the path turns once downwards at a C_j, proving the lemma. $\qquad \square$

6.2.4 Duplicate Notifications

Between any two nodes of the directed acyclic scope graph there may exist zero, one, or more different delivery paths—the scope graph is not a tree (Fig. 6.4). This may lead to duplicate notifications in certain implementations. The specification of scoped event systems does not consider delivery paths but demands notifications to be delivered at most once. So, concrete systems may violate the specification. However, there are two reasons for not eliminating duplicates in the scope model itself. First, duplicates generation and handling is highly implementation dependent. And second, in some applications delivery along different paths leads to different semantics of notifications so that they are not really duplicates.

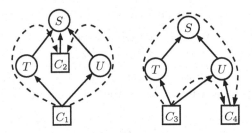

Fig. 6.4. Two ways of generating duplicates

The utilized implementation of scoping determines whether the conceptual replication really results in duplicate deliveries. A broad range of possible implementations of scoping exist,[7] and in some of them different delivery paths have no effect. For example, an explicit, externally available scope graph data structure can be used in a centralized implementation to infer all destinations before delivery is commenced. Furthermore, available countermeasures for duplicate detection are also highly dependent on the underlying implementation technique.

From an application point of view, there are several reasons for not eliminating duplicates in the scoped event system itself. First of all, in some applications notification processing is idempotent so that duplicate delivery does

---

[7] Please refer to Sect. 6.7.1 for an overview.

not influence the function of an application. On the other hand, if duplicates are not wanted, it is often easier to handle the elimination in the application layer, or at least as an additional layer on top of simple notification dissemination. In fact, the scope boundaries themselves offer a platform to install such logic.

The most interesting point, however, is that on application level different delivery paths may connote different notification semantics. Consider the left example of Fig. 6.4, where two different delivery paths connect C_1 and C_2, and assume that $C_1 \overset{n}{\rightsquigarrow} C_2$ results in two notifications n' and n'' being forwarded by T and U, respectively. Are the two notifications really equal? Are these notifications really duplicates if they originate, at least from the consumer's point of view, from different components T and U? Within S, these two notifications were published from different producers in the first place. The base event notified with n' may have a different meaning in the context of T than the event notified with n'' in U. Scope interfaces and mappings presented in the next section will enable administrators to control notification forwarding in a finer way.

In summary, there is no generic solution to handle duplicate notifications in a scoped event-based system. The many available choices of possible implementation techniques offer all sorts of corresponding duplicate handling capabilities, which are too divergent to be included in the general scope model. Note that duplicate notifications are forbidden in the specification of simple event systems but are possible in scoped systems. Different delivery paths conceptually deliver different notifications, even if triggered by the same base event.

6.2.5 Dynamic Scopes

The above definition assumed a static scope hierarchy to provide a basic definition that can be adapted and refined based on further requirements. In the case of dynamic scopes, four additional operations have to be offered: $cscope(S)$ and $dscope(S)$ to create and destroy a scope S, $jscope(X, S)$ and $lscope(X, S)$ to join X to scope S or leave it, respectively. These operations are typically available to the administrator role only, for individual components do not necessarily need to know about their scope membership.

A system with static scopes can then be simulated by having the administrator set up the scope hierarchy with the appropriate operations before clients start. However, dynamic scopes are not directly covered by the above specification. A changing scope graph may conflict with the safety condition, which is ambiguous in dynamic asynchronous system models. A notification n is only allowed to be delivered to Y if the producer X is visible to Y. But because delivery cannot be instantaneous, X may leave the scope in which n was published before it is delivered, and so $v(X, Y)$ may hold at time of publication but not on delivery, rendering the specification ambiguous. The specification does not cover systems that allow traces of the form

$$\sigma_4 = pub(X,n), \ldots, lscope(X,S), \ldots, notify(Y,n),$$

where scope graph reconfigurations and notification publication and delivery are mixed.

Several approaches to this problem exist. First of all, the assumed system model may require delivery to be instantaneous so that notification dissemination and scope reconfiguration cannot interleave. Any form of centralized implementation is able to achieve this guarantee. A second approach is to allow producers to leave a scope only if all their published notifications have been delivered, preventing the interleaving in σ_4 so that the resulting traces are equivalent to the static case with respect to the safety condition. In effect, this results in a type of synchronization similar to that of a global transaction: scope joins and scope leaves must be reliably acknowledged by all other brokers before the action is performed. Obviously, this type of dynamic scope semantics is unfavorable since it incurs a high synchronization overhead. However, scope reconfigurations may be so infrequent in practice that this is tolerable for medium-size systems. At least these semantics have the advantage that the safety part of Def. 6.3 can be used in the simple unmodified form. Interestingly, this restriction resembles an object-oriented programming approach where new subclasses and new methods are readily added, but modifying the inheritance hierarchy is complicated.

A different approach would be to not hide scope graph changes but to explicitly consider them in the specification. For the safety condition the visibility restriction $v(X,Y)$ would have to reflect time delays in notification delivery. On the other hand, the liveness part of Def. 6.3 does not consider dynamic scopes at all. By including $\square v(X,Y)$ in its precondition, only static graphs can fulfill liveness in the current definition. This specification is intentionally restricted because it is intended to specify only basic functionality. It currently covers a broad range of system models, and it can be refined (safety) and extended (liveness) to incorporate dynamic scopes in more specific system models. So, currently the following trace complies to the specification:

$$\sigma_5 = sub(Y,F), jscope(X,s), jscope(Y,s), pub(X,n_1), lscope(Y,s), \ldots,$$
$$jscope(Y,s), pub(X,n_i), lscope(Y,s), \ldots$$

In σ_5 components X and Y start off in the same scope and X publishes an "infinite" sequence of notifications n_i. However, since Y leaves the scope again after every publish operation, there is no point in time from which on X and Y remain in the same scope. Therefore, delivery is not required and σ_5 satisfies the liveness requirement. Of course, without knowing future traces a notification service has to try to deliver any pending notifications.

So, dynamic changes of a scope graph can be supported if changes and publications are serialized, or the safety condition has to be relaxed to cover only durations in which the visibility of producer and consumer remain unchanged.

6.2.6 Attributes and Abstract Scopes

The layout of a scope graph carries information on system structure. Annotations of scopes allow the administrator to associate further information on system operation, which will be done in the next subsections. Or annotations are simply used to add application-specific data into the structure. Technically, the notion of *scope attributes* is introduced. Attributes associate data to a specific scope according to a simple name/value pair model.

For example, a scope S is named and stores its time of creation in two attributes:

$$S.name = \text{``ItsMe''} \qquad S.creation = \text{``2004-12-20 12:22''}$$

How attributes are set and used is described in Sect. 6.6.

Attributes may carry information about system configuration and management. Section 6.7.1 introduces alternative implementation approaches, and attributes can store such annotations that refine the model expressed in the scope graph. However, these kinds of information are typically valid for more than one component of the graph. An obvious way to assign this information to a group of components is to use a scope, which bundles the components in question, just as a container carrying configuration data. This scope would be a special type of scope, termed abstract scope.

Abstract scopes group components, but there is no communication within. They are created for descriptive purposes and not to control communication of their members. They are used for system management (cf. Sect. 6.6).

6.2.7 A Correct Implementation

The following presents a possible implementation of Def. 6.3 as a proof of concept. The implementation uses a simple event system as specified in Sect. 2.5.2 as basic transport mechanism. This modular approach underlines the system's structure and shows the possibility of implementing the specification. But as before, it does not concentrate on efficiency issues, and any available notification service satisfying the simple event system specification can be used instead.

The architecture of the implementation is sketched in Fig. 6.5. The interface operations of the scoped event system are local library calls, which are mapped to appropriate messages of the underlying simple event system. Again, this part of the client process is the *local event broker* of the client. Conceptually, for every client an additional process at the interface of the simple event system is generated, the client's *proxy*. Practically, the proxy will be part of the local event broker. Note that the clients' proxies are the only components accessing the underlying simple service; no complex components are instantiated in this implementation scenario.

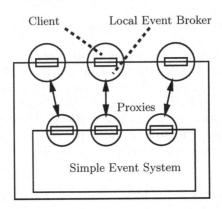

Fig. 6.5. A possible implementation of a scoped event system

Although dynamic scoping is not considered in the specification, the presented algorithm includes dynamic scopes in the style of Sect. 6.2.5. To simplify the implementation, changes to the scope graph $G = (C, E)$ are restricted: only components with no incoming edges may join or leave scopes. This restriction prevents individual brokers from having to store G completely.

As noted above, the scope graph describes a transitive partial order $\leq$ on C with $X \leq X' \Leftrightarrow (X, X') \in E$. The maximal elements of C have no outgoing edges, i.e., they have no superscopes. These elements are termed *visibility roots*, as the recursive definition of $v(X, Y)$ is terminated by common superscopes. The maximal elements that are visible from a component are used to determine visibility of notifications.

Data Structures

For every client X, its proxy $Prox_X$ holds a list V_X of its visibility roots. In a system with static scopes, V_X is initialized to the set of its visibility roots in the given scope graph. With dynamic scopes where changes are limited to the addition of new leaves—nodes with no incoming edges—V_X is set at the time of addition. In both cases, it remains constant and is not changed until the whole systems stops or X is deleted.

Algorithm

If a client invokes $pub(X, n)$, a message (pub, X, n) is sent to the client's proxy. At the interface of the simple event system, the proxy then invokes $pub(Prox_X, (n, R))$, where R is set to the constant value V_X.

Calls to $sub(X, F)$ and $unsub(X, F)$ are sent in a similar way to $Prox_X$. Using F, the proxy derives a filter $\tilde{F}$ that matches all notifications $\tilde{n} = (n, R)$ for which n matches F, and subsequently calls $sub(Prox_X, \tilde{F})$.

Whenever the simple event system notifies the proxy of Y about a notification $\tilde{n} = (n, R)$, the proxy checks whether $V_Y \cap R \neq \emptyset$. If the test succeeds, a message is sent to the local broker of Y to invoke $notify(Y, n)$. Otherwise the notification is discarded.

Correctness

In order to show that Def. 6.3 is satisfied, the presented implementation must obey the visibility $v(X, Y)$ of the safety condition and the additional precondition $\Box v(X, Y)$ of the liveness condition. The remaining part is satisfied by using the simple event system which satisfies Def. 2.5.

Lemma 6.2. *For every pair of clients X and Y and for the set of visibility roots V_X and V_Y stored at the proxies, the following holds:*

$$v(X, Y) \Leftrightarrow V_X \cap V_Y \neq \emptyset$$

Proof. We need to show two implications. The first implication ($\Rightarrow$) is proved by induction over the "visibility" path from X to Y. The second implication ($\Leftarrow$) is shown as follows: If $V_X \cap V_Y \neq \emptyset$, there exists a maximal element Z of $\leq$ such that $X \leq Z$ and $Y \leq Z$. By the definition of $\leq$ this implies $v(X, Y)$. □

Now, the correctness of the sketched implementation can be proved in terms of the safety and liveness conditions of scoped event systems.

Proof of Safety

Assume that $notify(Y, n)$ is invoked at client Y. It must be shown that this implies validity of the three conjuncts of the implication in the safety property of Def. 6.3.

The first conjunct follows directly from the safety property of the simple event system.

To prove the second and the third conjuncts, assume that the local broker issues $notify(Y, n)$ at client Y. This means that (a) the proxy of Y has previously received a notification $\tilde{n} = (n, R)$ and that (b) the test $V_Y \cap R \neq \emptyset$ succeeded.

From (a) and the safety property of the simple event system follows that $\tilde{n}$ was previously published by some proxy $Prox_X$. From Lemma 6.2 and (b) follows that $v(X, Y)$ holds. This proves the second conjunct.

From (a) and the safety property of the simple event system follows that $\tilde{n}$ matches some transformed filter $\tilde{F}$ of $Prox_Y$. This together with the algorithm proves the third conjunct. This concludes the proof of the safety property.

Proof of Liveness

Assume a client Y invokes $sub(Y, F)$ and never unsubscribes to F. From the algorithm it is implied that an "equivalent" subscription $\tilde{F}$ is issued into the simple event system. Since scope reconfigurations are restricted to occur at leaves, the values of V_X and V_Y of existent components are constant. From Lemma 6.2 this implies that $v(X, Y)$ is always true for all clients X and Y for which $V_X \cap V_Y \neq \emptyset$.

From the liveness property of the simple event system and the algorithm follows that there is a point in time after which every published notification $\tilde{n} = (n, R)$ that matches $\tilde{F}$ is delivered to every client proxy. So assume that after this point in time some client X publishes a notification n matching F. From the algorithm we have that $\tilde{n} = (n, V_X)$ is published within the simple event system. Its liveness property gives us that $\tilde{n}$ is eventually delivered at the client proxy of Y. From the algorithm and because $v(X, Y)$ holds, the test $V_X \cap V_Y \neq \emptyset$ will succeed and Y will eventually be notified of n.

6.3 Event-Based Components

6.3.1 Component Interfaces

So far, visibility is an only two-level hierarchy induced by the topmost super-scopes, the visibility roots of the graph G. Any two components are either able to see all of their published notifications or none at all. In order to overcome this problem and to improve the structuring abilities, visibility is refined by assigning input and output interfaces to scopes.

Input and *output interfaces* for simple components are subscriptions and advertisements, respectively. Both include filters that describe the set of notifications allowed to cross a component's boundary. As defined in Sect. 3.1, a notification n is either mapped on itself or to ϵ, indicating that n is either matched or blocked. In the following, similar filter sets are associated with scopes to make interfaces a feature of all components.[8]

6.3.2 Scope Interfaces

Scope input and output interfaces describe the set of notifications that are allowed to cross the scope boundary. Only those notifications that match one of the scope's output filters are forwarded up into its superscopes as outgoing notifications, and only those matching at least one of its input filters are treated as incoming notifications that are forwarded to scope members. Filters of scope interfaces are expressed in the same filter model used for subscriptions and advertisements of simple consumers and producers.

---

[8] The relationship between scopes and simple components is shown in the UML class diagram in Fig. 6.1.

The base interface I_C of a component C contains two sets of filters, ^iF_C and ^oF_C, representing the input and output interfaces of the currently active subscriptions and advertisements of the component. This base interface is associated with every component of the event-based system with the known function of letting notifications pass if they match one of the filters in ^iF_C for incoming notifications or ^oF_C for outgoing notifications.

Formally, the interfaces are bound to edges of the scope graph. Depending on the conceptual placement of filters with respect to the starting or ending node of an edge, two refinements and the resulting combination of filters are distinguished: selective, imposed, and effective interfaces (Fig. 6.6). While the next paragraphs discusses the different forms of interfaces, the formal definition of a scoped event system with interfaces is given in Sect. 6.4.1.

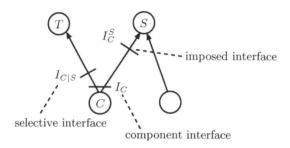

Fig. 6.6. Different scope interfaces

Selective Interfaces

According to the preceding definition a component has an interface independent of its scopes; it does not distinguish between superscopes. This conforms to the intended loose coupling of event-based interaction. However, the administrator knows the configuration of scopes and as part of this role it is possible to distinguish superscopes.

A *selective interface* $I_{C|T}$ controls the communication between a component C and a specific superscope T. It functions in the same way I_C does, but governs communication only between C and T. It is applied in addition to the base component interface. In Fig. 6.6, for instance, some of the notifications published by C are forwarded to S but not to T. If, in a type-based scheme, $I_{C|T}$ contains an output filter that accepts notifications of type A but not B, and if C happens to publish notifications n_A and n_B of type A and B, n_A would be visible in T but n_B not. Communication with S is not affected by $I_{C|T}$.

So, notification forwarding depends on the destination scope. A component may now exhibit different interfaces toward different superscopes. From an engineering point of view, this offers a fine control of interaction, which is especially important when composing existing subsystems. Furthermore, the functionality of the selective interfaces may be used to mitigate problems of duplicate notifications by blocking certain delivery paths. On the other hand, the administrator must be aware of possible effects of discriminating interfaces. If the distinguished superscopes share a common visibility root two different delivery paths may exist that preclude duplicate notifications but break causal order of messages. Consider S and T in Fig. 6.6 having a common superscope Z, then a short path exists connecting C and T directly, and a longer one crossing S and Z to reach T. A first notification n_1, which is blocked by $I_{C|T}$, may reach T after a second notification n_2 that matches $I_{C|T}$. Although the specification of simple event systems does not assume a specific ordering, many concrete systems provide a sender FIFO ordering that would be broken in this way.

Imposed Interfaces

A converse refinement of interface definition is to install filters at the "other" end of the scope graph edge. An *imposed interface* I^S is specified within a scope and wraps all of its members with an extra interface. It allows only those notifications that match the imposed interface to be exchanged within this scope, dedicating the scope to a specific kind of data. This interface does not influence the communication of the affected component in other scopes. Furthermore, interfaces can also be imposed on individual components. I_C^S in Fig. 6.6 restricts the interaction of C with S, without affecting the other children in S. If I_C^S contains an output filter that accepts notifications of type B but rejects A, the above-mentioned notification n_B published by C would be forwarded into S, but n_A is rejected by the imposed interface. Note that notifications of type A may published by other members of S, which are not affected by I_C^S.

Imposed interfaces are a means to control communication within a scope. Especially when an administrator integrates existing preconfigured components, not all of their provided interfaces are of interest within the new scope, or on the other hand, not all of the scope's internal traffic shall be visible to all components. As such, imposed interfaces are a security mechanism, too. They enforce predefined filters on scope members and thus control what is published and consumed within the scope. For instance, depending on security credentials, different interfaces may be imposed on newly connected scope members.

Effective Interfaces

The *effective interface* of a component concatenates the previously introduced base interface with the selective and imposed interfaces. It is given with respect to a specific outgoing edge of the component and describes the set of notifications that are effectively allowed to cross the respective edge of the scope graph. A notification matches the effective interface $\hat{I}_C^S$ of a component $C \lhd S$ iff it matches I_C and $I_{C|S}$ and I_C^S and I^S.

6.3.3 Event-Based Components

Scopes are a composition mechanism that facilitates creating new, more complex event-based components, showing essential characteristics of component frameworks in the flavor of Szyperski [369]. They encode the interactions between components and act themselves as components on a higher level of abstraction. The composed function is provided through a defined interface, thus facilitating the reuse of the bundle while abstracting from its internal configuration. Scopes are distributed event-based components (Sect. 6.6).

6.3.4 Example

The example stock trading application introduced in Sect. 5.1.2 is expanded to illustrate the use of scopes (Fig. 6.7). There are two main scopes, M1 and M2, denoting two different stock markets. Within each market customers are grouped into subscopes distinguishing private and professional customers. Each customer is permanently represented by one of the scopes C1, C2, etc., which remain connected in the graph of scopes even if customers are not personally logged in. They group a customer's PCs, cellular phones, or agents running on a remote server. An example "agent" would be a limit watcher which continuously monitors a share's price and issues a notification when a specific share deviates from the overall market performance. Such agents can be installed within a customer's scope without changing existing components—one of the obvious benefits of event-based systems—and without affecting other parts of the system, which is the prime attribute of scoping.

For the sake of simplicity, interest for at most one share is indicated below the rectangles representing the customers' PCs. The figure illustrates the scenario when the trading floor TF participates in the stock market M1 and issues a notification concerning SAP quotes. Although both consumers C3 and C4 have subscribed for notifications on SAP quotes, this notification will only reach C3, because C4 is not visible from the trading floor and C1 has subscribed to a different share. On the other hand, consumer C3 listens to both markets and may receive "duplicate" SAP quotes.

To illustrate how scope interfaces help in structuring event-based applications, let us consider the interfaces of the components in our running example as summarized in Fig. 6.8.

Customers send out notifications of type *Order* which contain a share identification, the number to be sold or bought, and potential price limits. The trading floor TF listens to these orders, issues acceptance notifications, and

---

[9] Delayed forwarding is discussed in Sect. 6.5.

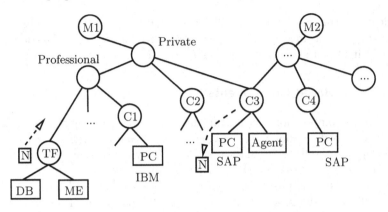

Fig. 6.7. The graph of the stock application

Component	Description	Input	Output
M1,M2	The Stock Markets	–	–
Private	scope of all private customers	–	Trade
Prof.	scope of all professionals	Order	Accept, Quote(delayed)[9]
C1,C2,...	Customer representation	Accept	Order
TF	Trading Floor	Order	Accept, Quote
ME	Matching engine	Order	Accept, Quote, OrderBook
DB	The logging database	Order, Quote	–

Fig. 6.8. Interfaces of the components in the example application

sends out *Quotes*, informing about successfully executed orders. The trading floor itself is composed of the matching engine ME and the database DB. While the database only logs all *Orders* and *Quotes*, the matching engine receives orders and issues *Quotes* of current prices. It maintains a list of open orders and executes the matching algorithm that leads to acceptance notifications (*Accept*) of matched orders. Additionally, the matching engine publishes an orderbook summary with prices and volumes of the ten best bid and ask orders. The summary is only visible within the trading floor, because the interface of TF prohibits further distribution. Based on this data, additional services may be integrated into the trading floor, like market makers ensuring that there is always at least one buy and one sell order open.

6.4 Notification Mappings

So far, uniform data and filter models were assumed, which prescribe syntax and semantics of notifications and filters throughout the whole system. In large systems, however, characteristics and demands of applications are likely to diverge and homogeneous models will not fit the needs, as pointed out in the discussion of the engineering requirements in Sect. 5.1. If all components are forced to agree on the same data and filter model, system integration and efficiency is impeded drastically.

The diverging requirements will best be met with tailored data and filter models—an idea which is obvious but hardly considered in the context of event systems. Different system parts will use different representations and semantics of events. With an appropriate support, one part of an application can exchange binary encoded notification while still being able to communicate with other parts of the system via serialized Java objects or XML encoded notifications. Efficiency considerations result in differentiating low-volume external representations in XML from more efficient, optimized internal representations.

An obvious implication of decomposing applications is that bundling of related components should not only encapsulate functionality but also delimit common syntax and semantics. Constraining the visibility of notifications is the basis for dealing with heterogeneity issues. Consequently, *notification mappings* are introduced as extensions of scope interfaces. They transform notifications at scope boundaries to map between internal and external representations, without interfering with internal notifications.

Scopes are an appropriate place to localize such transformations because bundled components are likely to agree on a common data and filter model, whereas the interaction with the remaining system is decoupled by the scope boundary. Notification mappings clearly address the heterogeneity requirements stated in Sect. 5.1 and facilitate construction and maintenance of large event-based systems.

6.4.1 Specification

Notification mappings transform notification from one data model to another. Mappings, however, do not primarily block notifications but transform them. *Notification mappings* are defined as binary, asymmetric relations on the set $\mathcal{N}$ of notifications. They are associated with scope graph edges, like scope interfaces, and two mappings $\nearrow_e$ and $\searrow_e$ are attached to every edge $e = (C, S) \in \mathsf{E}$. Let n_1 and n_2 be two notifications. For any edge e and its associated relation $\nearrow_e$, the mapping $n_1 \nearrow_e n_2$ means that when "traveling" upwards along the edge (i.e., in direction of the superscope) n_1 is transformed into n_2. The relation $\searrow_e$ is defined analogously for the reverse direction. Note, in order to support heterogeneous data models the relations map between two sets

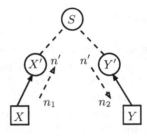

Fig. 6.9. Recursive definition of the relation $(n_1, X) \rightsquigarrow (n_2, Y)$

of notifications used in C and S, respectively, i.e., $\nearrow_e \subset \mathcal{N}_C \times \mathcal{N}_S$, but it is implicitly assumed that $\mathcal{N}$ contains the different models for simplicity.

Now, the general visibility of notifications can be defined using these relations.

Definition 6.5. *The* visibility of notifications *in a scope graph* $G = (C, E)$ *is defined by the relation* $\rightsquigarrow$ *on* $\mathcal{N} \times \mathcal{K}$, *where*

$$(n_1, X) \rightsquigarrow (n_2, Y) \quad \text{or shorter} \quad X \stackrel{n_1}{\rightsquigarrow}_{n_2} Y$$

means that n_1 *visible to* X *is also visible to* Y:

$$(n_1, X) \rightsquigarrow (n_2, Y) \Leftrightarrow$$
$$(X = Y \wedge n_1 = n_2)$$
$$\vee \left(\exists e = (X, X') \in \mathsf{E}. \ \exists n' \neq \epsilon. \quad n_1 \nearrow_e n' \right.$$
$$\left. \wedge \left[(n', X') \rightsquigarrow (n_2, Y) \right] \right)$$
$$\vee \left(\exists e = (Y, Y') \in \mathsf{E}. \ \exists n' \neq \epsilon. \quad n' \searrow_e n_2 \right.$$
$$\left. \wedge \left[(n_1, X) \rightsquigarrow (n', Y') \right] \right)$$

The recursive definition of $(n_1, X) \rightsquigarrow (n_2, Y)$ is illustrated by Fig. 6.9. Intuitively, notification n_1 "flows" from X to Y and, after potentially being transformed several times, it is received as notification n_2. The path on which n_1 flows to n_2 is the same as for the visibility relation defined in Sect. 6.2, i.e., it can be characterized by a path from X up to a common superscope and then down to Y. But in addition the notification is subject to any mappings assigned to the relevant edges.

The semantics of scoped event systems with mappings are derived from those of scoped event systems by the refined visibility definition. With like arguments the graph of scopes and the relations $\nearrow$ and $\searrow$ are assumed to be static in the sense that a component's mappings are not allowed to change until all of its published notifications are delivered; otherwise the visibility clause may corrupt the safety condition in the specification.

Definition 6.6 (scoped event system with mappings). *A scoped event system with mappings $ES^{\mathcal{M}}$ is a system that exhibits only traces satisfying the following requirements:*

- *(Safety)*

$$\square\Big[notify(Y,n') \Rightarrow \big[\bigcirc\square\neg notify(Y,n')\big]$$

$$\wedge \big[\exists n.\,\exists X.\,n \in P_X \,\wedge\, ((n,X)\rightsquigarrow(n',Y))\big]$$

$$\wedge \big[\exists F \in S_Y.\,n' \in N(F)\big]\Big]$$

- *(Liveness)*

$$\square\Big[sub(Y,F) \Rightarrow$$

$$\Big(\lozenge\big[\square((n,X)\rightsquigarrow(n',Y)) \Rightarrow$$

$$\square\big(pub(X,n) \wedge n' \in N(F) \Rightarrow \lozenge notify(Y,n'))\big]\Big)$$

$$\vee \Big(\lozenge unsub(Y,F)\Big)\Big]$$

The difference between this definition and that of scoped event systems (Def. 6.3) is that the term $v(X,Y)$ is replaced by the term $(n,X)\rightsquigarrow(n',Y)$ and that the published notification n is not necessarily equal to the delivered n'. This formulation extends the system to not only obey the visibility of components but the visibility of individual notifications. The delivered notification n' is the result of repetitive applications of the mappings $\nearrow$ and $\searrow$ along the path implicitly defined by $\rightsquigarrow$. The present definition is even a generalization of the scoped delivery. This is because a scoped event system can be regarded as one with event mappings where all mappings are the identity relation, i.e., they do not change anything along the delivery paths. In such a system, $v(X,Y)$ is implied by the existence of a notification n such that $(n,X)\rightsquigarrow(n,Y)$.

Interfaces as Mappings

Notification mappings are a generalization of and subsume scope interfaces. The relation $\nearrow$ might be undefined for an outgoing notification n_1 so that there is no n_2 such that $n_1 \nearrow n_2$. This blocks the notification just as a nonmatching filter does. In order to seamlessly extend scope interfaces, $\nearrow$ and $\searrow$ are constrained to always map to some notification, with the empty notification ϵ as default.

Definition 6.7 (notification mappings). *A notification mapping is given by a function in $\mathcal{M} = \{m \mid m : \mathcal{N} \to \mathcal{N}\}$.*

$$n_1 \nearrow n_2 \Rightarrow \exists m \in \mathcal{M}.\,m(n_1) = n_2$$

Whenever a notification is mapped to ϵ it is considered to be blocked so that filters are but special mappings: $\mathcal{F} = \{f \in \mathcal{M} \mid f(n) = n \vee f(n) = \epsilon\} \subset \mathcal{M}$. With this definition, a uniform way of filtering and transforming notifications is accomplished so that, conceptually, interfaces and mappings can be concatenated at scope boundaries, e.g., $F_1 \circ F_2 \circ M_1 \in \mathcal{M}$.

Next, interfaces and their concatenation are defined more formally to define $\nearrow$ and $\searrow$ as concatenated interfaces and mappings.

Definition 6.8 (interface). *An* interface I *consists of an input mapping* ^iI *and an output mapping* ^oI: $I = (^iI, {}^oI) \in \mathcal{M} \times \mathcal{M}$. *The base interface* I_C *of a component* C *represents the sets of open subscriptions and advertisements of* C:

$$I_C = (^iI_C, {}^oI_C) \in \mathcal{M} \times \mathcal{M}$$
$$\triangleq (^iF_C, {}^oF_C) = \{\{F_1, F_2, \ldots, F_k\}, \{F'_1, F'_2, \ldots, F'_l\}\} \in P(\mathcal{F}) \times P(\mathcal{F})$$

where ^iI_C *and* ^oI_C *are defined as*

$$^iI_C(n) = \begin{cases} n & \exists F \in {}^iF_C.\, F(n) = n \\ \epsilon & \textit{otherwise} \end{cases}$$

$$^oI_C(n) = \begin{cases} n & \exists F \in {}^oF_C.\, F(n) = n \\ \epsilon & \textit{otherwise} \end{cases}$$

Selective interfaces $I_{C|S}$ *and imposed interfaces* I^S *and* I_C^S *are defined likewise.*

According to this definition an interface can transform notifications for the seamless concatenation of filters and mappings.

Definition 6.9 (concatenation of interfaces). *Two interfaces* I_1 *and* I_2 *are concatenated by*
$$I_1 \circ I_2 = (^iI_1 \circ {}^iI_2, {}^oI_2 \circ {}^oI_1).$$

Note that the resulting interface evaluates the composed input and output interfaces in inverse order. This is not necessary if only filters are considered, but by incorporating mappings the sequences are no longer commutative. The effective interface between two components $C \lhd S$ describes the notifications transmitted along this edge in the scope graph and combines the aforementioned interfaces *and* notification mappings assigned to this edge, extending the informal description given in Sect. 6.3.2.

Definition 6.10 (effective interface). *The* effective interface $\hat{I}_C^S$ *between two components* $C \lhd S$ *is given by concatenating base interface, selective interface, mapping, and imposed interface:*

$$\hat{I}_C^S = I_C \circ I_{C|S} \circ M_C^S \circ I_C^S \circ I^S$$

Finally, the interfaces between two components $C \lhd S$ are correlated to the mapping relations $\nearrow$ and $\searrow$ as follows:

$$n_1 \searrow n_2 \Leftrightarrow (I_C \circ I_{C|S} \circ {}^i M_C^S \circ {}^i I_C^S \circ {}^i I^S)(n_1) = n_2$$
$$\Leftrightarrow {}^i \hat{I}_C^S(n_1) = n_2$$
$$n_1 \nearrow n_2 \Leftrightarrow ({}^o I^S \circ {}^o I_C^S \circ {}^o M_C^S \circ {}^o I_{C|S} \circ {}^o I_C)(n_1) = n_2$$
$$\Leftrightarrow {}^o \hat{I}_C^S(n_1) = n_2$$

The rules of notification forwarding in the scope graph given by the publishing and delivery policies in Eqs. (6.1) and (6.2) can be refined corresponding to the above discussion:

$$\textbf{PP}: \quad X \overset{n_1}{\rightsquigarrow} S \wedge X \lhd S \lhd T \wedge {}^o \hat{I}_S^T(n_1) = n_2 \Rightarrow X \overset{n_1}{\rightsquigarrow}_{n_2} T \qquad (6.3)$$
$$\textbf{DP}: \qquad \overset{n_1}{\rightsquigarrow} T \wedge S \lhd T \wedge {}^i \hat{I}_S^T(n_1) = n_2 \Rightarrow \overset{n_2}{\rightsquigarrow} S \qquad (6.4)$$

Despite the integration of interfaces and mappings, the scope overview in Fig. 6.1 still distinguishes interfaces and mappings to underline their different intentions, and also because their implementations are apt to diverge.

Some Further Comments

The already mentioned issue of duplicate notifications has to be reconsidered here. A notification is duplicated if it travels along different paths from producer to consumer, but it may now be subjected to different mappings so that different versions of the same original notification are created. The specification cannot rule out this case since it is highly application-dependent whether this is an unwanted situation or not. The mappings may help handling alternative delivery paths as they can annotate passing notifications, e.g., to include information about the delivery path in the notification.

Trying to offer a sophisticated concept of heterogeneity support in event-based systems is beyond the scope of this book, and thus notification mappings are presented as a starting point for including appropriate enhancements. The mappings underline the extensibility of the scoping concept and open it to integrate existing works in the area of syntactic and semantic transformations that are applicable here [46, 79, 232]. Furthermore, the current if implicit assumption that notifications are mapped one-to-one is used for simplicity only. Scope boundaries may turn out as the appropriate place to implement more sophisticated event composition [146, 406].

6.4.2 A Correct Implementation

The following presents an implementation sketch of the scoped event system with mappings. The implementation of a scoped event system with mappings

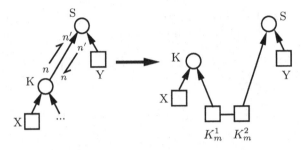

Fig. 6.10. Transformation of mappings into components

$ES^{\mathcal{M}}$ is based on a scoped system $ES^{\mathcal{S}}$ and a transformation of the graph of scopes G that essentially follows the idea of adding activity to edges. Figure 6.10 sketches the transformation that creates G' by replacing every edge (K, S) that does not apply the identity mappings $n \nearrow n$ and $n \searrow n$ for two extra mapping components K_m^1 and K_m^2. *Two* mapping components are taken to constrain the visibility of the transformed notifications to the appropriate scopes. If only *one* K_m would be inserted, additional measures had to be taken to distinguish the superscopes.

Figure 6.11 describes the architecture of the implementation for the example system in Fig. 6.10. A component X connected to $ES^{\mathcal{M}}$ is also directly connected to an underlying scoped event system $ES^{\mathcal{S}}$. Calls to $pub(X, n)$ of $ES^{\mathcal{M}}$ are forwarded to $ES^{\mathcal{S}}$ without changes, and vice versa, calls to $notify(X, n)$ of $ES^{\mathcal{S}}$ are forwarded to $ES^{\mathcal{M}}$.

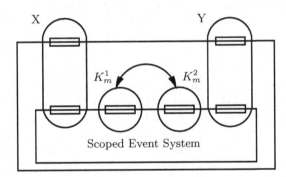

Fig. 6.11. Architecture of scoped event system with mappings

In general, if a scope K is to be joined to a superscope S by calling $jscope(K, S)$, two mapping components K_m^1 and K_m^2 are created that communicate directly via a point-to-point connection. K_m^1 joins K, subscribes to all notifications published in K, and transforms and forwards them to its peer. Furthermore, subscriptions in K have to be transformed before they are forwarded. The implementation relies on externally supplied functions that map notifications and filters/subscriptions between the internal and external representations in K and S, respectively. K_m^2 joins S and republishes all notifications it gets from its peer K_m^1. It subscribes in S according to the subscriptions forwarded by K_m^1, transforms any notifications received out of S, again with externally supplied functions, and forwards them to K_m^1, which republishes them into K.

Correctness

The algorithm from the previous section has to satisfy the requirements given in Def. 6.6 of $ES^{\mathcal{M}}$, i.e., safety and liveness conditions. The correctness proof largely depends on the correctness of the underlying scoped event system $ES^{\mathcal{S}}$. The next lemma relates the graph transformation to the structure of delivery paths.

Lemma 6.3. *If $(n, X) \rightsquigarrow (n', Y)$ holds, then in the implementation of $ES^{\mathcal{M}}$ exists a sequence $\rho = C_1, C_2, \ldots, C_m$ of components for which holds:*

1. *$C_1 = X$ and $C_m = Y$.*
2. *for all $1 < i < m$ holds that C_i is a mapping component.*
3. *for all $1 \leq i \leq m-1$ holds that C_i and C_{i+1} either share a communication link or reside in the same scope of $ES^{\mathcal{S}}$.*

Proof. Assume $(n, X) \rightsquigarrow (n', Y)$ holds. From the definition of $\rightsquigarrow$ follows that there exists a delivery path $\tau = (X, S_1, S_2, \ldots, S_l, Y)$ in the scope graph G. Since visibility is recursively defined by having common superscopes, all S_i must be scopes.

The construction method of building G' from G implies that every consecutive pair of scopes (S_i, S_{i+1}) in τ where mappings are applied is enhanced with two mapping components K_i^1 and K_i^2, which are joined by a direct communication link. The mapping components K_i^2 and K_{i+1}^1 of neighboring edges reside in the same scope S_{i+1} or are visible to each other. The projection of τ to mapping components (and X and Y) results in a sequence $X, K_1^1, K_1^2, K_2^1, K_2^2, K_3^1, \ldots, K_l^2, Y$, which is the witness for the sequence ρ of the lemma. □

Proof of Safety

Assume that Y is a simple component and that $notify(Y, n')$ of $ES^{\mathcal{M}}$ is called. It must be shown that the three conjuncts of the implication in the safety property of Def. 6.6 hold.

From the algorithm description follows that $notify(Y, n')$ of $ES^{\mathcal{S}}$ was called before, implying that n' is notified at most once and that n' matches an active subscription of Y. This proves the first and the third conjuncts.

The second conjunct is proved by a backward induction on the path guaranteed by Lemma 6.3. The fact that Y is notified about n' implies that there is a component Z that has published n' which resides in the same scope. If this Z is not a mapping component, Z plays the role of X in the formula, $n' = n$, and the second conjunct follows immediately (this is the base case of the induction). The step case of the induction is as follows: Assume that a component Z'' along the path has published some notification n'' which from backward notification mappings resulted from n'. Then there exists a component Z''' which is either in the same scope or connected by a communication link to Z''. In the first case, the step follows from the properties of ES^8, and in the second case from the algorithm. This implies that $n \in P_X$ and that $((n, X) \rightsquigarrow (n', Y))$, giving the second conjunct.

Proof of Liveness

The liveness property is proved by forward induction on the path guaranteed by Lemma 6.3 in a similar way as in the proof of the safety property. Assume that Y subscribes to F and never unsubscribes. Then assume that after subscribing, $(n, X) \rightsquigarrow (n', Y)$ begins to hold indefinitely. Then Lemma 6.3 guarantees a path between any publisher X of a relevant notification n and Y. A similar way of reasoning as in the safety proof implies that n is forwarded and transformed along the path resulting in n', which Y is eventually notified about.

6.4.3 Example

Returning to the stock exchange example, mappings can be exploited to convert between different currencies.[10] Quotations are typically given in a local currency which needs to be transformed at the boundary of the local scope in order to achieve comparability. As another example for the usefulness of mappings, consider XML languages like FIXML [273] that standardize financial data exchange. These languages are used to connect external partners, but they are typically too expensive for internal representations due to efficiency reasons. Also, most likely, different representations of events will be used inside the consumers, within the market, and within the trading floor, e.g., Java objects, XML financial data, and EBCDIC mainframe text fields. Notification mappings are installed at the consumers and at the trading floor to map between serialized Java objects and their XML representation and between XML and EBCDIC, respectively.

6.5 Transmission Policies

The discussion of engineering requirements in Sect. 5.1 argued not only for the heterogeneity of data models but also emphasized the necessity to adapt noti-

---

[10] At least from a technical point of view, disregarding varying exchange rates.

fication delivery semantics. The ability to accommodate diverging application needs improves the utilizability of the event service. It helps to provide tailored and efficient implementations, and it avoids a one-size-fits-all approach, which is not appropriate for a communication substrate targeted at evolving networked systems.

The next paragraphs distinguish *transmission policies* to describe how notifications are forwarded in the scope graph. Transmission policies are a way to influence notification dissemination beyond filtering on notifications. While filters operate independently on independent notifications, i.e., they are stateless, transmission policies may have their own state and they exploit additional information not available in filters and interfaces. They refine the visibility definition both within a scope and with respect to its superscopes. Changing it affects the functionality of the overall system in a fundamental way. However, once delimited by scope boundaries, such modifications are the means that allow administrators to customize the interaction within and the composed functionality of specific scopes.

Conceptually, notification forwarding at a node in the scope graph first determines a set of eligible next-hop destinations according to the effective interfaces and then applies the policies to refine this set before transmission. Default policies implement the known semantics of notification delivery, and by explicitly binding them to individual scopes in the specification of event systems, they are subjected to modification on a per-scope basis. This gives the administrator a tool to not only compose but to program scopes. Three different policies are involved in notification transmission: publishing, delivery, and traverse policies.

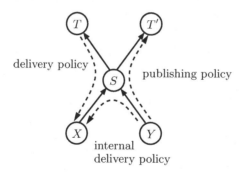

Fig. 6.12. Three important transmission policies in scope graphs

6.5.1 Publishing Policy

A *publishing policy* is associated with a component and controls into which superscopes an outgoing notification is forwarded. In Fig. 6.12, a publishing policy at S can prevent a notification $Y \overset{n_1}{\leadsto} S$ from being forwarded to T, even if the notification conforms to the effective output interface $^o\hat{I}_S^T$. Out of the set of eligible superscopes the publishing policy selects the subset to which a notification is actually forwarded. One might reject the idea of manually selecting the scopes into which data is published as contradicting the event-based paradigm. However, the same arguments as for selective interfaces apply here, too. The selection is part of the administrator's role and is not interwoven with application functionality in simple components. It can be seen as an additional way to control interaction of components outside of the components themselves.

In general, a publishing policy of a component C is a mapping of notifications to a subset of its scopes:

$$pp_C : \mathcal{N} \to P(\mathcal{S})$$

The mapping relation $\nearrow$, which determines the visibility of notifications, can be extended to respect publishing policies. For an edge $e = (S, T)$ of G let

$$n_1 \nearrow_e n_2 \Leftrightarrow {}^o\hat{I}_S^T(n_1) = n_2 \wedge T \in pp_S(n_1)$$

The general rule of forwarding outgoing notifications in the scope graph is implied as follows. Assume Y made a notification n_1 visible in its scope S, $Y \overset{n_1}{\leadsto} S$, and S is a subscope of T, $S \lhd T$, then the notification shall be visible in T if n_1 matches the effective output interface between S and T and the publishing policy (PP) does not object to T. That is,

$$\mathbf{PP}: \quad \underbrace{Y \overset{n_1}{\leadsto} S \wedge Y \lhd S \lhd T}_{\substack{\text{component} \\ \text{visibility}}} \wedge \underbrace{{}^o\hat{I}_S^T(n_1) = n_2}_{\substack{\text{interface} \\ \text{mappings}}} \wedge \underbrace{T \in pp_S(n_1)}_{\substack{\text{publishing} \\ \text{policy}}} \Rightarrow S \overset{n_1}{\leadsto}_{n_2} T \quad (6.5)$$

This definition of PP refines the previous one of scoped delivery with mappings given in Eq. (6.3). It can be reduced to the former definition by setting $pp_S(n_1) = \mathcal{S}$, which always validates $T \in pp_S(n_1)$ and makes Eqs. (6.5) and (6.3) equivalent. Note that the equation also implies $Y \overset{n_1}{\leadsto}_{n_2} T$.

A publishing policy might be used to check for attributes not available in filters and interfaces. Since it is implemented as part of the administrator role, it possibly has access to the scope graph layout and associated metadata. If the availability of security credentials can be checked by the policy, a scope may thus mandate that *its* notifications are only delivered if a certain privilege level is held by the destination scope. But this simple definition leaves room for any form of implementation. In the stock exchange example a market was divided into a professional and a private market. The former gets undelayed stock quotations and is modeled as a subscope of the private market. A publishing policy at the boundary between these two scopes may be used to delay each notification for a certain amount of time. Such implementation-specific issues are not excluded by the above definition.

6.5.2 Delivery Policy

A *delivery policy* is associated with a scope and guides notifications that are to be delivered to scope members. They may either be published in a superscope or by some other constituent component. The delivery policy determines to which members of the scope a notification is forwarded. In Fig. 6.12, a delivery policy at S might direct a notification $T \overset{n}{\leadsto} S$ to X, prohibiting the delivery to Y even if the notification conforms to the effective input interface ${}^i \hat{I}_Y^S$. Out of the set of eligible children the delivery policy selects a subset to which the notification is actually forwarded.

Similar to publishing policies, a delivery policy of a scope S is a mapping of notifications to a subset of components:

$$dp_S : \mathcal{N} \to P(\mathcal{K})$$

The mapping relation $\searrow$ can be refined so that it obeys scope interfaces and reflects delivery policies on incoming notifications. Consider $e = (X, S)$ as given in Fig. 6.12 and a notification visible to S in T, $T \overset{n_1}{\leadsto} S$. The visibility of the notification within S is then determined by

$$n_1 \searrow_e n_2 \Leftrightarrow {}^i \hat{I}_X^S (n_1) = n_2 \wedge X \in dp_S(n_1).$$

Please note that this equivalence not only guides forwarding of incoming notifications but also of internal notifications published by scope members; in the example, $T \overset{n_1}{\leadsto} S$ and $Y \overset{n_1'}{\leadsto} S$ would go down the same edge $e = (X, S)$. However, since internal and external communication is typically treated differently, an additional *internal delivery policy* idp_S is introduced to facilitate this differentiation. The definition of $\searrow_e$ has to distinguish between applying dp_S and idp_S. In the first case, n_1 is an incoming[11] notification that is made visible by a superscope T, i.e., $X \lhd S \lhd T$ and $T \overset{n_1}{\leadsto} S$. In the second case n_1' is an internal notification that is made visible by a member of S, i.e., a sibling of the considered consumer X, $X \lhd S \rhd Y$ and $Y \overset{n_1'}{\leadsto} S$.

$$n_1 \searrow_e n_2 \Leftrightarrow \begin{cases} {}^i \hat{I}_X^S(n_1) = n_2 \wedge X \in dp_S(n_1), & \text{if } X \lhd S \lhd T \wedge T \overset{n_1}{\leadsto} S \\ {}^i \hat{I}_X^S(n_1) = n_2 \wedge X \in idp_S(n_1), & \text{if } X \lhd S \rhd Y \wedge Y \overset{n_1}{\leadsto} S \end{cases}$$

The rule of downward notification delivery (p. 173) is thus given as follows:

---

[11] The term "internal" and "incoming" notifications are also discussed on page 157 in Fig. 6.3a.

$$\mathbf{DP}: \quad T \overset{n_1}{\leadsto} S \wedge \underbrace{X \lhd S \lhd T}_{\substack{\text{component visibility} \\ \text{incoming notification}}} \wedge \underbrace{{}^i\hat{I}_X^S(n_1) = n_2}_{\substack{\text{interface} \\ \text{mappings}}} \wedge \underbrace{X \in dp_S(n_1)}_{\substack{\text{delivery} \\ \text{policy}}} \Rightarrow S \overset{n_1}{\leadsto}_{n_2} X$$

$$(6.6)$$

$$\mathbf{iDP}: \quad Y \overset{n_1}{\leadsto} S \wedge \underbrace{X \lhd S \rhd Y}_{\substack{\text{component visibility} \\ \text{internal notification}}} \wedge \underbrace{{}^i\hat{I}_X^S(n_1) = n_2}_{\substack{\text{interface} \\ \text{mappings}}} \wedge \underbrace{X \in idp_S(n_1)}_{\substack{\text{internal} \\ \text{delivery policy}}} \Rightarrow S \overset{n_1}{\leadsto}_{n_2} X$$

$$(6.7)$$

Again, from the equations and the definition of $\leadsto$ also follows that $T \overset{n_1}{\leadsto}_{n_2} X$ and $Y \overset{n_1}{\leadsto}_{n_2} X$.

An example of a delivery policy is an 1-of-n delivery where an incoming notification is forwarded to only one out of a group of possible receivers. In this way load-balancing characteristics may be implemented in a specific scope. Internal delivery policies are pertinent whenever the data flow within a scope shall be controlled in addition to the established filters. An internal delivery policy is able to arrange multiple consumers into a chain. Consider a sequence of exception handlers, each subscribed to the same type of failure, which it tries to solve, and if not possible it republishes the received notification. An internal delivery policy can forward each published error notification to the next hop in the preconfigured list of consumers/handlers.

6.5.3 Traverse Policy

The last, only informally presented policy is the *traverse policy*, which is associated with a scope S and controls the downward path of incoming notifications in a scope. In contrast to the preceding policies, the traverse policy does not select destinations within a certain scope but selects the scope into which to descend first. It searches at different levels in the scope hierarchy below S for a scope with eligible consumers, and if one is found it will stop searching and refer the notification to the respective scope.

Actually, this policy allows a notification to deviate from a default path through the graph of scopes. In a top-down traverse policy eligible receivers, i.e., simple components with a matching subscription, are searched in the current scope first. If no consumer is found at this stage, the search is continued in the next lower level of scopes if the policy still applies there (same administrative domain). The bottom-up traverse policy starts the search in the deepest subscopes. "Broadcast" is the default policy, which does not inhibit descending the scope graph and delivers to all eligible consumers $C \overset{*}{\lhd} S$ below the current scope S, subject to interfaces and delivery policies, of course.

This kind of dissemination control is apparently inspired by dynamic binding and method lookup in object-oriented class hierarchies. Multiple consumers of the same notification, which are located at different levels in the inheritance/scope hierarchy, can be considered to implement some form of generalized method overriding. While traditional programming languages like

C++ and Java use only one static policy to resolve calls to overridden methods, traverse policies draw ideas from metaobject protocols [221] to determine what kind of method lookup is used. The bottom-up policy resembles a virtual method call in Java in that the implementation of the most derived class is used. Other policies are possible that implement other kinds of method lookups.

6.5.4 Influencing Notification Dissemination

Transmission policies are a means to adapt the event-based dissemination within scopes, i.e., to tailor the quality of service (QoS). They make the interaction in the graph programmable.

To some extent transmission policies bear similarities to metaobject protocols (MOP) known in object-oriented programming [221]. Metaobject protocols offer the ability to redirect or transform messages sent as method calls, and this control allows one to influence object interaction outside of the objects' implementation. Here, notifications are selected, transformed, ordered, or queued, to manipulate the default visibility of notifications and to adapt event-based interaction within the bounds given by the scoping structure to which the policies are associated.

As for the expressiveness and possible implementations of transmission policies, note that the above definition is not intended as an algorithmic description. It integrates with the specification of scoped event systems with mappings given in Def. 6.6, and since the specification relies on linear temporal logic, it only describes valid traces of system execution. In particular, any implementation that exhibits such traces conforms to the specification. So, even if the rules PP, DP, and iDP might connote an algorithm for notification forwarding, possible implementations covered by the definition of pp_S, dp_S, and idp_S, and of $\nearrow_e$ and $\searrow_e$, can be Turing-complete. For instance, delaying notification as part of a transmission policy is sanctioned as long as any later delivered notification still adheres to the visibility definition and the safety condition of the specification.

The decision made by a transmission policy is based on additional data not available in filters and interfaces. Various characteristic approaches to decision making can be distinguished. There are policies that essentially implement filters on notifications like component interfaces, but which are able to exploit additional metadata. Notifications carry management information, which is annotated by the event system and stripped off before delivery, and as a tool of the administrator policies might access this data. So, they would be able to differentiate producers, e.g., to check security credentials. Furthermore, transmission policies probably have (limited) knowledge about the current scope graph layout and of a notification's (partial) path through the graph.

The second, more complex form of transmission policy does not filter any data contained in notifications, but compares all eligible destinations, ranking them to do a top-k selection. The ranking may be random, based on lowest utilization, etc. And finally, when the policy implementation maintains its own state, it might keep a record of the last sent notifications in order to limit the maximal bandwidth toward a consumer by rejecting too frequent notifications. Or it might realize a round robin 1-of-n delivery. With its own state the policy is capable of delaying notifications for a certain amount of time or until a specific condition becomes valid, i.e., a "releasing" event occurs. This opens a venue to bind event composition to scope boundaries, or to implement a form of acknowledged notification forwarding where acknowledgment is given components other than the original producer.[12]

6.6 Engineering With Scopes

Scopes are an engineering abstraction for event-based systems. To some extent they are comparable to classes and objects in object-oriented design and programming. They can be used to model system entities and their relationship and, on the other hand, they provide the basis for system implementation in form of a specific object/component model.

So far, there was no clear distinction made between using the scope graph as a modeling tool or as means of implementing system structure. In order to reflect the different objectives, two types of scope graphs are distinguished. *Descriptive scope graphs* describe a set of components, their relationships, and visibility constraints as expressed by the scope features annotated in the graph. An *instantiated scope graph* scoped event system, describes a running which contains instances of various descriptive scope graphs. The former can be seen as a collection of scope types and classes, while the latter constitutes the runtime environment. Interestingly, both can be combined in one graph. If the descriptive graph is treated as abstract scopes (cf. Sect. 6.2.6) in a combined graph, instantiated components are members of their respective descriptive scopes. This combination does not affect communication within the instantiated scope graph, but allows for instance grouping and runtime reflection [245].

In the remaining subsections a development process is described that shows how scope graphs are created and how they are deployed. A language for specifying and programming scopes and scope graphs is introduced afterwards.

6.6.1 Development Process

The development process for scoped event systems consists of four stages:

1. **Component design.** Individual simple components and preconfigured scopes are created and put into repositories for later use. The design at this stage specifies required and provided interfaces and employed scope

---

[12] Let us call the releasing notifications *commit* and *abort* and you see the link to transactions.

features. Larger descriptive scope graphs can be built up from these pre-configured components.

2. **Scope graph design**. From a selection of existing and newly created components a descriptive scope graph is created. This step concentrates on orchestrating preconfigured components, resolving open interface constraints. No implementation issues are handled.

3. **Scope graph deployment**. An existing descriptive scope graph is translated into a running system. Implementation techniques are chosen, integration code to bridge with existing systems is generated, infrastructure code is deployed to selected nodes of the network, etc.

4. **System management**. A running system is monitored and adapted at runtime. This is necessary to react to failures, to install new components, and to evolve the system where necessary.

6.6.2 Scope Graph Handling

Component Definition

From the engineering point of view, a scope can be considered as *a module construct for event-based systems*, being an abstraction and encapsulation unit at the same time. As an abstraction unit, a scope provides the rest of the world with common higher-level input and output interfaces to the bundled subcomponents, eventually mapping these interfaces to the interfaces of the individual constituents. As an encapsulation unit, a scope constrains the visibility of the notifications produced by the included components. It hides the details of the composition implementation. The engineering of single scopes is about building new event-based components.

Generally, programming of scopes has two sides. First, it is about arranging and orchestrating a set of components; this is the structure of the scope. Second, programming is about specifying the dependencies on other components that are not part of the predefined scope. At runtime a certain environment of available producers and consumers might be required, which are essential for the operation of this scope, but not part of its definition; this is the context of the scope.

How are these two tasks accomplished? Three ways for specifying and programming scopes are considered here: scope API, XML description, and SQL-like language. A basic programming API, e.g., in Java, is easily conceivable. A scope class is the base class with a default implementation of scope, which can be specialized in subclasses. On the programming language level, scope classes are part of the descriptive scope graph and objects constitute the instantiated scope graph. However, the scope concept is too generic to come up with exactly one API proposition; an example can be found in [135].

The context of a scope is a list of requirements that is better encoded in a descriptive language, like XML or SQL. An XSchema definition of scope graphs defines the entities that compose a descriptive scope graph in form of an XML document. It includes descriptions of single scopes and their dependencies in a scope graphs, but may also contain information about network layout and broker networks; an example is available in [268].

The specification of dependencies to other points are called *coupling points*, which is a variation of UML (Unified Modeling Language) ports and interfaces. A coupling point is a description of what other components are needed at deployment. It contains an expression on scope attributes, required interfaces, and the roles eligible components must play. Roles are introduced as a suggestion to describe functionality on a level more abstract than interfaces. Technically, roles involve only string matching on a well-defined attribute. However, they enable system engineers to distinguish components even if they have identical interfaces. As an example consider two components subscribing for temperature events. One component calculates the average, the other one logs all published temperatures. Both would use the same interface and a role annotation could help distinguish them. Roles are used to name sets of interfaces and/or semantics of interfaces. A meaningful interpretation of the names relies on agreements made outside of the notification service.[13]

The SQL-like language presented in later in this section facilitates the definition of scopes and their features, and includes coupling points to express dependencies, rules for modifying scopes, and their position in the scope graph.

Who is responsible for setting up and maintaining the scope graph? In order to not impair the loose coupling of application components, they should not be forced to interact with the scope graph. For this reason, they may access the graph structure through the Java API, but typically programming and configuration is done by the administrator, who knows the included components and is able to govern their interaction. Of course, different administrators may be responsible for different scopes. We can use abstract scopes to define different administrative domains [268].

As a result of component design a repository of components, i.e., a descriptive scope graph, is created for later composition in bigger scope graphs and for later deployment.

Scope Graph Composition

This second stage of the development process creates the descriptive scope graph. From a selection of existing and newly created components a graph is designed, typically for a specific application. This task includes the resolution of dependencies on interfaces, attributes, and roles, and the specification of application-specific implementation requirements.

The graph describes the relationship between components and stores predefined configurations on a larger scale than single components. Similar to

---

[13] The use of ontologies like in concept-based publish/subscribe [79] is an example of such externally provided agreements.

class hierarchies, the scope graph offers a way to statically describe system structure. The graph is created for a specific application, and so the question is raised, what can be modeled with a scope graph? Since scopes are a generic concept to partition applications and control their interaction, this question asks for a methodology and design guidelines. Unfortunately, there are no general guidelines available so far.

The question by what means an administrator creates this graph is answered, though. Scope graph design must comprise tools and primitives to compose scope graphs from given specifications, to create and configure connections in the graph, and to resolve open dependencies. The Java API can be used to wire specific components or to resolve dependencies by application-specific rules. Existing scope specifications based on the XSchema grammar can be joined, whereby unambiguous dependencies can be resolved with a simple search on the available component definitions. The SQL-like scope language also facilitates this step by altering existing definitions, substituting descriptions of coupling points with lists of concrete components.

However, it may happen that not all dependencies can be resolved before deployment, especially if the runtime environment consists of instances of different descriptive scope graphs. They must be resolved at deployment time or even at runtime. The scope language offers event–condition–action (ECA) rules for this purpose.

Finally, the descriptive scope graph can carry annotations that have no immediate meaning in this step, but are interpreted in later on, similar to stereotypes in the Unified Modeling Language (UML, [154]). For example, annotations of required quality of service attributes may govern the following deployment step, hinting at appropriate implementation techniques.[14]

Scope Graph Deployment

Scope deployment creates or extends an instantiated scope graph, which contains all scopes currently running in the system. This step deploys pre-configured scopes of one or more descriptive scope graphs, it resolves open dependencies, and chooses and parameterizes the implementation techniques for the deployed scopes.

The remaining context dependencies of the descriptive scope graph are resolved at deployment time. Often multiple descriptive graphs are used to describe different applications and subsystems. Their models evolve independently and they only rely on some of the services provided by others. So the deployment step is also an integration step that combines (independently administered) systems at a high level of abstraction.

Some dependencies are not resolved once and for all at deployment. They do not pertain to the static structural layout of the system, but rather depend

---

[14] Obviously, the stepwise transformation and deployment of the scope graph resembles the ideas of model-driven development [155].

on the execution of the event-based system. This is described as part of the *management* paragraph below.

An important point, not only in this step but for the scope concept in general, is the fact that the choice of a concrete implementation technique is postponed until now. The implementation of a scope and its communication facilities is determined here based on annotations made in the descriptive scope graph and/or based on decisions made by the administrator. This approach allows for a model-driven implementation, which fits the needs of the application to the services available in the system. Requirements on causal ordering or security considerations can be part of the application model, and the administrator decides how these things are implemented using available group communication protocols and encryption and key management schemes. Consequently, scopes are the appropriate place to customize specific parts of a system, as demanded in Sect. 6.1.

Management

Scope graph management comprises tools and primitives to maintain and update the instantiated scope graph. All features of scopes are subject to updates and even the layout of the scope graph can be changed, establishing and destroying edges by joining and leaving scopes. It also covers the manual creation of new scopes, and thus deployment is part of scope graph management. These tasks must be available in the API of the publish/subscribe service.

It gets interesting when considering automatic updates. As mentioned above, scope graph layout can be dynamic depending on the execution of the system. Automatic updates of the graph use the management functions to react to events and conditions observed in the system. The scope language presented below allows ECA rules to be associated with scopes. Each rule reacts to arbitrary notifications visible to the respective scope, and if an optional conditional expression is fulfilled arbitrary management commands are executed. Binding these rules to scopes uses the visibility constraints of the scope graph to apply them only in limited areas of the graph. As for the scoped communication, this controls the execution of rules and reduces the complexity of rule analysis [29].

Such rules can be used to define scopes that automatically include all components conforming to a certain condition. One example is mobile systems, which are an apparent application domain of scoped notification delivery. The geographic vicinity to a reference location groups all components within this area.[15] In fact, whenever location models do not strictly correlate to the topology of the network infrastructure, some form of application-specific scoping is necessary [142].

---

[15] Grouping always implies a common context, and scoping thus may contribute to the discussion about context in mobile systems [335].

6.6.3 Scope Graph Language

In order to support the development process a specification language for scope graphs is defined next. Corresponding to the generic nature of the scope concept, the language definition is intended to be open for further refinements, which are probably domain dependent. A Backus–Naur form is used to specify the syntax in form of production rules like

```
rule1 ::= ( "A" | rule2 ) [ rule3 ] rule4-commalist
```

Here, rule `rule1` is expanded to either the literal "A" *or* the result of `rule2`, followed by zero or one expansion of `rule3`, followed by one or more comma-separated expansions of `rule4`.

The next paragraphs introduce a grammar for defining scopes, their features and dependencies. It includes the rule "..." at places of possible future extensions.

Component References

In order to identify any specific component, a reference scheme for components must be defined. For the sake of simplicity only symbolic names are considered here.

```
simple-component-name ::= symbolic-name
simple-component-ref ::= simple-component-name

scope-ref ::= symbolic-name | ( "MEMBERS(" scope-ref ")" )

component-ref ::= simple-component-ref | scope-ref
```

Wherever a scope is referenced by its name, e.g., *scope1*, the scope itself is meant, that is, the node in the scope graph. The *MEMBERS(scope1)* expression is used to refer to the members of the scope, that is, the set of nodes $C_i \triangleleft scope1$ of the scope graph.

Names are not globally unique; they are scoped. A component is part of some scope and its name is, at first, only valid within its scope. A reference to a scope is always resolved from a specific node in the scope graph. If for a given name no component exists in the current scope, all superscopes are considered recursively. This approach is similar to references to overloaded methods in object-oriented programming languages, where the "nearest" definition is used up the inheritance hierarchy. Of course, it may happen that a name cannot be resolved or a name is ambiguous. For a concrete system, rules may be established to devise globally unique names.

Scope Definition

The definition of a scope consists of several parts: component selection, interface and attribute definitions, actions, and update rules. Implementation issues are not specified here. Defining a scope makes it part of the descriptive scope graph; deployment is a second step described later in this section.

```
scope-definition ::=
     "DEFINE SCOPE" component-name "AS"
     component-selection-clauses
     scope-feature-clauses

component-selection-clauses ::=
   component-selection-clause [ component-selection-clause ]

component-selection-clause ::=
     [ component-identifier ":" ]
     ( super-selection | member-selection )
     [ "WHERE" boolean-expression ]
     [ ":" selection-property-clause ]

super-selection ::= "SUPERSCOPE"
     selection-qualifier
     "FROM" ( scope-ref-commalist | "*" )

member-selection ::= [ "MEMBER" ]
     selection-qualifier
     "FROM" ( component-ref-commalist | "*" )

scope-feature-clauses ::=
     scope-feature-clause [ scope-feature-clause ]

scope-feature-clause ::= (
     ( component-identifier ":"
               selection-property-clause ) |
     interface-clause |
     role-clause |
     set-clause |
     action-clause |
     update-clause )

selection-property-clause ::= "{"
     [ interface-clause ]
     [ action-clause ]
     "}"
```

```
boolean-expression ::=
    ( attribute-test | interface-test | role-test | ... )
    [ ( "OR" | "AND" ) boolean-expression ]

attribute-test ::=
    attribute-name
    ( numerical-comparison | string-comparison | ... )

numerical-comparison ::= numerical-operator number
string-comparison ::=
    ( string-comparison-op | string-matching-op ) string
```

Component selection determines superscopes and members of the defined scope if it starts with SUPERSCOPE or MEMBER, respectively. Any selection consists of two steps. First, a base set of components is given after the *FROM* keyword, and the *where* clause selects in a second step those satisfying a boolean expression.

The base set can be given as an enumeration of specific components, for example

```
DEFINE SCOPE example AS ALL FROM prod1, prod2, scope1
```

which defines a scope *example* that contains exactly the components *prod1*, *prod2*, and *scope1*. Or specific components and members of other scopes can be mixed.

```
DEFINE SCOPE temp AS
ALL FROM MEMBERS(world), A, B
WHERE has-temp-sensor = 1
```

defines a scope *temp* containing those components of the predefined scope *world* plus A and B, which have an attribute *has-temp-sensor* set to one. The star * is a special scope name available for template definitions. It is later replaced with the superscopes and siblings of the current scope when it is deployed. It denotes all components visible at deployment time.

The *where* clause is a boolean expression on component attributes and acts as filter. The expression tests individually each of the components given in the from clause; no pairwise comparisons of components are done here. If the where clause is omitted, a scope is defined containing exactly the specified list of components.

The same syntax is used for selecting superscopes. The following definition additionally specifies the superscopes *S1, S2* of *temp*.

```
DEFINE SCOPE temp AS
m: ALL FROM MEMBERS(world), A, B
    WHERE has-temp-sensor = 1
s: SUPERSCOPES ALL FROM S1, S2
```

A component identifier is a name that is valid only within the scope definition. It denotes each component included by the selection it is prepended to. The names do not correspond to nodes in the scope graph; they rather identify selections for later references, for example, when updating or refining a scope definition. In the above example, s refers to S1 and S2 and m refers components selected by the first selection clause.

Selection Qualifier

So far, *all* components matching the where clause are selected for the new scope. However, sometimes a comparison and ranking of eligible components is necessary. For example, the administrator may want to select those that are nearest to a specific location or have the most free computing resources. A *selection qualifier* is part of the selection:

```
selection-qualifier ::=
    ( ALL |
      ( TOP "(" attribute-name "," number ")" ) |
      ( "[" [ number ] ".." [ number ] "]" ) )
```

The default qualifier is ALL (as in the previous examples). TOP performs a top-k selection of all components satisfying the where clause. It sorts components by the given attribute and chooses the first k of them. A qualifier of the form $[n..m]$ specifies the size of the respective selection. A minimum of n matching components up to a maximum of m are chosen here. Either boundary can be omitted, denoting a cardinality of zero and as many as possible, respectively. Omitting both is like choosing ALL.

Many extensions are conceivable at this point. The top selector may take a predicate as argument that evaluates expressions like "*fixed-location - location-attribute*", which sorts according to a distance metric. Other domain-dependent functions may be added in specific implementations.

Interfaces

Component interfaces are defined as part of the scope feature clauses after all selections. Selective and imposed interfaces are specified in selection property clauses, which are either appended to the respective selection clauses or are also given after all selections (see below). The *interface* clause begins with the keyword "INTERFACES" and then includes a comma-separated list of interface specifications. There is no specific filter model preset in the language (cf. Sect. 6.3.1), and so a syntax corresponding to the available filter model must be chosen.

```
interface-clause ::= [ "INTERFACES" interface-commalist ]
interface ::= ( "INPUT(" | "OUTPUT(" )
        [ "0" | "1" | channel-interface | topic-interface |
```

```
      typebased-interface | content-interface | ... ]
")"
```

```
channel-interface ::= channel-name-commalist
topic-interface ::= topic-commalist
topic ::= "/" topic-name [ topic ]
typebased-interface ::= notification-type-name-commalist
content-interface ::=
        boolean-attribute-expression-commalist
```

The two special interfaces "0" and "1" denote filters rejecting and accepting all notifications. The following snippet defines a scope that outputs temperature alarm notifications, but it does not receive any input from its superscopes S1 or S2.

```
DEFINE SCOPE temp AS
ALL FROM MEMBERS(world)
WHERE has-temp-sensor = 1
SUPERSCOPE ALL FROM S1, S2
INTERFACES OUTPUT(AlarmNotification)
```

The next example is an extension that also includes imposed interfaces on the components of *temp* that allow them only to send temperature notifications. All other kinds of input or output traffic of members is prohibited.

```
DEFINE SCOPE temp AS
m: ALL FROM MEMBERS(world)
   WHERE has-temp-sensor = 1
SUPERSCOPE ALL FROM S1, S2
m:{
   INTERFACES OUTPUT(TempNotification), INPUT(0)
}
INTERFACES OUTPUT(AlarmNotification)
```

or alternatively

```
DEFINE SCOPE temp AS
ALL FROM MEMBERS(world)
   WHERE has-temp-sensor = 1 : {
      INTERFACES OUTPUT(TempNotification), INPUT(0)
   }
SUPERSCOPE ALL FROM S1, S2
INTERFACES OUTPUT(AlarmNotification)
```

Note that omitting a component interface is like setting it to "0", whereas omitting a selective or imposed interface is like setting it to "1" (cf. Sect. 6.3.2).

Coupling Points

Coupling points generalize component selection. Coupling points are queries on available components and their properties. They are half-edges in the scope graph that describe dependencies on other components based on properties like component interfaces, roles, or attributes.[16] The dependencies must be resolved at deployment by creating the necessary edges in the scope graph.

A coupling point either provides or demands a specific property. If it demands, the coupling point of matching components must provide the required properties, and vice versa. So far, where clauses request for attributes and interface clauses provide interfaces. What is still needed are means to set attributes, to require interfaces, and to set and require roles. References to the following grammar rules are already part of where clause and scope feature clause:

```
interface-test ::= [ "HAS" ] interface
role-test ::= "IS ROLE(" role-name ")"

role-clause ::= "ROLES" role-name-commalist
role-name ::= symbolic-name

set-clause ::= "SET" set-attribute-commalist
set-attribute ::= attribute-name "="
     ( value | notification-attribute | component-attribute )
```

The *set* clause supports setting scope attributes to constant values as well as to values of notification or components declared in the update clause of the scope (see below).

The next statements define two scopes *admin* and *company*. The latter includes one instance of the former due to its role definition. It imposes an output interface so that only notifications conforming to the holidayAnnouncement type can be passed into *company*. The latter also includes the top ten components, termed *worker*, that either produce or consume other important notifications.[17]

```
DEFINE SCOPE admin AS
ALL FROM c1, c2, c3
INTERFACES INPUT(something), OUTPUT(else)
ROLES boss

DEFINE SCOPE company AS
b:[1..1] FROM world
```

---

[16] Dependencies on attributes can subsume the other two if a sufficiently rich data model is available.

[17] Actually, two distinct clauses should select producers and consumers to avoid getting only one kind of components.

```
WHERE IS ROLE(boss)
INTERFACES OUTPUT(holidayAnnouncement)
worker:TOP(experience,10) FROM world
  WHERE OUTPUT(necessaryInformation) OR
        INPUT(furtherProcessing)
SET name = "Acme, Inc."
```

Actions

Scopes put components into groups for visibility purposes, but they can also perform actions on notifications and components. Scope features like mappings and transmission policies are functions executed on notifications.

```
action-clauses ::=
      ( map-clause | policy-clause | do-clause )
      [ action-clauses ]

map-clause ::= "MAP" ( "INWARD" | "OUTWARD" )
      ( "{" set-attribute-commalist "}" |
        external-code-ref )

policy-clause ::=
      ( delivery-policy | publication-policy | ... )
      [ policy-clause ]
```

The *map* clause defines a mapping which is either inward or outward, transforming incoming or outgoing notifications, respectively. If only one direction is specified, the other one must be derivable or prohibited by interface. Mappings may be defined within the specification language, but most likely externally provided functionality will be used as implementation. So, the map clause includes a reference to external code, which could be a symbolic name that refers to a repository of the notification service or a URL to an external code repository. For the same reason there is no syntax for defining transmission policies; they are supposed to be externally provided, too.

```
do-clause ::= "DO" command
```

The *do* clause is included as hint for future extensions, but is not used so far. It may provide a way to customize scope functionality or even to apply code to all members of the scope. The latter is sketched in [349] for a scenario of wireless sensor networks: application code is assigned to network nodes based on scoped definitions.

Updates

The update clause defines ECA rules to adapt instantiated scopes. Any kind of (application-specific) event visible to the scope can be used in these rules.

There are special event types like pub(F(n)), which is the publication of a notification n conforming to filter F, and sub(F), which is the event of some component subscribing to the filter F, etc.

```
update-clause ::= "UPDATE ON" event
      [ condition ]
      DO action-commalist

event ::=
      ( ( "pub(" | "con(" )
         notification-identifier ":" interface ")" |
      ( "sub(" | "unsub(" | "adv(" | "unadv(" )
         interface ")" |
      join(C,S) | leave(C,S) | ... )

condition ::= "IF" boolean-attribute-expression

action ::= scope-change | create-clause

create-clause ::=
      "CREATE NOW"
      [ INCLUDE COMPONENT [ component-identifier ] ]
```

The notification identifier is a symbolic name valid within the scope definition. It is bound to the actual notification triggering the action and can be used in other parts, e.g., in the set clause to update scope attributes.

Actions comprise the alter scope statement explained below and creation rules. The create clause is a powerful tool to control the dynamics of scope graphs. It defines rules to automatically create predefined scopes when specific events occur. Because this automatic creation can be combined with join actions, new scopes can be created with the publisher of the triggering notification as first member of the scope. "INCLUDE COMPONENT" joins the component that triggered the action. This is the producers or the consumer of a notification (consuming a notification is considered as an event here), the component changing its interface, etc.

In this way session scopes can be defined. They include the initial publisher, all consumers, and consumers of subsequently produced notifications. The condition of the ECA rule controls the extension of such a dynamic scope—a precondition to implement spheres of control or transaction contexts in event-based systems.

Deploying Scopes

Scope definitions extend the descriptive scope graph of the system. It is like defining a class or type in a programming language; it does not create an instance of the subject. An instance of a scope is created and deployed with the following statement:

```
scope-deployment ::= "DEPLOY SCOPE" scope-ref
  [ component-selection-clauses ]
  [ scope-feature-clauses ]
  architecture-clause

architecture-clause ::=
  ( brokerscope-clause | intergrated-routing-clause | ... )

brokerscope-clause ::= "BROKERSCOPE(" host ")"
```

To deploy a scope, an existing definition and an implementation is necessary. The architecture clause lists scope architectures, which are introduced in Sect. 6.7.1. Essentially, it refers to a scope implementation available in the system. It carries implementation-specific parameters, like a host name for a brokerscope implementation.

```
DEFINE SCOPE temp AS
a: ALL FROM *
   WHERE has-temp-sensor = 1

DEPLOY temp
SUPERSCOPE ALL FROM S
a:{ INTERFACES OUTPUT(TempNotification) }
BROKERSCOPE(localhost)
```

This example defines a scope containing all members of S that have temperature sensors. The scope is deployed in an existing scope S using a brokerscope implementation on host *localhost*. It also adds imposed interfaces on selection a permitting only temperature notifications.

Changing Scopes

An ALTER SCOPE statement is introduced to change any part of a scope. It may refer to a definition as well as to an instantiated scope.

```
scope-change ::= "ALTER SCOPE" scope-ref
      [ "ADD" | "DEL" ]
      [ component-selection-clause ]
      [ scope-definition-clauses ]
```

The statement adds new selections or features to an existing scope, or deletes or replaces existing parts of it.

```
ALTER SCOPE temp ADD
ALL FROM c
SUPERSCOPES ALL FROM S
```

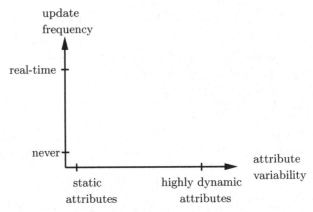

Fig. 6.13. Scope definition accuracy

The above statement adds a component c to the scope *temp* and joins it to S, i.e., $temp \lhd S$.

Maintenance and Definition Accuracy

The where clauses of component selections are rules that determine to which of the available components edges are established in the scope graph. But when are these rules evaluated? Once at deployment? Every t seconds? Or if attributes deviate by more than 20%? Fig. 6.13 sketches alternative views on the accuracy of scope definitions.

The degree of correlation between the rules expressed in the where clauses and the currently established connections in the scope graph is called *scope definition accuracy*. It depends on the variability of attributes and the frequency with which rules are reevaluated.

We assume that queries are evaluated at deployment time only and that their result is not automatically updated afterwards. This corresponds to the lower left point in Fig. 6.13. However, the update clauses in scope definitions allow system engineers to install custom ECA rules to maintain accuracy.

6.7 Implementation Strategies for Scoping

The concept of scopes can be implemented on top of a variety of techniques. In fact, the ideas underlying the scope concept are quite common, but visibility control is often implemented only partially and in an ad hoc manner.

This section investigates a number of approaches for implementing scopes. They differ in the characteristics of the communication media used to convey messages and in the strategies for scope graph distribution. The resulting *scope architectures* are the blueprints of the implementation. All the architectures implement the visibility constraints defined by scopes, but they diverge

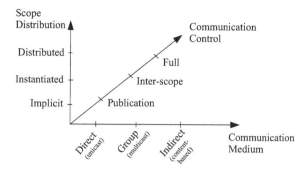

Fig. 6.14. Design dimensions of scope architectures

in their support of other quality of service parameters, like communication reliability and performance, and they also influence system extensibility and adaptability. They emphasize different aspects of the visibility abstraction and are therefore eligible for different application environments.

6.7.1 Scope Architectures

The concept of scopes can be implemented to target any of a wide range of diverse requirements. The implementation influences the functionality and quality of service an application can count on. The architectures presented in this chapter cannot be ranked in general; they may fit the needs of an application or not. There is no best architecture.

Two architectural dimensions are distinguished (Fig. 6.14): communication medium and scope implementation. The combination of these dimensions gives rise to a number of *scope architectures* that determine the principal layout of the scoped event service (cf. Fig. 6.16). The third dimension turns out to classify the architectures' ability to control communication. This section details the architectural choices and defines a metric for comparing the architectures presented later.

Communication Medium

The notion of a *communication medium* denotes any technology that is used to convey notifications between nodes of the scope graph. The communication medium is the basic building block of scope implementation and determines which scope features are supported directly, which features can be implemented efficiently on top, and which features are hardly achievable at all. Any means of data sharing and transport can act as communication medium, ranging from shared memory and TCP [370] connections to IP multicast [106] and peer-to-peer networks [310, 374]. Moreover, existing publish/subscribe services, database management systems [172], and tuple spaces [174] are also eligible candidates for implementing scope graphs. They offer different quality of service and determine the flexibility and functionality of a scoped event system beyond visibility rules.

Although within a single scope different kinds of traffic might be conveyed on top of different communication media, a single medium per scope is assumed for simplicity here. Please refer to Sect. 6.8 for a general discussion on combining media and scopes.

In the following, communication media are differentiated according to their support for unicast/multicast delivery and their addressing capabilities. These are not orthogonal dimensions, rather they highlight different technical aspects that affect scope implementation.

Unicast vs. Multicast

The basic distinguishing feature of communication media is whether they forward data point-to-point or point-to-multipoint, i.e., unicast versus multicast. *Unicast media* send data directly to a specific, identified receiver. In order to reach a number of recipients the send operation must be repeated. Examples include TCP, RPC, and messaging systems. Perhaps surprisingly, unicast media are viable implementation techniques for certain classes of event-based systems; they are considered as a medium to implement scopes, while the producer's and consumer's view (API) on the notification service remains unchanged. *Multicast media* send data to groups of receivers. Multicast media like shared memory, IP multicast, existing notification services, and database tables are common implementation techniques that intuitively correspond to the characteristics of notification distribution.

Obviously, multicast media distribute notifications more efficiently than unicast media. On the other hand, multicast limits the ability to distinguish recipients and control the actual set of receivers. Scope features like delivery and security policies, which are meant to re-introduce control, cannot be implemented directly on top of multicast media without additional filtering (cf. *client-side filtering* later in this section). Exploiting the knowledge about scope members enables system engineers to shape traffic, implement advanced transmission policies, encrypt data, etc. At the cost of multiple send operations and the need to maintain the current set of scope members, unicast media are more flexible than multicast media. In practice there are applications for both unicast and multicast media, and the main issue is a tradeoff between efficiency of data distribution and addressing granularity.

Direct, Group, and Indirect Addressing

Communication media can be further distinguished according to their addressing schemes. While unicast media use *direct addressing*, which identifies an individual receiver uniquely in the network, multicast media can be subdivided into group addressing and indirect addressing. In *group addressing* data are sent to a named group of recipients. The name of the group is specified by the sender, and all members of the group get messages sent within. Group membership is handled separately via membership protocols. IP multicast and group communication protocols [319] are examples of this form of communication.

In *indirect addressing*, the second form of multicasting to a set of receivers, no destinations are specified. Instead of naming groups of receivers, the set of receivers is determined indirectly with the help of information given in messages and by potential receivers. For instance, content-based routing delivers notifications according to consumer-provided filters that test notification content. Another example is proximity group communication [258, 320], where messages are sent only to receivers that are physically close by, i.e., addressees are implicitly determined by location metadata.

Communication Media, Publish/Subscribe, and Visibility

The choice between unicast and multicast media is mainly a tradeoff between efficiency and control, as described above. But what media are good candidates to implement a publish/subscribe service, and do some of them even offer a visibility mechanism comparable to scopes? What are the characteristics of group and indirect addressing that influence the implementation of scopes?

As for the general applicability to implement a publish/subscribe API, group and indirect addressing is related to the discussion on filter models (channel-, subject-, and content-based filtering) given in Sect. 2.1.3. Group addressing is like channels in that a name representing a set of receivers is used by the sender to disseminate data. Subject-based addressing is an extension that allows for subgroups [289, 380], which is, to some extent, also supported by IP multicast [259].

Group-based multicast media establish visibility constraints in that they encapsulate intragroup traffic. Notifications published within a multicast group, or under a specific subject, are a priori not visible to outside consumers. However, groups classify messages either based on content (all notifications of type A) or based on application structure (all database servers in a company's back-end infrastructure). Furthermore, groups are often not able to reflect the acyclic scope digraph, because they are mostly arranged in trees, as in IP multicast and subject-based addressing. Even if one tries to model different viewpoints with the help of subgroups, the exponentially growing number of necessary groups limits practical applicability (see Sect. 2.1.3).

Scopes, on the other hand, are orthogonal to consumer subscriptions. They handle interfaces (i.e., subscriptions, group names, etc.) and system structure (the organization of scopes in the scope graph) independently. Thus, groups do not directly implement scopes.

Indirect addressing media can avoid many of the problems of group addressing. They are typically more flexible, but less efficient as they do not easily map to hardware-supported multicast mechanisms. In the generic form, like in content-based publish/subscribe, implementations based on database management systems (DBMS), and tuple spaces, they are able to carry different viewpoints (content vs. structure) simultaneously. Available products/ prototypes are able to offer only a few of the features of scopes, but they are an ideal basis for their implementation.

Scope Distribution

Considering individual scopes, there are three basic choices of how a scope can be realized: *implicit* with all the control in the local event brokers of members; *instantiated* with an explicit administrative component that represents the scope and is responsible for membership control, transmission policies, and mappings; and finally, the implementation of a single scope can be *distributed* on multiple administrative components residing in different nodes of the network. Note that similar alternatives exist for the scope graph. Implicit scopes imply an implicit scope graph, administrative components can either be centralized in a single node or run on different nodes of the network (centralized or distributed scope graph), and distributed scopes imply a distributed scope graph.

Implicit Scope Implementation

The first approach is to collocate scoping with application components. The implementation is shifted into the communication library used to connect application components to the notification service, i.e., into the local event brokers in REBECA terminology (cf. Sect. 2.4). The local event brokers use the addressing and filtering capabilities of the underlying communication medium to implement scope boundaries. The main idea is to annotate notifications to carry scope graph data. Extended subscriptions then exploit these annotations to filter not only on the original consumer's interest, but also on visibility constraints imposed by the scope graph. Consider, for instance, a scope graph with unique scope names, local event brokers that annotate notifications with scope names ($n.scope =$ *"MY-SCOPE"*) and modify each original subscription F to $F' = F \wedge n.scope =$ *"MY-SCOPE"* + *interfaces*.

The extended subscriptions F' must be mapped to the medium's filter capabilities, which is possible if expressive filter models are available like in the Java Message Service or in REBECA. If this mapping is not possible, client-side filtering must enforce the visibility constraints to guarantee that all requirements of the safety condition of scoped event systems are met, cf. Def. 6.3 in Sect. 6.2.2.

For example, consider the members of a scope forming a group that communicates notifications via a group-addressing medium like subject-based publish/subscribe to all scope members. This floods all notifications to all members of this scope, postponing original subscription processing to the

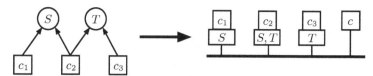

Fig. 6.15. Implicit implementation shifts visibility control into application components

client side. If content-based filters are available, processing of both client subscriptions and scope interfaces can be shifted into the medium; the former F' could be supplied to JMS or REBECA.

In an implicit scope implementation the structure modeled by the scope graph is transformed into a flat implementation, as illustrated in Fig. 6.15. Every component is connected to the same medium, and conventions must determine how visibility constraints are implemented on top of the addressing mechanisms offered by the medium. In order to meet the safety and liveness conditions, each component must maintain the necessary management information about the layout of the scope graph and the current scope interfaces. So, scoping structure can be transparently implemented in the local event brokers without modifying application code, but scope graph changes require update processing in potentially many of the components.

The problem of shifting scope control into local event brokers is that components not adhering to these conventions may bypass visibility constraints, both as consumer and as producer. Since the scope structure exists only implicitly in the components of the system, no external entity controls and enforces scope boundaries, giving rise to both reliability and security concerns. Consumers might arrange to listen to notifications they are not intended to receive, and even worse, they may send notifications to any component, disrupting correctness in other parts of the system as well. Moreover, more advanced features of scopes, namely transmission policies and mappings, are even harder to implement using an implicit implementation.

Instantiated Scope Implementation

To exert more control on notification dissemination the scope graph must be managed within the notification service infrastructure. A basic approach is to explicitly instantiate administrative components to represent scopes. They are generated and controlled by the notification service itself and contain an implementation of scopes outside of application components.

This scenario is further subdivided into a centralized graph and a centralized scope form. The former implements the whole scope graph in a single node of the distributed system and amounts to a central information hub. This is a widely used approach for implementing unscoped event systems, because it simplifies notification routing and access control, but comes at the expense of scalability and diversity support. Examples range from centralized databases [172, 292] (see later in this section) to content delivery networks [333], which can be seen as logically centralized nodes optimized for one-way delivery efficiency. In the centralized scope form, each scope is represented by one administrative component, but each such component may run on a different node in the network.

Administrative components make the scope structure explicit and accessible to the system engineer, who is now able to customize (parts of) it to the local needs of an application. This approach facilitates configuration and integration of heterogeneous components on a per-scope basis as each administrative component may act as bridge between different implementations (different data/filter models, communication medium, etc., see Sect. 6.8). In contrast to an implicit solution, instantiated scopes make it easier to control adherence to a specific scope graph and it relieves clients from management tasks.

Distributed Scope Implementation

A single, distributed scope consists of multiple administrative components that together constitute this scope. Each scope member is assigned to one administrative component. The same type of communication medium is still assumed for delivery to scope members, but communication between the administrative components may be based on a different technique. Scalability is obviously improved since multiple administrative components share and subdivide the load to distribute intrascope notifications; they may even exploit effects of locality when notifications are only forwarded within one administrative component.

For example, consider two groups of application components belonging to the same scope, but located at two different border brokers of the underlying network, e.g., an Internet of two LANs connected by a WAN. Instantiating a scope implementation solely in one LAN would diminish the benefits of locality for the other side. But if administrative components are available on both sides, they may draw on a local broadcast medium and connect each other using a point-to-point link.

Example Architectures

Figure 6.16 shows possible architectures that are defined as specific combinations of scope distribution and communication medium. They are sketched in the following and two of them are detailed in Sects. 6.7.3 and 6.7.4.

Static Deployment

The combination of implicit scope implementation and point-to-point communication leads to a *static deployment* where every scope member knows its siblings and communicates directly with them. When subscriptions are known

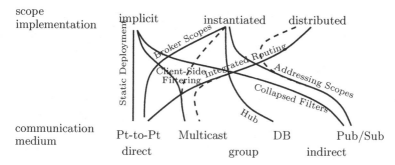

Fig. 6.16. A comparison of scope architectures

to all members, notifications are sent to subscribed consumers only. Otherwise, notifications are sent to all scope members, which evaluate their own filters on any notification published in the scope. Output interfaces toward superscopes and input interfaces of sibling subscopes must be known as well so that cross-scope notifications can be sent to a consumer in the destination scope, which in turn relays them within. This scenario is called static deployment since it is an eligible architecture option if the scope graph is static, rather small, and does not change at runtime. System configuration can then be compiled into the local event brokers without affecting the publish/subscribe API, as it is for any of the presented architectures. In such a situation even remote procedure calls are a suitable implementation technique to convey notifications. If the system is not static the necessary configuration data in the components must be kept up to date. Examples of this approach are data-driven coordination languages (e.g., Manifold [296, 297]), which connect input/output ports of coordinated entities, and even an implementation using TCP/IP to connect the participants is eligible, particularly if the system footprint has to be kept small. Interestingly, the JavaBeans programming model [84, 359] and component-oriented programming in general [234, 369] are related to this approach in that they facilitate the wiring of interfaces and ports.

Another application of static deployment is to wrap a callback-based system with a publish/subscribe API. That is, undirected subscriptions are resolved and directly registered at corresponding callback handlers that are visible according to the locally stored scope graph. Although somewhat unusual, this might offer a way to draw from existing request/reply or directed messaging systems, when possible, and from their established benefits, for instance, in security and transactional data management.

Client-Side Filtering

The *client-side filtering* architecture also utilizes an implicit scope implementation but is built on a multicast medium that provides group-based addressing, like IP multicast. Each scope is assigned a multicast group address and all members of a scope are reached with only one call to the medium. Compared to static deployment, the required network bandwidth is considerably reduced. However, since there are still no administrative components the visibility constraints defined by the current scope graph must be enforced on producer and/or consumer side. As described earlier in this section, the local event brokers may annotate notifications and must select appropriate destination group addresses on producer side. And on consumer side incoming notifications must be filtered out so that in combination only matching notifications are delivered that comply with the scope graph and satisfy the safety condition of the scoped event systems definition.

A different way of using group-based multicast here is to group according to content instead of structure. In such a scenario multicast groups might be used to group subscriptions, which is the common use of multicast in publish/subscribe systems [87, 291]. Consumers would have to determine the visibility of incoming notifications by evaluating the interfaces of the scope graph as part of their client-side filtering. Thus, in the first approach producers have to know the current scope graph layout to select the correct destination scopes, while in the second approach consumers are in charge of this. The two approaches differ mainly in the selectivity of the grouping and the implied costs of keeping the graph information up-to-date.

Another extension is to instantiate administrative components within scopes that are responsible for relaying incoming and outgoing notifications. In this way the need to store the full scope graph in local event brokers is removed, since these relaying components have to know their adjacent nodes only.

Client-side filtering is obviously applicable when scope graphs are rather static and of limited size. For instance, if scope graph changes are just induced by moving simple components the assignment of group addresses to scopes remain unchanged. The moving components have to join the respective groups, but the scope graph information need not be updated elsewhere. Scope graph management is thus reduced to group membership management, which is provided by the communication medium. Nevertheless, this architecture is left out of consideration in favor of more flexible solutions.

Collapsed Filters

In the *Collapsed Filters* architecture, the visibility constraints expressed in the scope graph are merged into the subscriptions issued by consumers. This leads to a flat notification service where enhanced subscriptions implement the scope graph implicitly, requiring an expressive subscription like in content-based publish/subscribe. Extra effort is necessary on both the producer and consumer sides. Producers, i.e., their local event broker, annotate notifications and add data necessary for visibility filtering. Consumers have to extend their subscriptions to test as much of the imposed visibility constraints as possible. If the filter model is not expressive enough, they must locally evaluate the remaining filters on every received notification.

The collapsed filters approach is a simple implementation of scoping as a layer on top of an existing communication infrastructure. But it does not provide the full control of visibility at runtime. Notification mappings and delivery polices are not always implementable. Furthermore, graph changes are difficult and costly to deploy, because application components are not easily reconfigurable and changes to the graph have to be consistently distributed to all affected components.

The system's functionality in a collapsed graph depends on the correct function of *all* participating components. It renders control of the visibility to the components. A corrupted or malevolent component may publish or eavesdrop in any scope. The discussion on combining different scope architectures in Sect. 6.8 leads to a possible solution when gateway components bridge two separated subgraphs and provide an explicit encapsulation of visibility constraints.

Central Hub

The "classic" data management approach of using a central database may also be beneficial in an event scenario. It is an alternative implementation of collapsed filters and it easily offers sophisticated quality of service guarantees in addition to the basic safety and liveness requirements of scoped event systems.

Using databases for implementation blurs the distinction between the collapsed filter and the central hub scope architectures. Similar to the content-based publish/subscribe medium assumed above, a database table can hold all published notifications, and subscriptions are merely queries to this table. In fact, database technology provides a wide spectrum of functionality [172] that may be exploited to extend the quality of service offered by the event system beyond the definitions given in Chap. 2. On the other hand, there are drawbacks like their maintenance complexity, resource consumption, and acquisition and operation costs.

Addressing Scopes

Addressing scopes is an extension of the client-side filtering approach that no longer relies on multicast but instead on content-based publish/subscribe. Each scope has a unique name that is appended to published notifications. Every subscription is extended to accept notifications only if they are issued in the consumer's scope. The scope address type of architecture introduces administrative components that localize the implementation of interfaces, publishing policies, and mappings. They offer a finer control of interscope communication than the collapsed scopes.

Scoping is still implemented on a shared multicast medium and the implementation is not aware of the underlying network layout. In fact, intrascope communication is not directly governed by the administrative components and relies on the filtering capabilities of the communication medium. The local event brokers of producers and consumers modify notifications and subscriptions before sending them out. With respect to intrascope communication, scope addressing is similar to collapsed scopes. Internal delivery policies, admission to scopes, and, in general, conformance to the visibility defined in the scope graph is achieved only if producers and consumers operate cooperatively and correct.

Compared to the collapsed scopes, which need only one access to the medium to reach every consumer, the administrative components repetitively access the medium to forward a notification along a delivery path in the scope graph. In situations where some consumers are connected via long delivery paths, this approach apparently induced a considerable communication overhead. But the indirection introduced by the administrative components relieves simple components from maintaining the current graph structure. Especially the last point touches on a well-known tradeoff between scalability and expressiveness [69]. In the collapsed scope graph approach lots of extended filters are issued, whereas with scope addresses the filter complexity is limited at the expense of increasing communication bandwidth.

Broker Scopes

Broker scopes are a one-to-one implementation of the scope graph in that each scope is explicitly represented by an event broker of the broker network (cf. Sect. 2.4). This approach is detailed in Sect. 6.7.3.

Integrated Routing

Integrated routing fully integrates scoped notification delivery into the routing infrastructure. The routing tables themselves are extended to reflect visibility constraints of the scope graph. This architecture is described in Sect. 6.7.4.

Scope Graph Distribution—Types of Architectures

While the choices described above consider individual scopes only, the following looks at scope graph implementation as a whole. The general processing steps of scoped notification delivery are described, which identify potential places to implement scoping functionality in the system. These steps serve as a basis to compare the preceding example architectures and to classify them in three types of architectures. These types differ in the degree they support scope graph reconfigurations, transmission policies, and, in general, any distribution control beyond scope interfaces.

Figure 6.17 sketches the delivery in a scoped event system. The numbered course shows the forwarding of a notification that moves along an exemplary delivery path $(p, S_2, \ldots, S_{n-1}, c)$ between producer p and consumer c in an arbitrary scope graph.

1. In the first step a notification is published by producer p.

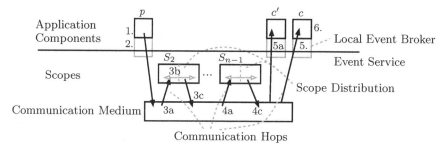

Fig. 6.17. Steps of scoped notification delivery

2. The access to the event notification service is provided by the local event broker, which is conceptually part of the application component. The broker may process the notification as part of an implicit scope implementation (cf. static deployment) before it is forwarded by accessing the communication medium.

3a. If scopes are instantiated in administrative components, the notification is delivered to an instance of S_2 of the example delivery path.

3b. If scopes are distributed, the notification is also sent to other instances of this scope if needed.

3c. Delivery in S_2 is completed when the notification is forwarded toward its members and superscopes, accessing the underlying medium for the second time.

4. The previous three steps are repeated for all other scopes.

5. The notification is received by the local event broker of the potential consumer, which may again process and filter the notification before it is delivered to the consumer.

6. Finally, the notification is delivered to the consumer c.

An implementation of scope graphs may stretch across up to three layers: On the lowest layer, the communication medium is parameterized to distinguish scopes or at least administrative components representing scopes. On the middle layer explicit administrative components implement scope features within the notification service. At the highest layer, code is collocated to application components in local event brokers to modify notifications and subscriptions.

The figure illustrates all possible steps although only a subset is relevant for a specific architecture. In an implicit scope implementation no scopes are instantiated within the event service and steps 3 and 4 are omitted. With a centralized scope implementation step 3b is not needed. Whether any processing is done in the local event brokers (steps 2 and 5) depends on the concrete implementation, but it is definitely required in implicit approaches. When group-based multicast is used to address all members of a scope additional client-side filtering is also needed in step 5.

	publication control	inter-scope control	full control
# accesses	1	$n-2$	$n-1$
possible medium	any	group or indirect	direct
scope distribution	implicit	central./ distributed	central./ distributed
data flow control	no explicit control	control inter-scope traffic	control every edge
examples	static deployment, client-side filtering, collapsed filters	addressing scopes	broker scopes, integrated routing

Fig. 6.18. Types of architectures, their characteristics, and examples

The different choices to partition scope implementation among these steps turn out to be a fundamental characteristic of scope architectures. It determines their ability to adopt scope graph changes and to implement any sophisticated control of communication beyond interfaces. For an assessment it is crucial to compare the amount of control residing within the notification service with the amount shifted into the communication medium and the application components, respectively. For this purpose the number of accesses to the communication medium that are necessary to forward a notification along a delivery path is taken as a measure to distinguish architecture types. These accesses are labeled as communication hops in Fig. 6.17, whereas communication between instances of the same distributed scope (step 3b) is not counted, since it does not leave the scope's sphere of control. Based on this consideration, three modes of notification forwarding are identified and depicted in Fig. 6.18.

1. **Publication control.** All consumers are reached with only one access to the medium. All interfaces and delivery policies bound to the scope graph must therefore be evaluated within the communication medium or as part of the local event brokers of producers and consumers. There is no control within the medium or the publish/subscribe service infrastructure once the message is sent. Accessing the communication medium means here that all eligible consumers in the whole system get the notification.

2. **Inter-scope control.** In this approach, scopes are represented by administrative components that govern the interfaces toward superscopes and relay incoming and outgoing notifications if they match the respective input and output interfaces. Within scopes, however, lists of members are

not maintained and notifications are not directed to specific addressees. A multicast medium is used that may reach all scope members in one step. Since producers do not distinguish any siblings, the consumers' subscriptions must either be completely handled by the communication medium or, if all scope members are indistinctively addressed as a group, consumer-side filtering must be applied.

Accessing the communication medium means here that a scope and all of its members get the notification. For an arbitrary delivery path, one access to the communication medium is needed for every edge, except for the root scope of the path where sending and receiving components are siblings. This leads to $n-2$ calls to the communication medium for a path of length n.

3. **Full control.** Each scope is represented by an administrative component, and notifications are forwarded strictly along the edges in the scope graph, resulting in $n-1$ accesses to the medium for a delivery path of length n. Each scope is implemented in one or more brokers in the routing network. Delivery is controlled even within a scope.

This is an one-to-one implementation of the scope graph, and accessing the communication medium means here that notifications are sent to the next hop node in the scope graph or only within one scope graph node that resides on multiple network nodes (e.g., integrated routing).

This classification describes what part of the scope graph is offered through the communication medium and the implicit implementation in application components, on the one hand, and what part is implemented in administrative components instantiated in the infrastructure, on the other hand. This distinction determines how the different number of accesses to the communication medium determines the ability of a scope architecture to adapt the current configuration of the system. While explicit administrative components are readily adaptable, it is far more difficult to update infrastructure code in a consistent and transparent way when it resides in local event brokers.

An even more important fact is that the granularity of the control exerted on notification distribution gets inevitably more coarse if fewer accesses to the medium are needed. With fewer accesses more consumers are reached in one step, which implies uniform delivery to larger sets of nondiscriminated components. However, any form of refining and controlling dissemination will have to differentiate subsets of these components. And the number of accesses to the medium characterize how much of the structure identified in the scope graph is reflected in the implementation.

6.7.2 Comparing Architectures

Scope architectures can be classified in the architectural dimensions given above. However, further criteria are necessary for comparing and assessing their functionality from an application point of view. The architectures presented in the next sections are compared according to the following criteria:

- Impact on infrastructure and components: What must be changed to implement scoping?
- Implementation overhead: What is the overhead implied by a given scope architecture? What are the communication costs compared to unscoped publish/subscribe and compared to other scope architectures?
- Reliability: How do failures of components affect single scopes or overall system correctness?
- Reconfiguration: What kinds of changes of the scope graph are possible in the running system? What are the costs of scope graph updates? Adaptability and flexibility to change system structure are the main issues here.
- Customization: While all scope architectures obey the visibility constraints expressed in a scope graph, which of the other features of scopes are supported? What kinds of mappings, transmission policies, security policies, etc. can be established?

The comparison of the scope architectures is summarized in Fig. 6.19.

| | impact on | | | | ability to | |
	infrastr.	components	overhead	reliability	reconfigure	customize
collapsed filters	+	−	○	−	−	−
hub	+	○	○	+	○	+
static deploy.	+	−	+	○	−	○
addressing scopes	+	○	○	○	+	○
broker scopes	○	+	○	+	+	+
integrated routing	−	+	+	+	+	+

Fig. 6.19. Comparison of scope architectures (+ means low impact and overhead, and high ability to achieve reliability, reconfiguration, and customization)

6.7.3 Implement Scopes as Event Brokers

The broker scope approach is the most general implementation of scopes. It uses administrative components representing scopes, as before, but relies on their forwarding even for intrascope communication. It directly implements the structure of the scope graph in the sense that publishing within a scope first requires accessing the communication medium to send the notification to the representing scope instance, which, in the second step, sends the notification to all its children and, after applying the output filters, to the eligible superscopes. In terms of Fig. 6.17, all the steps are explicitly implemented. With brokering each notification individually, even the delivery of notifications to separate consumers could be distinguished in steps 5 and 5a. The existence of step 3b depends on the internal implementation of each scope representative, of course.

The characteristics of this approach are the independently operating administrative components that represent each scope and have full knowledge about adjacent subcomponents and superscopes. And, in principle, a point-to-point communication between the nodes is assumed so that arbitrary delivery can be implemented in scopes. In practice, a number of different communication media and schemes for implementing and locating administrative components are possible.

One Scope, One Broker

The simplest form is a one-to-one implementation of the scope graph, which instantiates exactly one administrative component per scope and uses point-to-point media to convey data as defined by the edges of the graph. The point-to-point communication to all children offers the full control of intrascope traffic. Any constraint bound to the scope graph is easily implemented at this explicit point in the infrastructure: no restrictions of applicable transmission policies, mappings, and security measures are imposed.

From a technical point of view, an implementation with scopes as brokers is similar to the architecture described in Sect. 2.4, only that a strict treelike network is no longer mandated. Instead, the undirected form of the directed acyclic scope graph constitutes the overlay network used to convey the data. The original restriction to trees was made to simplify analysis and implementation of general routing protocols, which is a reasonable initial assumption for a research prototype. Here, this restriction is removed. However, the problems inherent to arbitrary graphs are not solved in general, rather scoping and the definition of visibility constrains the possible routing configurations in the graph. The network layout is no longer an infrastructure independent of the application components; the administrator of the system is provided with means to shape its layout and control the distribution of notifications. Routing is the implementation of visibility, and the responsibility of ensuring sensible routing is now partially transfered to the administrator.

A possible drawback of this approach might be its degradation of communication efficiency. To convey data along a given delivery path of length n, $n-1$ accesses to the underlying medium are necessary, which is only one more than in the scope address approach. But if only intrascope traffic is considered, which may dominate in many systems anyway, the necessary accesses are doubled. However, even if other implementation approaches may be more efficient for certain system configurations, broker scopes provide the most general implementation of scope graphs, and the ones most adaptable to any kind of reconfigurations. So, the alleged inefficiency has to be compared with the indirection of the scope brokers and the enhanced control they introduce thereby.

Distributed Scopes

The above discussion assumed a single administrative component per scope, which is responsible for filtering incoming and outgoing traffic and internal forwarding. With distributed scopes, this task is performed by multiple instances, that is, by distributed administrative components of one scope. Whenever the instances are not independent, they have to communicate with each other and thus implement step 3b of Fig. 6.17. For the communication between these instances a communication medium can be used that is different from the one conveying data between the scope graph nodes. However, the same arguments regarding addressing capabilities, scalability, and flexibility hold as before.

A number of objectives are achievable with distributed scopes. An obvious improvement is to instantiate multiple administrative components for each scope to prevent single points of failure. The instances may be identical replicas using a primary/backup approach [11] or operating in parallel independently of each other. Alternatively, each of the instances may be responsible for a different subset of the scope's components so that in case of failure only one subset is affected, but not all components of the scope. In these cases, a point-to-point communication within a known set of scope representatives is indicated.

Furthermore, scope distribution facilitates adaptation. For example, if one administrative component is instantiated per superscope, each instance handles the interfaces, mappings, and transmission policies with respect to one superscope. The addition of edges simply requires adding the respective administrative components. And if a multicast medium is used to forward notifications from scope members to all the administrative instances, edge configuration does not even influence any other parties in the scope. Another option is to provide specialized services by different scope representatives for certain types of notifications, such as internal delivery policies or encryption for specific notifications. This implementation partially backs off the initially stated assumption that only one communication medium is used per scope. The same result could be achieved if each of the specialized administrative components is created as a full scope in the scope graph.

The above examples employ separate administrative components to facilitate the implementation and reconfiguration of a scope graph, but they do not consider distribution with respect to the actual layout of the physical network. A very important aspect of distributed scopes is their ability to bridge between the structure of the application given in the scope graph and the structure of the underlying network. Consider a scope that groups physically dispersed members located in two different subnetworks. With a single administrative component all traffic would be centralized, whereas distribution helps exploit locality. If an instance of the scope is present in each of the subnetworks, notification forwarding is decoupled and done locally in each network. And the bandwidth necessary between the networks can be reduced once the connected administrative components remember the remotely published subscriptions, i.e., they maintain a routing table.

The previous description shows clearly that multiple explicit scope instances constitute a distribution network by itself. When several scopes are distributed, several of these overlay networks coexist. In this situation scoping and routing are mixed, which is investigated in Sect. 6.7.4.

Collocating Broker Scopes

A special solution is to collocate all administrative components at one node in the network. Scope-internal traffic still needs two accesses to the underlying medium, but all interscope communication is done locally. Although closely related to the central hub approach, cf. Sect. 6.7.1, the scope graph is explicitly instantiated here, only that interscope communication is implemented by interprocess communication (IPC). Separate administrative components can still evolve independently, they just happen to be collocated, so to speak, to improve efficiency, auditability, or other global constraints.

Evaluation

Scopes as brokers are the most flexible implementation of the scope graph. They offer all features of the scoping concept and the flexibility to adapt all aspects of the one-to-one realization of the scope graph. Every feature is localized in the infrastructure. Apart from this configuration viewpoint, broker scopes make the infrastructure itself visible and adaptable, for it provides administrators with means to map application structure to infrastructure components, that is, to event brokers.

This scope architecture is possibly not the most efficient implementation of a certain scope graph, but it is the most generic one. It is not a service of the publish/subscribe infrastructure, but instead a way to define and adapt the infrastructure itself, and it will serve as a basis for refining the implementation of subgraphs, as discussed in Sect. 6.8. However, it is not always acceptable to have such a close correlation between the application structure supposedly encoded in the scope graph and the implied, dependent layout of the network infrastructure.

6.7.4 Integrate Scoping and Routing

The explicit instantiation of administrative components described in the previous section makes the full range of scope features available to system engineers, i.e., administrators. However, it also determines the layout of the underlying network infrastructure, which is no longer independent of the applications. In contrast, the following integrates scoping into the routing infrastructure. Visibility control becomes an inherent service of the event notification service and is no longer implemented as a layer above the underlying broker network.

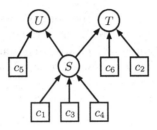

Fig. 6.20. An exemplary scope graph

Scopes as Overlays

Given a network of brokers and a scope graph, the simple components of a specific scope are in general connected to arbitrary border brokers, irrespective of their scope membership. They are reachable via a subset of the border brokers, and the notification service must ensure that notifications are forwarded to these brokers if they match one of the subscriptions of the respective simple components. Consider the exemplary scope graph and the broker network depicted in Figs. 6.20 and 6.21. Brokers B_1, B_2, and B_4 are part of scope T, that is, they are *scope brokers*[18] of T. Together with B_3, they are also scope brokers of S. B_2 is in both cases an intermediate broker that currently does not have any directly connected scope members. B_1 and B_5 are scope brokers of U.

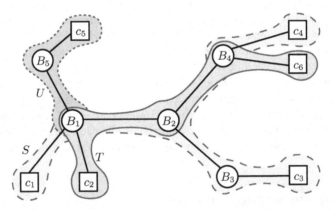

Fig. 6.21. Scopes as overlays within the broker topology

---

[18] Mind the difference between scope brokers and broker scopes. The former are part of an independent broker network and sustain a specific scope, whereas the latter is a scope architecture and a different way to implement the scope graph (Sect. 6.7.3).

Filter	Destination
$^iI_{c_1}$	c_1
$^iI_{c_2}$	c_2
$^iI_{c_3}$	B_2
$^iI_{c_4}$	B_2
$^iI_{c_6}$	B_2
$^iI_{c_5}$	B_5

Fig. 6.22. A flat routing table for broker B_1

The main idea is to rely on any of the existing routing schemes, e.g., those offered by REBECA (Sect. 2.4), as before, but to use it for intrascope traffic only and for each scope separately. Still, the same broker network is used to route all notifications and a connected subset of brokers routes the traffic internal to a given scope without heeding other scopes. The separate routing for each scope effectively establishes *scope overlays* in the broker network, which are sketched in Fig. 6.21. On the other hand, the separation of scope-internal routing necessitates a special handling of interscope transitions. In Fig. 6.21, B_1 is scope broker of both S and U to bridge between the overlays of the two scopes.

Consequently, two kinds of routing are utilized to integrate scoping into the broker network: intrascope within a specific scope and interscope routing between scopes adjacent in the scope graph. In intrascope routing each scope overlay maintains its own routing tables so that each broker has a routing table per scope it supports. The employed routing scheme maintains the independent routing tables and handles advertisements and notifications as before. Hence, brokers constituting a scope overlay behave like a traditional flat publish/subscribe service in which no visibility constraints exist. In interscope routing brokers must arrange for the transition of notifications between scope overlays according to the scope graph and the assigned interfaces and mappings. The current assumption is that two scopes $S \lhd T$ have to share at least one common scope broker to implement the scope graph edge at this point. In the previous example both B_4 and B_5 support scopes S and T, and both are able to let notifications cross the respective boundaries; the same holds for B_1 and S and U.

Enhancing Routing Tables

The original flat routing tables maintained in each broker contain filter-destination pairs that list issued subscriptions and the next-hop nodes from which they were received, describing the paths to consumers. Figure 6.22 shows the flat routing table RT_{B_1} of broker B_1 of the previous example. The *enhanced routing tables* subdivide these entries and group them in separate scope-specific tables $\mathsf{RT}_{B_1}^S$, $\mathsf{RT}_{B_1}^T$, and $\mathsf{RT}_{B_1}^U$, sketched in Fig. 6.23. From the point of view of a specific scope S, both simple and complex components are entries in a scoped routing table $\mathsf{RT}_{B_1}^S$. Although technically equal, entries of subscopes are distinguished from entries of superscopes, which is necessary to correctly implement the visibility of components as described in the next subsection.

The "Filter" and "Destination" columns have still the same semantics as before: an entry indicates that notifications are to be forwarded to the given destination if they match the respective filter. In distinction to the original flat table, however, the new tables store arbitrary mappings instead of just filters. In this way the effective interfaces between components can be tested, including any mappings assigned in the scope graph. Of course, any implementation is free to still store simple filters separately from more complex notification processing functions. For instance, the filter–link pairs of the original routing tables may be transformed into triples of filter sequences and links plus mapping sequences.

The destinations stored in the enhanced tables are either network links or locally stored data structures. The former represents an implementation to communicate with next-hop brokers and clients, the latter are the routing tables of next-hop nodes in the scope graph. They mix and integrate the two levels of routing between physical brokers, on the one hand, and between scope overlays, on the other hand.

The scoped routing tables $\mathsf{RT}_{B_i}^{S_i}$ govern notification forwarding both within and between scopes, once set up properly. But in order to establish new edges in the scope graph and to create and link the respective routing tables, additional information must be maintained in the broker network. Each broker keeps a *scope lookup table* ST^{B_i} that contains pairs of scope identifiers and network links, indicating in which direction scope brokers of the specified scope can be found. These tables are updated upon scope creation and deletion, as discussed below. For the previous example they look like in Fig. 6.24.

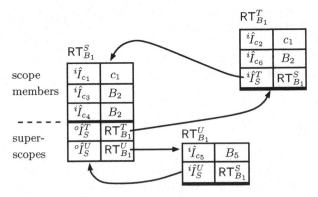

Fig. 6.23. Enhanced routing tables of B_1 incorporating scopes

ST^{B_1}	
S	B_1
T	B_1
U	B_1

ST^{B_2}	
S	B_2
T	B_2
U	B_1

ST^{B_3}	
S	B_2
T	B_3
U	B_2

$\ldots$

Fig. 6.24. Scope lookup tables

Setting Up Routing Tables

Once created, the routing tables are filled when consumers subscribe, and the underlying routing algorithm must forward and register these subscriptions. Chapter 2 described simple routing and covering and merging, which may be applied to accomplish this task. The scoped routing tables themselves and the references between them are set up as reactions to scope graph reconfigurations. In addition to the plain publish/subscribe primitives *pub*, *sub*, and *notify*, Sect. 6.2.5 on dynamic scopes introduced four new operations: *cscope(S)*, *dscope(S)*, *jscope(X, S)*, and *lscope(X, S)*, which create and destroy a scope S, and join X to scope S and remove it, respectively. While the network of brokers is still assumed fixed, the following describes how routing tables are adjusted to reflect these operations. Section 6.6 has suggested tools that support system engineers in this task.

Adding and Removing Scopes

The primitive *cscope(S)* creates a new scope S if invoked by the system engineer at a specific broker B. If no scope of this name is known before, a new routing table RT_B^S is created and the scope lookup table ST^B is updated. By default the creation is announced as unscoped notification and every broker listening to these kinds of notifications updates its scope lookup table accordingly. If a new scope shall not be made publicly available but only as a member of a specific superscope T, the initial announcement can be postponed until it has joined T. The announcement is then sent within T and its visibility is governed by the installed interfaces. Without such restrictions the full list of all scopes instantiated in the system would be listed in all lookup tables, as is the case for advertisements or subscriptions in flat publish/subscribe systems. Applying scope interfaces to restrict the distribution of scope announcements helps limit the amount of management information kept in the system.

To complete the scope configuration, additional data about its interfaces, transmission policies, or security policies is necessary. This information is also provided by the system engineer and is stored as an extension of its routing tables $\mathsf{RT}_{B_i}^S$ in all scope brokers.

A scope is removed from the system by calling *dscope(S)* at one of its scope brokers. Following the entries in its routing table a message is sent to all of its scope brokers to remove its routing tables and any references from routing tables of adjacent scopes. Its members are notified with a corresponding notification.

Joining a Scope

An arbitrary component C is joined to a scope S by calling *jscope*(C, S) at the local or border broker of C. The scope lookup table is used to route a ScopeJoin message to the first scope broker of S. These special messages leave a trail of temporarily stored source-pointers in the visited brokers that allows a response to be routed backwards to C. A scope broker of S that receives a ScopeJoin message takes two steps. It includes the border broker of C as scope broker of S and forwards the interface of the new component to existing scope brokers to get the routing tables updated. The first step requires that the current routing table is forwarded along the stored trail toward C so that each visited broker creates an initialized routing table for scope S. If security policies are installed in the scope brokers, a join request may be denied, which results in a rejection sent toward C.

A simple component leaving a scope is similar to just unsubscribing to all issued subscriptions. Scope brokers may regularly test if any members are locally connected, and if other scope brokers are reachable via only one link, this scope broker is an unused border broker of the scope overlay and may be shut down. If a scope leaves one of its superscopes, i.e., *lscope*(S, T), an appropriate message is distributed to the scope brokers of both scopes and the references to the respective other scope are removed from all involved $\mathsf{RT}_{B_i}^S$ and $\mathsf{RT}_{B_i}^T$ routing tables.

Scoped Routing

Scoped routing uses the enhanced routing tables to forward notification in accordance with the current scope graph. The algorithm basically extends the plain REBECA algorithm of flat publish/subscribe routing. It is executed in each broker B and operates on a set of enhanced routing tables $\mathsf{RT}_B^{S_i}$ of scopes S_i, of which B is currently a scope broker.

Notification Layout

The algorithm needs some additional management information to operate properly. This information is annotated to notifications by the routing implementation and is not accessible to applications.

To prevent loops and infinite forwarding, notifications must not be sent back on links they were received from, both network links and scope graph edges. As in the original REBECA routing, notifications are annotated in each broker with an identifier of the source network link to prevent it from being sent back in the direction from where it was received. Additionally, each notification carries an identifier of the current scope and of the source component, which are accessed by get_scope(n) and get_source(n), respectively. These identifiers signify the scope in which the notification is currently visible and the (last) component from where it was forwarded into this scope. Note that the latter does not name the original publisher but the last node in the scope graph visited before the current one. The local event broker of the original producer is responsible for setting the identifiers initially.

These component identifiers must be unique with respect to the current scope and its adjacent nodes in the scope graph so that they identify its components or superscopes unambiguously. However, such edgewise distinct names may not suffice, because many scopes may be hosted in one broker and naming must be unambiguous within a broker. So, besides the simple but restrictive assumption of globally unique identifiers, a scheme similar to the mappings of virtual channel identifiers in Asynchronous Transfer Mode (ATM) networks [233] might be devised that maps identifiers on both sides of a network link to guarantee uniqueness.

The next paragraphs introduce different states of routing that are accessed by get_state(n). Of course, all get-functions are accompanied by the respective set methods.

Routing States

Following the discussion about delivery paths in scope graphs and transmission policies, three states of routing are distinguished:

- scope internal routing: A notification is forwarded to siblings in the same scope.
- downward routing: An incoming notification is forwarded to scope members.
- upward routing: An outgoing notification is forwarded to superscopes.

A notification published by a simple component is initially handled in the internal routing state. It may alternate between internal and upward states, but once in downward routing it may not switch back. Adherence to this sequence is mandatory to not break the bipartite nature of delivery paths, that is, notifications are always first sent up in the scope graph before they solely travel down against the edges of the graph. Internal routing is expressly distinguished to facilitate the respective transmission policy, cf. Sect. 6.7.4.

The Algorithm

Figures 6.25 and 6.26 illustrate the algorithm, which basically extends the plain REBECA algorithm of flat publish/subscribe routing and is executed in each broker B. The main control loop main_loop is triggered whenever new data is appended to the receiving queue, which may either be due to incoming network traffic or via cross-scope traffic. The expected pair (n, l) contains the notification to be forwarded and a link from which it was received. The latter

```
        procedure main_loop
          loop
            // the queue is fed from network links
  4         (n, l) = get_next (recvQ)
            scoped_routing(n, l)
          end
        end


  9     procedure scoped_routing (n, l)
        Input n:  notification
              l:  source link


          s := get_state(n)
          S := scope(n)
  14
          --- internal routing
          D := destinations(n, remote_components(RT_B^S))
          foreach (n', l') ∈ D
            if l ≠ l' then send (n', l')
  19      end


          --- downward routing
          D := destinations(n, subscopes(RT_B^S))
          cross_scope(S, D, "downward")
  24
          --- upward routing
          if not s = "downward" then
            D := destinations(n, superscopes(RT_B^S))
            cross_scope(S, D, "upward")
  29      fi
        end
```

Fig. 6.25. Overall routing algorithm

may be either a network link or a local routing table, i.e., a routing destination in the enhanced routing tables.

The procedure scoped_routing determines the next destinations of a notification currently visible in a scope S. It interprets the current routing state and accordingly queries different parts of the routing table RT_B^S. The function subscopes(RT_B^S) returns a routing table that contains all entries that point to a locally stored routing table of a subscope of S. Similarly, superscopes(RT_B^S) contains entries of local superscope routing tables. Conversely, remote_components(RT_B^S) returns the remaining entries, which are reachable via network connections. In the case of $RT_{B_1}^S$ of Fig. 6.23, the three functions return entries of $\{\}$, $\{RT_{B_1}^T, RT_{B_1}^U\}$, and $\{c_1, B_2\}$, respectively. First, eligible destinations within the considered scope and then the locally available routing tables of

```
function destinations (n, T)
Input n:  notification
       T:  routing table
Output D: list of notification-destination pairs

5      foreach (I, d) ∈ T
           n' := I(n)
           if n' ≠ ε then
               D := D ∪ (n', d)
           fi
10     end
end
```

Fig. 6.26. The naïve matching algorithm with mappings

subscopes are determined; both must be done for all routing states. A distinction of states is at this point only necessary when transmission policies are applied, cf. Sect. 6.7.4. Last, the upward direction is examined to find all locally available routing tables of eligible superscopes, which is only done if routing is not in downward state. Taken together, these steps follow the default delivery and publishing policies of Sect. 6.4.1 that describe visibility in the scope graph.

The above procedures rely on the function **destinations** to determine all eligible destinations in the specified routing table. The naïve matching algorithm, extended with mappings, is given in Fig. 6.26 for illustrative purposes. It returns pairs of destinations and notifications to send there, allowing for a seamless integration of mappings in the routing decision. Of course, in practice more efficient matching algorithms, e.g., [133, 404], and a more sophisticated handling of notification copies may be applied.

Crossing Scopes

The scoped routing algorithm relies on **cross_scope** to forward a notification between scopes (Fig. 6.27). It is responsible for relaying the current notification to other routing tables stored in the same broker. In fact, an underlying assumption is that scope transitions take place only within a broker. Routing tables of a super- and subscope pair $S \lhd T$ must be collocated at the same broker to enable interscope routing. In the above example B_1 is a scope broker of all scopes and may route between S, T, and U, whereas B_2 and B_4 can route between S and T only.

cross_scope takes a list D of pairs of eligible destination scopes, whose interfaces match, and notifications that shall be sent there. In this way, the current notification may be forwarded in different representations. With the help of the reference to the source component (**get_source**(n)) the algorithm prevents notifications from being sent back along the scope graph edge they

```
      procedure cross_scope (S, D, s)
      --- forward all notifications to next routing tables
      Input S:   current scope
            D:   list of notification-routing table pairs
            s:   routing state

4
         foreach (n, RT_B^{S'}) ∈ D
            if get_source(n) ≠ S' then
               set_source(n, S)
               set_scope(n, S')
9              set_state(n, s)
               put_in_front(recvQ, (n, RT_B^S))
            fi
         end
      end
```

Fig. 6.27. Interscope forwarding

were received from. This does not preclude duplicates because of alternative paths in the scope graph, but it rules out erroneous duplication because of repeated processing, at least in one broker. How to prevent this repetition in different brokers is detailed below.

The procedure sets the source component to the current scope and the intended destination as new current scope and then puts the relayed notification into the incoming queue $recvQ$. This eventually triggers the main loop and starts routing of n in the destination scope. The routing state recorded in each notification is updated according to the specified parameter s that is supplied by the main scoped routing algorithm.

Crossing at Different Locations

Although interscope routing is not possible at arbitrary brokers, there still may be multiple brokers where two scopes $S \lhd T$ coincide. And thus a notification might cross a scope boundary repetitively at different brokers, duplicating notifications even along a single edge of the scope graph. Furthermore, security considerations or the implementation of advanced ordering schemes might necessitate a designated broker that bridges all traffic between the respective scopes. In the previous example, a notification published by c_1 is distributed in its scope S and may enter superscope T at B_1, B_2, or B_4.

Three choices for placing interscope routing are distinguished according to the following criteria. First, are the scope-crossing functions applied at only one broker or at several different brokers? Second, if only at one, is it a designated gateway broker or an arbitrary broker that conveys the traffic between the respective two scopes? The following alternatives are available:

1. Transition at designated central gateway: All interscope traffic of a scope S is handled by a single gateway broker B_i of that scope. Only at this gateway the routing table $\mathsf{RT}_{B_i}^S$ contains an entry pointing to sub- and superscopes.
2. Transition anywhere, but only once: Interscope traffic is transfered into its destination scope at the first possible broker, and nowhere else.

The first approach of having a *designated gateway* is the simplest solution. It instantiates the respective scope graph edge at a single point in the broker network. Only at this gateway broker a routing entry for the specific superscope is stored, say $({}^o\hat{I}_S^T, \mathsf{RT}_{B_1}^T)$ as part of $\mathsf{RT}_{B_1}^S$ if B_1 is the gateway broker of $S \leadsto T$. All other scope brokers of S register an entry ${}^o\hat{I}_S^T$ that points toward this gateway broker, e.g., $({}^o\hat{I}_S^T, B_2)$ is stored in B_4. This is necessary to get published notifications matching the output interface forwarded to the gateway broker. Within T all routing table entries pointing to the subscope S are similarly adapted to direct downward traffic to B_1 as well. Each gateway broker links a specific pair of scopes, but generally system engineers may decide to group all gateway brokers at one network node, to group them for each scope, or to place all gateways independently.

A drawback of this strict separation of inter- and intrascope routing is wasted network bandwidth. Consider c_4 and c_6 connected to broker B_4 in the previous example. Notifications from c_4 to c_6 are routed through broker B_1 to enter T there and go back to B_4 again. The adequate placement of gateway functionality has a major influence on network utilization. On the other hand, the centralized gateway offers full control of the incoming and outgoing traffic at a designated broker. This allows trusted software modules to be employed for cross-scope communication at a single trusted broker, for example, to authenticate all outgoing notifications or to link separate security domains without disclosing other scope brokers. The implementation of transmission policies is simplified, too, as pointed out in the next subsection. In general, if the placement of scope brokers corresponds to the physical layout of the underlying network, gateway brokers may also represent the physical gateway between different networks hosting the adjacent scopes.

The second approach allows notifications to cross-scope boundaries between two two scopes $S \lhd T$ at the first possible broker that sustains both scopes. When c_1 publishes n, it is forwarded into S, T, and U at B_1, assuming matching interfaces, of course. An appropriate countermeasure must be provided to prohibit repeated scope transitions in B_2 and B_4. This is achieved by testing whether the destination scope T was already seen in the last broker from which n is received, in which case the transition has already happened in a previous broker. Notification forwarding in `cross_scope` is denied if an entry in RT_B^T exists that points toward $\mathrm{link}(n)$. In the example, B_2 has stored an entry $({}^i I_{c_2}^T, B_1)$ in $\mathsf{RT}_{B_2}^T$ and does not forward n into T again.

Unfortunately, so far each scope transition generates a new notification and the transition at the earliest encountered broker leads to messages being sent on the network that differ only in the annotated current scope they are visible in. In the example, two messages are sent to B_2 and B_4, one visible in S and one in T. A possible improvement is a combined delivery to all eligible superscopes, which are identified by a list of scopes annotated on the notification instead of just one identifier. The multiplicity of messages is replaced by a list of scopes, at least as long as no mappings transform the notification. The routing decision is evaluated as before, only that `scoped_routing` is called multiple times to fill the list of next-hop destinations. At each broker, the available routing tables are checked, and whenever additional scopes are detected and entered the list of visible scopes is updated. In the example, a notification forwarded from S to T is annotated with both scopes and is transmitted only once between B_1, B_2 and B_4.

Transmission Policies

The distinguished routing states directly correspond to the delivery, internal delivery, and publishing policy. The policies are encoded as part of the enhanced routing tables, even if they include general mappings in the routing decision. As discussed in Sect. 6.5, the policies operate on sets of notifications and must be evaluated after the eligible destinations are determined by the matching algorithm in `destinations`.

The three policies can be inserted into the three parts of the sketched `scoped_routing` algorithm. Internal routing is refined by evaluating

$$D := idp_S(D)$$

on the set of eligible consumers before it is processed in the foreach loop. Delivery and publishing policy are intended to be applied at scope boundaries, and so they are evaluated in `cross_scope`,

$$D := pp_S(D)$$

for upward routing and

$$D := dp_S(D)$$

for downward routing, again just before sending the notifications in the foreach loop.

Scope Multicast

So far, intrascope routing has stuck to strict routing where notifications are forwarded only if a matching subscription is available. This prevents notifications from being always sent to all scope brokers of a scope, but induces multiple point-to-point messages and repeated routing decisions. An alternative strategy for routing in a scope S is to send all notifications to all of its scope brokers irrespective of any subscriptions. In a second step, the so-called *fan-out* of the broker network to the consumers is implemented via point-to-point communication. The routing tables of S are evaluated in every scope broker of S and each matching and locally connected consumer is notified separately.

If implemented as part of the broker implementation, an application layer multicast scheme is established within the broker network. This approach does not avoid multiple point-to-point messages between the scope brokers, but is readily applicable in most networks. On the other hand, IP multicast offers an established, well-known facility to speed up communication to a group of receivers. The original decision of using point-to-point communication in the broker topology is partially inspired by the assumption that the sets of consumers are rather volatile and vary frequently. A multicast solution that directly communicates to consumers requires frequent group changes, and the explicit control of individual delivery is lost. However, IP multicast is a convenient technique to connect scope brokers. The broker topology can be supposed to change less frequently than the consumers and thus does not overwhelm multicast group management. So, intrascope routing is reduced to a notification being conveyed to all scope brokers with one multicast datagram before it is explicitly directed to any matching consumers. This approach combines multicast efficiency with the full control of notification delivery.

Evaluation

The integrated routing architecture is possibly the most generic scope architecture. It combines the efficiency of a distributed solution, incorporates multicast delivery, and still offers the flexibility to control the hop of notification delivery to consumers. It extends the known routing tables and can build on various existing routing protocols, such as covering- or merging-based routing provided by REBECA and other notification services. Scoping is here offered as a service of the event infrastructure. The layout of the publish/subscribe network is independent from the actual application structure given by the scope graph.

On the other hand, the option to connect scope members to arbitrary brokers may increase network utilization, and the dispersion of components and traffic may increase the complexity of the system. But this is essentially always the case for distributed solutions.

6.8 Combining Different Implementations

The preceding discussion assumed the same type of architecture for all scopes in the system, which is, obviously, a severe limitation of potential application domains. In fact, one of the primary benefits of the scoping concept is its ability to facilitate the customization of the infrastructure. Once groups of components are identified, their scopes can be based on those architectures

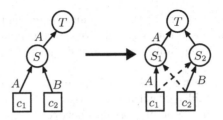

Fig. 6.28. Duplicate scopes to separate QoS requirements

that fit their respective needs best. The special requirements of their interaction are addressed by employing appropriate implementations of scoped
notification dissemination. But yet, the different implementations must be
seamlessly integrated.

6.8.1 Architectures and Scope Graphs

In the first place, scopes model application structure. But they are also a
tool for determining notification semantics within the application structure.
Different types of notifications may demand different quality of service (QoS)
even within a specific scope. For example, consider noncritical timer information sent in bulk (type A in Fig. 6.28) and personnel record updates (type B)
that are supposed to be encrypted and delivered to authenticated consumers
only. While both are consumed in the same part of the application, i.e., in
the same scope, these two data types obviously ask for different architectures
and communication media that facilitate scalable delivery of the former and
secured delivery of the latter.

In principle, several different communication media might be used in one
scope to facilitate different QoS. Alternatively, a scope with complex semantics
is duplicated in Fig. 6.28, and each instance is tailored for a different kind
of QoS supported. The interfaces are split so that the same notifications are
forwarded into T as before. Publishing policies and imposed interfaces assigned
to c_1 and c_2 ensure that the traffic within S_1 and S_2 is separated and directed
to the scopes that offer the necessary quality of service. In the above example,
the timer notifications would be distributed via scope S_1 operating on top of
a scalable messaging system, and S_2 would employ encrypted point-to-point
connections to meet the security requirements of type B. The edges (c_1, S_2)
and/or (c_2, S_1) are necessary if the c_1 and c_2 shall get the same notifications
as before, but additional interfaces are necessary to prevent messages from
leaking with wrong QoS.

Instead of dealing with arbitrary combinations of communication media,
dissemination semantics, and scopes, the following assumes a specific scope
architecture per scope. For implementation purposes, bridging takes place
between connected subgraphs of the graph of scopes that share a common

architecture. However, to simplify the discussion, only pairs of scopes and the bridging in between are investigated next.

6.8.2 Bridging Architectures

Combining different scope architectures requires a gateway between the different implementations of two scopes $S \triangleleft T$. The simple components of a scope have as part of their local event brokers an architecture-specific implementation for accessing the underlying communication medium (cf. Fig. 6.17). The gateway relies on two local event brokers to bridge the respective implementations of the architectures. Gateway functions are assigned to the considered subscope, S, and enforce the input and output interfaces of S, its publishing and delivery policies, and any mapping applied on the edge (S, T).

Collapsed Filters

The collapsed filters architecture does not instantiate administrative components and so gateway functions must reside in all members of the scope. Because of the required duplication of code in simple components, an extra gateway component is preferable. Such a component would not interfere with the internal delivery of notifications. It is similar to the mapping components used in Sect. 6.4.2 to sketch the feasibility of scoped systems. It acts as an additional producer/consumer in scope S and manually implements the edge to superscope T, being a regular member there as well.

The distinguished *scope destination* and *visibility roots* approaches to annotate notifications and extend subscriptions are hardly different regarding the implementation of the gateway. In both cases a gateway component is instantiated for each bridged scope–superscope pair or for all bridged superscopes collectively. Only their subscriptions must reflect the differences in the lists of annotated scope identifiers: the former lists all reachable scopes while the latter lists only visibility roots on upward paths. Upon receiving a notification, the gateway component tests which of the edges it controls is eligible, applies the assigned output interfaces and publishing policies, and forwards the data, if appropriate.

In the same way, the gateway registers in the superscope(s) and, upon receiving a notification from there, evaluates the assigned input interfaces and delivery policies. Since delivery policies need cooperative filtering in all consumers, the gateway's functionality depends on the filtering supported in the present implementation of the collapsed scope graph.

Scope Address

In the scope address architecture there are administrative components available to execute gateway functions. Cross-scope traffic is matched against the interfaces, while publishing and delivery policies are applied as before. The same implementation can be used, only a second local event broker to bridge the different architecture's implementations must be present.

Broker Scopes

Broker scopes are administrative components that represent a specific scope and explicitly control all internal and external traffic. Thus, they may directly implement any gateway to other architectures.

Integrated Routing

Although no individual representatives of scopes exist, scopes and transitions between scopes are explicitly recorded as routing tables with entries referencing other routing tables. Instead of pointing to other tables, the entries may refer to a second local event broker to access another's scope architecture. Interface and transmission policies are handled as before—they are always explicitly applied. The discussion about locating cross-scope transitions (cf. Sect. 6.7.4) holds for gateways as well.

6.8.3 Integration With Other Notification Services

The gateway of a scope may not only bridge different scope architectures, but may also facilitate coupling of a scoped system with other notification services. The gateway functions simply have to implement another service's API to act as a regular producer/consumer within that service. The traffic flowing between the scoped and the external system is controlled by the gateway functions, i.e., interfaces, mappings, and transmission policies. By creating an "outside" scope and a gateway that connects other communication services, external data is incorporated into the scoped system without impairing visibility control. On the other hand, this gateway retains the component characteristic of scopes with respect to the outside system. The flow of notifications leaving the scope follows the definition of the scope graph.

Of similar importance is the coupling of scoped and unscoped applications, which are likely to coexist. Consider the integrated routing approach, for instance, and two applications, one scoped and one unscoped. The scoped routing tables are used in addition to a traditional implementation, which is nothing more than a further routing table not connected to the scope routing tables. All scoped clients are assigned to some $RT_{B_i}^{S_i}$, and the nonscoped ("legacy") clients are still maintained in separate old-style routing tables RT_{B_i}. In fact, the overlay of all RT_{B_i} constitutes a *default scope* to which every newly created simple component may be assigned. In this way, scoped and nonscoped clients can interact in a controlled way.

6.9 Further Reading

This chapter has introduced the concept of scoping in event-based systems. It offers a module construct; it is an extension point for the integration of different communication techniques, for handling quality of service, security, and data heterogeneity; and it facilitates the management of event-based systems. Accordingly, related work is very broad and comes from many areas of computer science [135].

The Common Object Request Broker Architecture (CORBA) provides a number of mechanisms to organize and structure distributed systems [283]. It includes the CORBA Notification Service [287]. Event management domains [282] support the federation of multiple notification channels in arbitrary topologies. However, applications have to select their channels and thus move information about application structure into the components—there is no support for an administrator to orchestrate channels and components. A generic solution to avoid static configurations is reflective middleware [96, 223].

The enterprise edition of Java (J2EE, [365]) specifies an execution environment that contains a component model and a number of standardized services, including a notification service (JMS), which is described in Sect. 9.1.3. The standard offers the plain publish/subscribe API (plus transaction support), but not the engineering features of scopes. Many JMS implementations exist, some offer extensions like topic hierarchies, e.g., [97]. In terms of managing application components, Java Management Extensions (JMX) defines a standardized Java way to management and monitoring [363].

If a database management system such as Oracle Streams Advanced Queuing (AQ), cf. Sect. 9.2.3, transports our notifications, we can exploit all the features offered by a database, like transactions, rules, consistency constraints, logging, high availability, authentication, access control, etc., and apply them to the publish/subscribe communication as well [172]. The focus is then more on advanced QoS features than on lean implementation.

Many domain-specific implementations of publish/subscribe run into the engineering issues addressed by scopes. Eder and Panagos [118] pointed out the problems that arise from missing structures in workflow systems. They connected workflow engines from multiple sites with the READY notification service [185]. The service introduced *event zones* to cluster components based on (either) logical, administrative, or geographical boundaries. Boundary brokers connect zones and control the communication between them. A component can belong to only one zone, which limits the structuring capabilities and prohibits composition and mixing of aspects [147, 188].

Wireless sensor networks (WSNs, [95]) also exploit eventing, and the need for structuring mechanisms was identified before, cf. [350, 397].

The field of software architecture is concerned with the overall organization of a software system [165]. Architecture definition languages (ADLs) are employed to describe the high-level conceptual architecture consisting of components, connectors, and specific configurations [256] of these. Typical, well-understood arrangements of connectors and configurations are identified as *architectural styles* [3], the patterns of software architecture, and events and implicit invocation are among them. Luckham [242] presented the RAPIDE language family. It includes event processing agents to encapsulate event processing rules behind input and output interfaces. The architecture definition language can be used to arrange a number of these agents, similar to scope graphs.

Sullivan and Notkin introduced mediators as a design approach that explicitly instantiates and expresses integration relationships and separates them from component function [356]. In a less general approach, Evans and Dickman defined *zones* to support partial system evolution [130]. Barrett et al. [31] proposed an event-based integration (EBI) framework that also covers scope features like transmission policies, mappings, and hierarchical grouping.

As event services are the basis for application integration and evolution, they cannot be expected to run in homogeneous environments. Heterogeneity issues can be handled in traditional request/reply systems, but they are rarely considered in event systems [32]. Database research contributes to the necessary syntactic and semantic data mappings [54, 80].

The field of *coordination theory* investigates techniques for managing the dependencies between a set of active components [246]. It differentiates computation from coordination [295] and localizes interaction in *coordination media* [64, 78]. Scopes event-based communication directly corresponds to this viewpoint.

A key point of scoping is that it does not imply a specific implementation per se. Depending on the intended semantics, adaptability, communication efficiency, etc., alternative implementations are applicable. The system engineer can incorporate existing work on group communication [319]. A wide variety of work exists in this area that supports nested groups [42] and reliable communication [43, 213]. Peer-to-peer systems are another candidate [311, 331, 374].

On lower layers, IP multicast is an obvious implementation candidate. Deering and Cheriton [106] introduce multicast scope control with the help of time-to-live fields (TTL). Administratively scoped IP multicast exploits hierarchical administrative boundaries [259]. Multicast scopes bundle network nodes, but do not support communication between scopes and require static configuration within the IP network routers. Interestingly, such multicast scopes allow us to implement publish/subscribe on IP multicast [26, 291, 357] only within restricted parts of the scope graph.

Composite Events

For certain applications, the expressiveness of subscriptions used by the local notification matching algorithms introduced in Chap. 3 is not sufficient. As a remedy, a service for *composite event detection* facilitates the management of a large volume of events by enabling subscribers to specify their interest more precisely. The composite event service supports the advanced correlation of events through the detection of complex *event patterns*. In this chapter we describe composite event detection services for publish/subscribe systems.

We start with two application scenarios that benefit from composite event detection in the next section. After that, we list the requirements for such a detection service (Sect. 7.2) and introduce composite events in more detail in Sect. 7.3. We then give an example of composite event detectors based on finite state automata (Sect. 7.4.1) and a corresponding language (Sect. 7.4.2). Composite events (CE) are detected by automata that support distribution and a flexible time model for composite events. Event subscribers of the composite event service use a core composite event language to specify event patterns using a series of operators. We also gave examples for three higher-level specification languages for composite events that are domain-specific. Section 7.5 has a discussion of centralized and distributed architectures for composite event detection. We also explain how distribution is controlled by distribution and detection polices. The design space for distribution polices gives rise to a variety of different strategies for distributing composite event expressions. We conclude the chapter with an overview of other composite event detection services in Sect. 7.6.

7.1 Application Scenarios

Many application scenarios for a publish/subscribe service benefit from a general-purpose composite event service that enhances the expressiveness of subscriptions. In the following we will consider how composite events can be

used in a ubiquitous computing environment and for network systems monitoring. For each application scenario, we provide two examples of composite
event subscriptions.

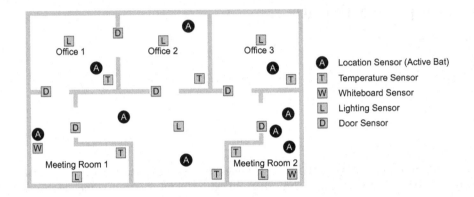

Fig. 7.1. The Active Office with different sensors

The Active Office

One example of a ubiquitous computing environment is the *Active Office*, a
sensor-rich environment inside a computerized building (Fig. 7.1). In this
building, sensors that are installed in offices provide information about the
environment to interested devices, applications, and users. The Active Office
is aware of its inhabitants' behavior and enables them to interact with it in a
natural way. The large number of sensors potentially produce a vast amount
of data. Building users wear *Active Bats* [5] that publish location information,
and static sensors gather data about doors, lighting, equipment usage, and environmental conditions. However, information consumers prefer a high-level
view of the primitive sensor data. Thus, a middleware used in this application scenario has to cope with high-volume data and be able to aggregate and
transform it before dissemination. Composite event detection can help process
the primitive events produced by the large number of sensors and provide a
higher-level abstraction to users of the Active Office.

1. A user may subscribe to be notified when a meeting with at least three
   people working in the messaging department takes place during working
   hours in one of the meeting rooms.
2. Building services may be interested in composite events about a drop in
   temperature by 15 degrees for at least 15 min in any occupied office.

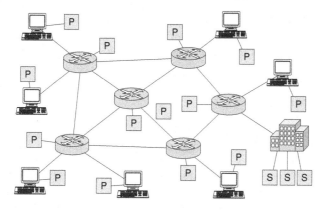

Fig. 7.2. A system for monitoring faults in a network

Network Systems Monitoring

When monitoring the operation of networks [49], network entities publish notifications that are related to fault conditions in the network, such as that shown in Fig. 7.2. In practice, millions of events may be published daily in respect of fewer than a hundred real faults that require human intervention. The task of network systems monitoring can thus be simplified by expressing patterns associated with real problems as composite event subscriptions.

1. The network management center may want to be notified when at least five workstations in different parts of the network detect a degradation in network bandwidth.
2. A network customer may be interested in composite events when none of its load-balanced Web servers are available to the outside world unless the downtime is part of scheduled maintenance work.

The XenoTrust Framework

Recent efforts, such as the XenoServer project [30], are building large-scale, public infrastructures for general-purpose, distributed computing with resource management and sharing. In such environments, reputation information about participants must be disseminated in a timely and scalable fashion so that entities can make trust-dependent decisions. Composite events can help participants receive notifications about changes in reputation of their resource providers or consumers, thus creating a global-scale trust management system [225].

1. A user may want to be notified when the reputation of any of its currently active resource providers drops below a certain threshold and there is an alternative provider that is capable of taking over the current resource contract.

2. A resource provider may submit a subscription causing a composite event when a new client receives a low reputation rating from at least three other providers within three days while requesting significant resources.

7.2 Requirements

From the above application scenarios for composite event detection, we derive several requirements for a composite event detection service.

- The composite event service must be *expressive* enough when it comes to the specification of composite events. Depending on the application domain, users will describe event patterns of varying complexity. The composite event service must naturally capture common use patterns.
- The composite event service must be *usable*. From a user's perspective, it must be easy to express complex event patterns in the composite event service. A too-expressive language for the specification of composite events may lead to poorly usability.
- The composite event service must be *efficient* in terms of the user's performance goals, such as low detection delay or bandwidth consumption. In particular, there must exist an efficient implementation technique for composite event detectors in the service. Often, a distributed implementation improves the service.

Obviously, there is a tension between these requirements: A very expressive composite event service may not result in an efficient or usable system. In contrast, a very efficient implementation of composite event detector may lead to a limited system with low expressiveness. In the following, we will describe a composite event service based on extended finite state automata that attempts to balance these trade-offs.

7.3 Composite Events

A composite event service is based on the notion of a *composite event*. Composite events prevent subscribers from being overwhelmed by a large number of primitive event publications by providing them with a higher-level abstraction. A composite event is published whenever a certain pattern of events occurs in the publish/subscribe system. This means that subscribers can subscribe directly to complex event patterns, as opposed to having to subscribe to all the primitive events that make up the pattern and then performing the detection themselves.

Often a subscriber is interested in the primitive events that caused a composite event. Therefore, when a composite event has been detected by the composite event service, it is published and contains all primitive events that contributed to its occurrence. Since composite events are build from primitive ones according to a well-defined set of rules, every composite event can be assigned a *composite event type*. It is built from the types of the included events and the relation between them in the composite event subscription. Note that primitive events can also be considered degenerate composite events, thus unifying primitive and composite event types within the publish/subscribe system.

Definition 7.1 (Composite Event). *Every composite event c has a composite event type τ_c and belongs to the composite event space $\mathbb{C}$,*

$$(c : \tau_c) \in \mathbb{C},$$

A composite event type τ_c corresponds to a valid expression $\mathcal{C}$ in a composite event language,

$$\tau_c \equiv \mathcal{C}.$$

A composite event c consists of an interval timestamp t_c and a set of composite subevents $\{c_1, c_2, \ldots, c_k\}$,

$$c : \tau_c = (t_c, \{c_1, c_2, \ldots, c_k\}).$$

A composite event is associated with a timestamp t_c that states when it has occurred. In a distributed system, there is no concept of global time [227], which is why the timestamps of composite events caused by distributed sources can be captured more accurately using partially ordered *interval timestamps* [238]. An interval timestamp has a start and end time so that it can express the local clock uncertainty at an event broker and also the duration associated with a composite event from the first contributing event to the last. To capture the temporal relations between composite events, we define a partial and a total order over interval timestamps that will be used by the weak and strong transitions in the detection automata described in the next section.

Definition 7.2 (Interval Timestamp). *An interval timestamp t_c,*

$$t_c = [t_c^l; t_c^h],$$

has a start time t_c^l and an end time t_c^h with $t_c^l \le t_c^h$. Interval timestamps are partially-ordered ($<$) and totally-ordered ($\prec$) as follows,

$$t_{c_1} < t_{c_2} \triangleq t_{c_1}^h < t_{c_2}^l,$$
$$t_{c_1} \prec t_{c_2} \triangleq (t_{c_1}^h < t_{c_2}^h) \vee (t_{c_1}^h = t_{c_2}^h \wedge t_{c_1}^l < t_{c_2}^l).$$

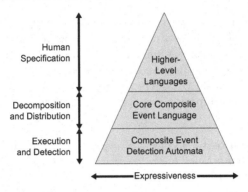

Fig. 7.3. The components of the composite event detection service

7.4 Composite Event Detection

From the description of the application scenarios, it becomes clear that it is challenging to design a single language for composite events that is both expressive and intuitive to use by, say, a human user in the Active Office environment. Therefore, an alternative approach is shown in Fig. 7.3, with several specification layers that have different powers of expressiveness. At the bottom layer, *composite event detection automata* provide maximum expressiveness and perform the actual detection of composite events described in the next section. Composite event subscriptions specified in the *core composite event language* presented in Sect. 7.4.2 can be decomposed for distributed detection. Finally, domain-specific *higher-level languages* constitute the top layer and only expose a subset of the core language—supporting a simpler definition of composite events for a given application domain. Expressions in higher-level languages are automatically compiled down to composite event detection automata by the composite event service.

7.4.1 Composite Event Detectors

In this section, we describe a composite event service that uses *composite event detection automata*, which are finite state automata [193] that are extended with support for temporal relationships and concurrent events, to analyze event streams. Basing composite event detection on extended finite state automata has several advantages. First, finite state automata are a well-understood computational model with a simple implementation. Second, their restricted expressive power has the benefit of limited, predictable resource usage, which is important for the safe distribution of detectors in the publish/subscribe system. Third, regular expression languages have operators that are tailored toward the detection of patterns, which avoids the risk of redundancy or incompleteness when defining a new composite event language. Finally, complex expressions in a regular language may easily be decomposed for distributed detection.

A detection automaton consists of a finite number of states and transitions between them. To ensure that each state only has to consider certain events for transitions, it is associated with an *input domain* Σ, which is a generalization of the concept of an input alphabet in traditional finite state automata. An input domain is a collection of *describable event sets* $A, B, C, \ldots$, which correspond to sets of events that are matched by a primitive or composite event subscriptions. In a given state, only these events need to be considered by the automaton because other events are not relevant for the composite event being detected. In practice, the automaton issues subscriptions for all describable event sets in the input domain of a state. The resulting incoming events are ordered according to the total timestamp order ($\prec$) (see Def. 7.2) into an *event input sequence* and are consumed by the automaton sequentially.

Fig. 7.4. The states in a composite event detection automaton

As shown in Fig. 7.4, a detection automaton has four types of state. Detection starts in a unique *initial state* and continues through a series of *ordinary states*. A *generative state* is an accepting state that also publishes the composite event that has been detected by the automaton. *Generative time states* deal with timing by publishing an internal *time event* when a timer (e.g., 1 min) associated with the state expires. The automaton treats time events like regular events, but they are not visible externally.

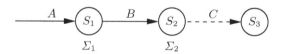

Fig. 7.5. The transitions in a composite event detection automaton

Each state can have two forms of outgoing transition that are labeled with the describable event sets of the events that trigger them. Note that since describable event sets can be defined by composite event subscriptions, our automata support the detection of event patterns involving concurrency. In the sample automaton in Fig. 7.5, the transition between states S_1 and S_2 is a *weak transition* that requires the timestamps of the events from the describable event sets A and B to be partially ordered ($<$). A *strong transition*, such as between states S_2 and S_3, mandates a total ordering ($\prec$) between events from B and C. Strong and weak transitions therefore allow the expression of different temporal orderings between events. When an event that is part of the input domain but without a matching outgoing transition is received, the detection in the automaton fails. Several matching transitions and empty ϵ-transitions are followed nondeterministically. Although the following presentation of the detection automata uses nondeterminism, standard techniques can be used to convert them into deterministic automata [193].

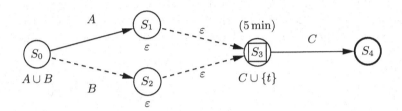

Fig. 7.6. A composite event detection automaton

In Fig. 7.6, we give an example of a composite event detection automaton. This automaton starts in state S_0 with an input domain of $A \cup B$. A strongly followed event from A causes a transition to state S_1; a weakly followed event from B leads to state S_2. Once the generative time state S_3 is reached, a timer starts that will expire after 5 min, publishing time event t. Since this event is part of the input domain for state S_3 but there is no corresponding outgoing transition, detection will fail unless an event from C is received before the timer expires, triggering a transition to state S_4. The generative state S_4 signals the successful detection of a composite event with a composite event publication.

7.4.2 Composite Event Language

In a composite event detection service, composite event subscriptions are expressed in a composite event language. Expressions in this language define the set of composite events in which an event client is interested. In this section we describe a *core composite event language* that corresponds to the extended finite state automata introduced in the previous section. We present the language's operators and the construction of corresponding composite event detection automata from subautomata. Some operators in this language, namely concatenation, alternation, and iteration, are influenced by those found in regular languages. However, other operators reflect the special features of our detection automata. We finish with examples of core language expressions and discuss three higher-level languages that can be built on top of the core language for domain-specific composite event specification.

Atoms. $[A, B, C, \ldots \subseteq \Sigma_0]$

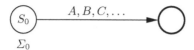

Atoms detect individual events in the input stream of all events that are in the input domain Σ_0. Here only events in the describable event sets $A \cup B \cup C \cup \ldots$ are matched and cause a transition to a generative state. Other events in Σ_0 result in failed detection, and events outside Σ_0 are ignored. The trivial atom $[A \subseteq A]$ is abbreviated as $[A]$.

Negation. $[\neg E \subseteq \Sigma] \triangleq [\Sigma \backslash E \subseteq \Sigma]$

Negation is shorthand for an atom that matches all events in the input domain Σ except for events in the negated describable event set E. Note that this semantics differs from more powerful negation operators found in other event algebras.

Concatenation. $\mathcal{C}_1 \mathcal{C}_2$

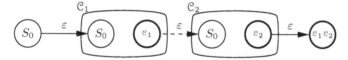

The concatenation operator detects a composite event matching expression $\mathcal{C}_1$ with a timestamp that *weakly* follows the timestamp of a composite event matching $\mathcal{C}_2$. The detection automaton for concatenation is constructed by connecting the generative state of $\mathcal{C}_1$ with a weak ϵ-transition to the initial state of $\mathcal{C}_2$.

Sequence. $\mathcal{C}_1 ; \mathcal{C}_2$

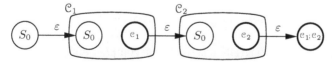

The sequence operator detects an event of type $\mathcal{C}_1$ *strongly* followed by an event of type $\mathcal{C}_2$. Unlike concatenation, this means that the interval timestamps of the events matching $\mathcal{C}_1$ and $\mathcal{C}_2$ must not overlap. The construction of the sequence detection automaton uses a strong transition for the ϵ-transition between the two subautomata.

Iteration. $\mathcal{C}_1^*$

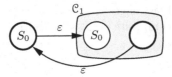

Any number of occurrences of $\mathcal{C}_1$ are matched by the iteration operator. Its detection automaton creates a loop from the generative state of $\mathcal{C}_1$ back to its initial state. If $\mathcal{C}_1$ receives an event that causes it to fail, then the composite expression $\mathcal{C}_1^*$ also fails.

Alternation. $\mathcal{C}_1 \,|\, \mathcal{C}_2$

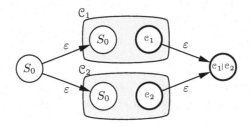

This composite event expression matches if either $\mathcal{C}_1$ or $\mathcal{C}_2$ is detected. The new automaton has an initial and a generative state with ϵ-transitions to both of the two subautomata introducing nondeterministic behavior.

Timing. $(\mathcal{C}_1, \mathcal{C}_2)_{T_1=tspec}$

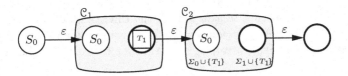

Timing relationships between composite events are supported by the timing operator that can detect event combinations within, or not within, a given time interval. This operator generates an event of type T_1 at the relative or absolute time specification *tspec* after a composite event of type $\mathcal{C}_1$ has been detected. The second expression C_2 may then use T_1 in its specification for atoms and input domains. Since time events are only locally visible, automata $\mathcal{C}_1$ and $\mathcal{C}_2$ must reside on the same node.

Parallelization. $\mathcal{C}_1 \parallel \mathcal{C}_2$

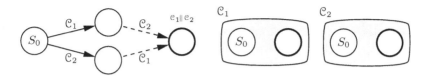

The final operator is parallelization and allows detection of two composite events $\mathcal{C}_1$ and $\mathcal{C}_2$ in parallel, only succeeding if both are detected. Unlike alternation, any interleaving of the two composite events is permitted. The detection automaton for parallelization is constructed by creating a new automaton that uses the composite events detected by $\mathcal{C}_1$ and $\mathcal{C}_2$ for its transitions.

Examples

The following examples illustrate valid expressions in the core composite event language. Let the describable event set A represent events corresponding to the subscription that "Alice is in the office", let $\bar{A}$ be "Alice has left the office", let B be "Bob is in the office", and let P be "anyone is in the office", as detected by an Active Bat.

1. $[A];[B]$. Alice enters the office followed by Bob.
2. $[A \subseteq \{A, B\}]$. Alice enters the office before Bob.
3. $([A], [B \subseteq \{B, T_1\}])_{T_1=1\,\mathrm{h}}$. Alice enters, and Bob follows within 1 h.
4. $[\bar{A}]\,[\neg A \subseteq P]\,[A]$. Someone else enters the office when Alice is away.

Higher-Level Composite Event Languages

We can now use the core composite language as a basic building block for other composite event detection languages. In general, when designing a language for composite event detection, we have two conflicting requirements. On one hand, the language should be *machine processable* so that it supports the efficient creation of composite event detection automata and the automatic decomposition of expressions for distributed detection. On the other hand, the syntax and semantics of the language should be high-level and intuitive, facilitating the task of writing expressions by programmers or end users. This means that the language should be *human processable*. To unify these two requirements, one can define higher-level composite event languages for the specification of composite events in a natural and domain-specific way. Expressions from higher-level languages are then translated automatically into the core language described above. The following are three possible higher-level composite event languages.

Pretty Language

The "pretty" language has a more verbose syntax compared to the core language and resembles rule-based specification languages found in active database systems. It has a redundant set of operators, and its specifications are close to English language statements. A composite event specification, such as

Event_A followed_by Event_B within 1 hour,

makes it easier for nonprogrammers to use composite events.

Programming Language Binding

This binding provides programming language-specific access to composite event specification. It avoids having to deal with a special composite event language by allowing the construction of composite event expressions from method calls, such as

eventA.after(eventB.repeated(3)).

At runtime these method calls are translated into core composite event language expressions for detector construction.

Graphical Composition Model

In a ubiquitous computing environment, a user-friendly way for composite event specification is needed that makes it easy for users to interact with the system at runtime. Composite events, such as "Turn the office light out after 7 pm when the office is empty", can be described using a graphical composition tool that is based on a simple model familiar to users. For example, composite event streams could be visualized as water flows with different forms of piping for the construction of composite event expressions [194].

7.5 Detection Architectures

In this section we present an architecture for a composite event service. The design requirements for the service can be derived from the above application scenarios. In general, the service should be applicable to a wide range of publish/subscribe designs and therefore should make few assumptions about the underlying publish/subscribe implementation. Ideally, it should only rely on standard interfaces provided by the publish/subscribe system and not require special extensions to the event model. For example, content-based routing and filtering support should be exploited for the dissemination of composite events. To satisfy the requirement of scalability, composite event detection can be distributed, decomposing complex composite event subscriptions into subexpressions and detecting them at different nodes in the system.

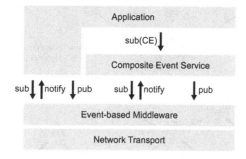

Fig. 7.7. The architecture for the composite event detection service

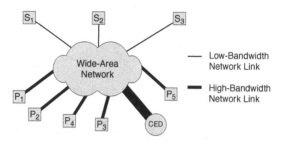

Fig. 7.8. Illustration of centralized composite event detection

The architecture for a general composite event service is shown in Fig. 7.7. The service uses the event client API of the publish/subscribe system so that composite event detectors can subscribe to primitive events and detect the occurrence of composite events. The publish/subscribe system is also used to coordinate the detection of decomposed composite event expressions and publish detected composite events. Note that the publish/subscribe system does not need to be aware of composite event types because composite event publications can be disguised using new primitive event types. Content-based routing and filtering of events is carried out by the publish/subscribe system. An application with event clients can either use the composite event service to submit composite event subscriptions and cause the instantiation of detectors, or interact directly with the publish/subscribe system for normal middleware functionality.

7.5.1 Centralized Detection

The most straightforward architecture for a composite event detection service is centralized, as shown in Fig. 7.8. In a centralized architecture, a single composite event detector (CED) is subscribes to all primitive events that may contribute toward the detection of composite events. When a composite event

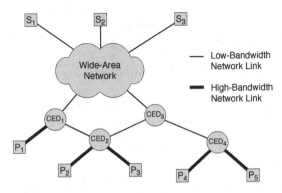

Fig. 7.9. Illustration of distributed composite event detection

has been detected, a new composite event notification is published by the detectors.

An obvious disadvantage of a such an approach is that the centralized event detector can become in a bottleneck in a large-scale system. This may happen if the network bandwidth or processing resources at the detection site are insufficient to keep up with the stream of incoming primitive events. In addition, a centralized detector wastes network bandwidth because many primitive events are sent to the detector over the network only to be discarded there. A distributed implementation of the composite event detection service is more complex but has the advantage that primitive events can be discarded close to event publishers.

7.5.2 Distributed Detection

A composite event service can also implement the detection of composite events in a distributed fashion. This is achieved by decomposing expressions from the composite event language into subexpressions that are detected by separate detectors distributed throughout the system. The support for the decomposition of composite event expressions allows popular subexpressions to be reused among event subscribers, thus saving computational effort and network bandwidth. In particular, the amount of communication is reduced because detectors for subexpressions can be positioned close to primitive event publishers that produce the events necessary for detection. Subexpressions can also be replicated for load balancing and increased availability, and computationally expensive expressions can be decomposed to prevent any detector from becoming overloaded.

A system that benefits from distributed composite event detection is shown in Fig. 7.9. The composite event detectors CED_{1-4} for subexpressions are located close to the primitive event publishers P_{1-5} that publish events at a high rate and therefore must be connected through high-bandwidth network links. Low-bandwidth links in a wide-area network are used to connect the composite event subscribers S_{1-3}. The traffic on these network links is significantly lower because fewer event publications need to be transmitted after composite event detection. Since each detector subscribes to at most two event streams, no detector can get overwhelmed by the event rate.

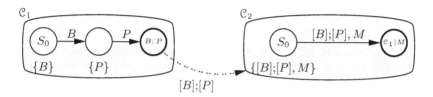

Fig. 7.10. Two cooperating composite event detectors for distributed detection

The detection automata described earlier directly support distribution because they can subscribe to composite events detected by other automata in the publish/subscribe system. In Fig. 7.10 the two automata $\mathcal{C}_1$ and $\mathcal{C}_2$ cooperate in order to detect the composite event expression $([B];[P]) \,|\, [M]$. The subautomaton $\mathcal{C}_1$ detects the expression $[B];[P]$, which is then used by $\mathcal{C}_2$ in the event input domain and transition of state S_0. When this composite event is received, it causes a transition to the generative state S_1. Next we present the capabilities of mobile composite event detectors.

Mobile Composite Event Detectors

A *mobile composite event detector* implements the distributed detection of composite events. Mobile composite event detectors are agentlike entities cohosted at event brokers that encapsulate one or more composite event detection automata for expressions from the core composite event language. They can subscribe to event publishers (and other mobile detectors) and publish the composite events detected by their automata. In addition, a mobile detector can move from one event broker to another in order to optimize the detection of composite events in the system.

When an event subscriber submits a new composite event subscription, a mobile detector is instantiated at an event broker and is then responsible

Fig. 7.11. The life cycle of a mobile composite event detector

for the detection of the new expression. The life cycle of a mobile composite event detector is summarized in Fig. 7.11. In the *construction phase*, the mobile detector establishes the detection of the new composite event subscription by cooperating with other existing mobile detectors. It then enters a *control phase*, during which the detection is optimized by adapting to dynamic changes in the environment and ensuring that it maintains compliance with distribution and detection policies described below. Finally, a *destruction phase* is reached when the mobile detector is no longer required because all event clients have unsubscribed or other detectors have made it redundant.

While in its control phase, a mobile detector can carry out several actions that are governed by distribution policies explained in the next section.

1. It can *instantiate* new automata for the detection of new composite event expressions or any subexpressions.
2. For distributed detection, it can decompose composite event expressions and *delegate* detection to other, already existing mobile detectors.
3. The mobile detector can *migrate* to another event broker that, for example, is closer to the event publishers that the detector has subscribed to.
4. Finally, it can *destroy* any of its composite event detection automata that are no longer required.

Distribution Policies

A remaining difficulty is the decision on an optimal strategy for the decomposition of composite event expressions and the placement of composite event detectors in the system. This is complicated by the fact that the requirements for distributing detectors are potentially conflicting. For example, to minimize usage of network bandwidth, existing detectors should be reused for subexpressions as much as possible. However, if low notification latency is important, detectors should be replicated in various parts of the network, thus leading to increased bandwidth consumption. An optimal solution is a trade-off that takes the static and dynamic characteristics of the application and the network into account.

To make these trade-offs explicit, we introduce the notion of a *distribution policy*, which is a set of heuristics that governs the actions of mobile composite event detectors in the control phase. Each composite event subscription submitted to the composite event service includes its own distribution policy for detection, depending on the application requirements of the event subscriber. During their lifetime, mobile composite event detectors attempt to comply with their distribution policy. Some distribution policies may require the aggregation of network or event broker statistics by mobile composite event detectors, such as communication latency or computational load. When defining distribution policies, three independent dimensions can be identified that help restrict the design space, as shown in Fig. 7.12.

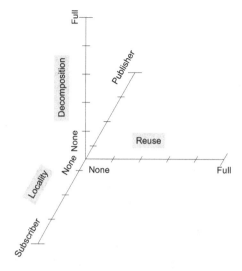

Fig. 7.12. The design space for distribution policies

Decomposition. The degree of decomposition of the composite event expression must be stated in the policy (with optional hints from the application). In order to reuse existing detectors in the system, an expression may have to be decomposed into subexpressions. Decomposition may increase the reliability of detection if multiple detectors are detecting overlapping expressions. For load-balancing reasons, a complex expression may be decomposed into manageable subexpressions. The degree of decomposition ranges from no decomposition to full decomposition, where every possible subexpression is factored out. Some policies allow decomposition only when there already exist detectors that can be reused for a subexpression.

Reuse. This dimension specifies to what extent already existing detectors are reused for a new composite event expression or any of its subexpressions. Not reusing existing detectors can result in more reliability, whereas maximum reuse will save bandwidth and computational effort. In situations in which detection latency is important, only local detectors that are in close proximity should be reused.

Locality. The location of new mobile composite event detectors must be determined. For certain scenarios, bandwidth usage can be reduced by moving detectors as close to primitive event sources as possible. Primitive events that constitute a composite event may be of interest only to the CE detector and should therefore not be widely disseminated throughout the entire system unnecessarily. This is called *publisher locality*. The opposite approach is to put new CE detectors close to application components that subscribe to them to improve reliability and detection latency. This leads to a policy with *subscriber locality*.

Table 7.1. Example of five distribution policies

Policy Name	Decomposition	Reuse	Locality
Minimum Latency	None	With locality only	Subscribers
Minimum Bandwidth	For reuse only	Max	Publishers
Minimum Impact	For reuse only	Max	None
Minimum Load	Max	Max	None
Maximum Reliability	For reuse only	At least 2	None

In practice, only certain combinations of these three dimensions will result in useful distributions policies. Table 7.1 summarizes five example policies that each attempt to optimize a different metric in the composite event framework.

Minimum Latency Policy. The detection latency is minimized by placing new detectors as close to subscribers as possible. Composite event expressions should not be decomposed into subexpressions as this would increase the detection latency. Similarly, an existing detector should only be reused if it is close to the subscriber and detects exactly the required composite events.

Minimum Bandwidth Policy. Bandwidth consumption is minimized by placing the detectors close to the primitive event publishers, leveraging the filtering aspect of composite event detectors. In addition, existing detectors should be used as much as possible so that no new traffic is generated. The reuse of subexpressions may lead to decomposition.

Minimum Impact Policy. This policy minimizes the impact that new detectors have on the entire system. This involves minimizing bandwidth, as before, but also means that computational load should be spread out evenly among detectors. Therefore, new detectors do not have locality, but existing detectors should be maximally reused.

Minimum Load Policy. The fourth policy minimizes the load on composite event detectors by decomposing an expression into the smallest possible subexpressions and distributing them evenly among detectors in the system. It attempts to reuse already existing detectors.

Maximum Reliability Policy. The last policy makes the composite event detection more resistant to node failure by instantiating redundant detectors for extra reliability. Old detectors are reused only when at least two already exist; new detectors are created otherwise. (This "at least 2" partial reuse policy lies between no reuse and full reuse in Fig. 7.12). To limit extra points of failure, detectors are decomposed for reuse only, and no locality restrictions are imposed on new detectors.

Note that a distribution policy is associated with a particular CE expression, so that every mobile CE detector can have its own policy. This enables event subscribers to specify a desired distribution policy at subscription time depending on application requirements. The effectiveness of distribution policies can be enhanced when mobile CE detectors are able to obtain network- and system-specific parameters such as the current load of a broker node or the communication latency to a particular publisher. A mobile CE detector may use this information to optimize detection in compliance with its distribution policy.

Detection Policies

In a distributed system, events from different event sources travel along separate network routes to a mobile CE detector. Even if we assume that the network itself does not reorder events, out-of-order arrival of events at the detector can occur because of the different associated network delays. Whenever a new event arrives, it has to be inserted at the correct position in the totally ordered event input stream before the stream is fed into the automaton.

The problem is to decide when the next event in the event input stream can be safely consumed by the automaton without risking that an event with an older timestamp is still being delayed by the network. Premature consumption could lead to an incorrect detection or nondetection of a composite event. Thus, each CE subscription is annotated with a *detection policy* that specifies when a detector can consume an event from an event input stream.

Best-Effort Detection. A best-effort detection policy states that events are consumed from event input streams without delay. Whenever an event is available, it will cause a state transition (or failure) in the automaton. Although this policy may lead to incorrect detection, it can be applied by applications that are sensitive to detection delay and are willing to ignore false positives.

Guaranteed Detection. Under a guaranteed detection policy, an event is consumed from an event input stream only once it has become *stable*[1] [238]. The consumption of only stable events ensures that no spurious composite events are detected. In our model, we assume that the network itself does not reorder events autonomously so that events coming from the same event source can be expected to arrive in chronological order at a detector. A detector knows that an event is stable and can be consumed after another event with a later timestamp from the same event source has been inserted in the event input stream. An event source that does not publish events at a high enough frequency can publish dummy *heartbeat events* that are used to "flush the network".

In an asynchronous distributed system, a guaranteed detection policy potentially introduces an unbounded delay at the detector. For instance, an event source might fail or decide not to cooperate by not sending heartbeat events. This could prevent the detector from consuming any events of that

---

[1] An event is stable if there is no other event with an earlier timestamp in the system that should be part of this event input stream and should thus be consumed instead.

type. To avoid this problem, we are currently investigating a *probabilistic stability* metric. As opposed to a simple binary stability measure, a detector attempts to model the probability that a particular event in an event input stream is stable and the event is only consumed if its stability metric is above a given threshold.

7.6 Further Reading

In this section we provide an overview of related work on composite event detection. A more detailed description of distributed composite event detection can be found in [314]. Composite event detection first arose in the context of triggers in active database systems. Other related application areas are network systems monitoring and the interaction with ubiquitous computing environments. In general, distributed publish/subscribe systems leave the detection of composite events to the application programmer. An exception is SIENA (described in Sect. 9.3.2), which includes restricted event patterns without defining their precise semantics or giving a complete pattern language. A service for the detection of composite events using CORBA is presented by Liebig [238]. Similar to the described composite event service, it uses interval timestamps to make the uncertainty of timestamps in a distributed system explicit. The notion of event stability is introduced to handle communication delays. A system and language for *complex event processing* is proposed by Luckham [242]. The *Rapide* language [243] supports the specification of event patterns in areas such as process management, network monitoring, and enterprise management. Event patterns are detected using event processing agents that have access to event histories and mine the event stream.

Active Database Systems

Composite event detection in active database systems is usually not distributed. Early languages for triggers follow an event–condition–action (ECA) model [105, 304] and resemble database query algebras with an expressive, yet complex, syntax. In the *Ode* object database [173], composite events are specified with a regular language and detected using finite state automata. Equivalence between the language and regular expressions is shown. Since a composite event has a single timestamp—that of the last primitive event that led to its detection—a total event order is established that does not deal with time issues. Composite event detectors based on Petri nets [307] are used in the *SAMOS* database [170]. Colored Petri nets can represent concurrent behavior and store complex event data during execution. A disadvantage is that even for simple composite event expressions, Petri nets quickly become complicated. SAMOS does not support distributed detection and has a simple time model. The motivation for *Snoop* [74] was to design an expressive composite event language with temporal support. A detector in Snoop is a tree that mirrors the structure of the composite event expression. Its nodes implement language operators and conform to a given *consumption policy*. A consumption policy determines the semantics of operators by resolving the order in which events are consumed from an event history. For example, under a *recent* consumption policy only the event that most recently occurred is considered and others are ignored. Detection then propagates up the tree with the leaves being primitive event detectors. A drawback of this approach is that detectors are Turing-complete, which makes it difficult to estimate their resource usage in advance. In addition, consumption policies influence the semantics of operators in a nonintuitive and operator-dependent way. For simplicity we have decided to only support a *chronicle* consumption policy.

Distributed Systems Monitoring

Similar to network systems monitoring in Sect. 7.1, composite events can be used for the monitoring of distributed systems. Schwiderski presents a distributed composite event monitoring architecture [339] based on the 2g-precedence time model. This model makes strong assumptions about the clock granularity that are not valid in large-scale, loosely coupled distributed systems. The composite event language and detectors are similar to Snoop and suffer from the same shortcomings. The work addresses the issue of delayed events in distributed detection by *evaluation policies*. *Asynchronous* evaluation allows a detector to consume an event without delay, whereas *synchronous* evaluation forces it to wait until all earlier events have arrived, as indicated by a heartbeat infrastructure. Although the detection can be made distributed, the placement of detectors in the system is left to the user. The *GEM* system [250] has a rule-based event monitoring language. It also follows a tree-based approach and assumes a total time order. Communication latency is handled by annotating rules with tolerable delays, which may not be feasible in an environment with unpredictable delays, such as a large-scale distributed system.

Ubiquitous Systems

Research efforts in ubiquitous computing have resulted in composite event languages that are intuitive to use by users of environments such as the Active Office. The work by Hayton [189] on composite events in the Cambridge Event Architecture defines a language that is targeted at nonprogrammers. Push-down finite state automata are used to detect composite events, but the semantics of some of the operators is nonintuitive. Although detection automata can use composite events for input, distributed detection is not handled explicitly and only scalar timestamps are used in the time model.

8

Advanced Topics

In this chapter we provide an overview of several areas of event-based systems that are still the focus of ongoing research. The entire space is too vast to be covered in this book, so we have chosen five topics that are of particular interest to designers of event-based systems instead. In Sect. 8.1 we discuss *security* in a publish/subscribe system and describe a secure publish/subscribe model that can be used as a foundation for access control using events. Section 8.2 investigates the issue of *fault tolerance* in event-based systems. The goal is to build systems that are robust in the face of failure, for instance, by designing self-stabilizing routing algorithms that are guaranteed to reach a correct state after a finite number of steps. The issue of *congestion* in publish/subscribe systems is addresses in Sect. 8.3. Here, we give an example of two congestion control algorithms that are targeted at the asynchronous, decoupled communication in a network of event brokers. Finally, in Sect. 8.4 we focus on *mobility* in event-based systems. The loose coupling of clients in a publish/subscribe system has natural advantages when applied to mobile clients that migrate through the systems, deattaching and reattaching at different points in the network.

8.1 Security

Security has received surprisingly little attention in publish/subscribe systems so far. Unlike composite event detection, it affects many different parts of a publish/subscribe system. In this chapter, we provide an example of a security service [34] for a distributed event system that uses role-based access control to provide three mechanisms: restrictions on the interaction of event clients with the publish/subscribe system, trust levels for event brokers, and the encryption of event data to control information flow in the publish/subscribe system on a fine-grained basis. An advantage of this approach is that it does not require separation of the overlay broker network into distinct trust domains but instead any broker can handle any potentially encrypted event.

The described security service is influenced by the security needs of two applications scenarios discussed in the next section. In Sect. 8.1.2 we define the requirements of a security service, showing how publish/subscribe communication impacts on security. After briefly summarizing existing access control techniques in Sect. 8.1.3, we introduce the secure publish/subscribe model implemented by the service in Sect. 8.1.4. It includes boundary access control using restrictions, different levels of event broker trust, and encryption of event attributes. We finish the overview of related work on security in publish/subscribe systems in Sect. 8.1.5.

8.1.1 Application Scenarios

In this section we look at two application scenarios and examine how they motivate the need for security in a publish/subscribe system. When considering security, we focus on issues of access control to the system and confidentiality of the event data being disseminated in the system.

The Active City

The *Active City* is an extension of theActive Office environment introduced in Sect. 7.1 to a geographically larger system covering an entire city. In an Active City, different city services, such as police and fire departments, ambulances, hospitals, and news agencies, cooperate using a shared event system for information dissemination. Since these city services are under separate management and have individual security implications, the event system must be flexible enough to accommodate a wide range of security policies and mechanisms to enforce them.

An excerpt of a sample event type hierarchy with event attributes that could be employed by cooperating services in an Active City is shown in Fig. 8.1. Information about a road traffic accident reported to the police in an AccidentEvent should be visible to the emergency services so that an ambulance can be dispatched if there are any casualties, but only anonymized data should be passed on to a news agency. The challenge is that some information may flow freely through the Active City, whereas other information has to be closely controlled. A simple solution would be for each city service to operate a separate, trusted event-based middleware deployment with controlled gateways between networks, forming an event federation [192]. However, this would result in complex policy management at the gateways, a significant waste of resources due to redundancy, and an increased event notification delay between services. It would also prevent event clients from one domain using the infrastructure of another while roaming. For this application scenario, a more complex solution is required.

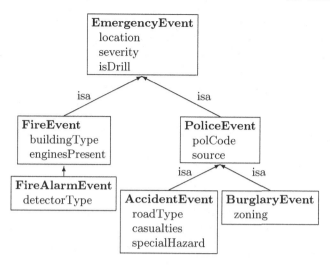

Fig. 8.1. An event type hierarchy for the Active City

News Story Dissemination

In an Internet-wide system for the dissemination of news stories, it is important that customers only receive the service that they are paying for. For example, a customer who has subscribed to a premium service should receive up-to-date news bulletins without delay, as opposed to a standard service subscriber that can only see events relating to older news reports. Moreover, subscribers should only be allowed to subscribe to the news topics that they are entitled to. To ensure this, it is not sufficient to merely rely on subscriptions in the publish/subscribe system because event brokers that perform content-based routing of news events may be under the administration of customers and thus not trusted to honor subscriptions correctly. Using partially trusted event brokers for event dissemination in customer networks is otherwise in the interest of news agencies because it reduces the resource requirements of their middleware deployments. When the service subscription of a customer changes, the event system should quickly adapt to the change in policy.

8.1.2 Requirements

Security mechanisms for an event system differ from traditional middleware security because of publish/subscribe communication semantics. Many-to-many interaction in a publish/subscribe system mandates a scalable access control mechanism. The anonymity of the loose coupling between event publishers and subscribers makes it difficult to use standard security techniques, such as access control lists, since principals can often not be identified beforehand. Content-based routing of events conflicts with the encryption of data because an event broker must have access to the content of an event for its routing decision [390]. Any access control mechanism should incur little overhead at publication time because event publications may have a high rate and thus routing should be carried out as quickly as possible.

Since event clients are not trusted, a security service should include perimeter security to control access of event clients to the publish/subscribe system. As seen in the application scenarios, event brokers are trusted to cooperate for the sake of event dissemination, but they may not be allowed to see all event data. Different levels of event broker trust are necessary and must come with mechanisms to remove compromised event brokers. The confidentiality of data stored in event attributes must be preserved even in the light of event matching and content-based routing. At the same time, as much as possible of the overlay broker network should be used for event dissemination so that a single infrastructure for both public and private information exists in order to improve efficiency, administerability, and redundancy in the publish/subscribe system.

8.1.3 Access Control Techniques

In this section, we describe different access control techniques and highlight their applicability to publish/subscribe communication. We assume that the system consists of a set of *objects*, a set of *principals*, and a set of *permissions*. The goal of an access control scheme is to define what principals have what permission to access what objects, as shown below. In a publish/subscribe context, principals correspond to event clients, objects are the event notifications, and permission are the standard operations, such as *publish* and *subscribe*.

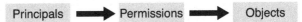

Our discussion will focus on *discretionary* access control, where users themselves set access control rights, as opposed to *mandatory* access control, in which rights are set by a centralized authority for the entire system. Discretionary access control is more suitable for large-scale distributed systems because it does not depend on a centralized entity.

Access Control Lists

A simple way to implement discretionary access control are *access control lists (ACLs)*. An ACL is associated with every object and specifies the access permissions of principals to that object. For example, an ACL for file *foo* may state that the file is writable and readable by user *Alice* and only readable by user *Bob*.

A drawback of ACLs is that they create a direct mapping between objects and principals. This is undesirable in a publish/subscribe context, in which the identities of event publishers and event subscribers are not globally known. In addition, events in a publish/subscribe system are short-lived, which makes them bad candidates to manage access permissions.

Capabilities

The opposite approach to an ACL is a *capability* that stores access permissions with the principal instead of the object. When a principal wants to access a given object, she must present the capability with the appropriate permissions first. For example, the capability owned by user Alice may state that she can read and write file *foo* and only read file *bar*. Of course, this means that capabilities need to be protected from tampering by principals through digital signatures or secure storage in memory.

Capabilities are more compatible with publish/subscribe communication because event clients manage their own capabilities. However, they are harder to manage because access permissions cannot easily be revoked. In addition, capability-based access control is often not scalable because principals may end up with a large number capabilities when the set of objects in the system changes dynamically. Finally, it also suffers from the problem that both principals and objects need to know about each other, which is not the case in a publish/subscribe system.

Role-Based Access Control

An access control model that extends capabilities and attempts to address some of its short-comings is *role-based access control* model (RBAC) [332]. RBAC simplifies security administration by introducing *roles* as an abstraction between *principals* and *permissions*, as shown below. Roles permit principals and permissions to be grouped intuitively in the system and addresses the anonymity of event clients in an event-based middleware. This grouping increases scalability of the access control mechanism because there are fewer roles than principals and permission in the system. The access control policy for the system focuses on the concept of a role, which is long-lived. To obtain privileges, a principal such as an event publisher or subscriber presents credentials that allow it to acquire a role membership that is associated with the desired permissions. In the rest of this chapter, we describe an access control model for publish/subscribe communication that is based on RBAC.

In our secure publish/subscribe model, we assume the decentralized implementation of a RBAC scheme, such as *Open Architecture for Secure Interworking Services (OASIS)* [22]. OASIS includes an expressive policy language to specify rules for role acquisition. It uses a session-based approach with

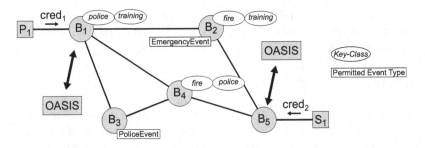

Fig. 8.2. Illustration of the secure publish/subscribe model

event communication to revoke currently active roles of principals in a timely manner after prerequisite credentials were revoked. Credentials in the OASIS implementation are protected with X.509 certificates [207] for authentication and proof of role membership.

8.1.4 Secure Publish/Subscribe Model

In this section we describe a secure publish/subscribe model for an event system. As a general design philosophy, the model couples access control with types of events. If the event space is already structured into event types, it is intuitive to leverage this for the specification of access control policy, but more fine-grained specification in terms of event attributes and content-based subscriptions is also supported by the model.

An example of an distributed publish/subscribe deployment with event brokers that implement the secure publish/subscribe model is given in Fig. 8.2. There are three mechanisms in the model to accomplish access control. First, boundary access control to the middleware, as described in the next section, is achieved by controlling access of event clients to local event brokers with an OASIS policy. Services requested by event clients can either be granted, rejected, or *partially granted* after imposing restrictions. As explained in Sect. 8.1.4, the second mechanism assigns an event broker to a particular trust category that prescribes the types from the event type hierarchy that the event broker is permitted to handle. Finally, confidential event attributes in event notifications are encrypted, limiting access to those attributes. A single event publication can contain both public and private data. Content-based routing decisions on encrypted event attributes can only be carried out by event brokers that possess the necessary decryption key, otherwise events need to be flooded. Event attribute encryption will be introduced in Sect. 8.1.4.

Boundary Access Control

Local event brokers that host event clients are OASIS-aware and perform access control checks for every request made to them. This ensures that only authorized clients have access to the publish/subscribe system in compliance with access control policies. As shown in Fig. 8.2, local event brokers delegate the verification of credentials passed to them by event clients to an OASIS engine. Four types of OASIS policy restrict the actions of event clients. The event client that creates a new event type becomes its *event type owner* and is then responsible for specifying policy.

Connection Policy. This policy states the required credentials for an event client to be permitted hosting by a given event broker. A client can only use the publish/subscribe system if it maintains a connection with at least one local event broker.

Type Management Policy. The creation, modification, and removal of event types in the event type hierarchy is controlled by a type management policy. Usually, credentials certifying that an event client has the role of event type owner for an event type allow it to perform type management. This also avoids conflicts between clients from different applications.

Advertisement Policy. For every event type in the system, an advertisement policy specifies the roles an event publisher must acquire in order to advertise events of this type. This policy is generally specified by the event type owner.

Subscription Policy. Similarly, a subscription policy lists the necessary roles for an event subscriber to subscribe to events of that type. The policy may also prescribe the content-based filter expressions that are permitted and is again defined by the event type owner.

When an event client violates the connection or type management policies, the event broker rejects the operation invoked by the client. For advertisement and subscription policies, certain requests may be partially accepted by imposing a *restriction* on the original event advertisement or subscription. An advertisement restriction limits the advertisement by restricting the events that the event publisher is allowed to publish. Likewise, a subscription restriction transforms the client-requested subscription into a different, less powerful one. The client may or may not be notified by its local event broker that a restriction has been imposed for privacy reasons. The secure publish/subscribe model supports two flavors of restrictions.

Publish/Subscribe Restrictions

This kind of restriction takes the original submitted advertisement or subscription and replaces it by a different, more limited one, as defined by a coverage relation. In the case of an event advertisement, the event type in the advertisement is replaced by a less specific parent type from the event type hierarchy. For event subscriptions, the publish/subscribe restriction specifies an upper bound on the event type and content-based filtering expression that the event subscriber is allowed to submit. If the submitted subscription is covered by the subscription restriction, the subscription is accepted without change, otherwise it is automatically downgraded to the restricted subscription.

Generic Restrictions

A generic restriction is not expressible by the publish/subscribe system since it can include any predicate evaluations permitted by OASIS. Although the original advertisement or subscription submitted by the event client is passed on to the publish/subscribe system, all later events are restricted according to the arbitrary predicate function in the generic restriction. For example, a generic advertisement restriction may reject the publication of events with certain content, and a generic subscription restriction may perform additional filtering of events on the message size of the event notification, which otherwise could not be expressed in an event subscription.

The advantage of publish/subscribe restrictions is that they do not incur an overhead during the dissemination of events. Since the original advertisement or subscription is replaced by a more limited version, the event-based middleware implicitly enforces policy and no events need to be dropped at client-hosting brokers. The same is not true for generic restrictions because their additional expressiveness comes with the price of having to evaluate arbitrary predicates at client-hosting brokers to decide whether an event client can publish or be notified of a given event publication.

Event Broker Trust

The previous mechanism for boundary access control using restrictions assumes that all event brokers are equally trusted to process data, which is not true in practice. When an event broker joins the publish/subscribe system, it authenticates with its credentials and is then believed to participate correctly in the routing of events according to a content-based routing algorithm. It maintains encrypted network connections with its neighbouring event brokers in the overlay broker network. However, an event broker may not be trusted enough to gain access to data in particular event notifications or subscriptions. To make these trust relationships explicit, event brokers are associated with event types from the event type hierarchy that they are permitted to handle. This is illustrated in Fig. 8.2. Event broker B_3 is only permitted to process events of type PoliceEvent. Event brokers may be authorized to handle all event types that are more specific or more general than a given type, in other words are sub- or supertypes of an event type.

When routing event advertisements, subscriptions, and notifications in the overlay broker network, an event broker only passes on a message to the next event broker after obtaining proof in the form of a role membership certificate that this event broker is authorized to handle that particular event type. Otherwise, the event broker is forced to make a different routing decision. This can be done by acting as though the untrusted event broker has failed, relying on the fault tolerance properties of event routing in the publish/subscribe system that will ensure a different routing path. Note that event broker trust encompasses the handling of entire event types only, but we relax this restriction by using of event attribute encryption, as described in the next section.

Event Attribute Encryption

The mechanism for event broker trust from the previous section excludes brokers that are not trusted to handle specific event types from routing. As mentioned before, this coarse-grained approach effectively splits the overlay broker network into several trust domains, thus weakening the reliability and efficiency of event routing. A better solution is to prevent an untrusted event broker from accessing confidential data but still enabling it to perform content-based routing on other attributes. We achieve the goal of a single event that can hold private and public information by encrypting event attributes in event publications with different cryptographic keys. Although this introduces a larger runtime overhead due to cryptographic operations during event routing, this is justifiable as it leads to more expressive access control specifications where event types no longer have to be strictly divided into private and public categories. Another advantage of this scheme is that access control policy can also associate event clients, which have access privileges, to event attributes.

In addition to event types, event brokers are also trusted with a number of *key classes*. A key class is a collection of cryptographic keys for encrypting event attributes, that supports key rotation and revocation. Access control to individual event attributes is achieved by signing and encrypting them with a key from a given key class so that only trusted event brokers can decrypt these attributes. An event broker can only read or write an event attribute if it has a role membership that includes access to the appropriate key classes. This also means that an event client can only submit an event subscription or notification that refers to encrypted attributes to its local event broker if it can prove that it possesses credentials for the required key classes. The event broker then performs the cryptographic operations on the client's behalf.

To include event attribute encryption in a typed event model, the event type hierarchy is extended with a description of the key classes that are necessary to access the content of event attributes, as shown in Fig. 8.3. Each event attribute is annotated with its key classes in disjunctive normal form. A conjunction of key classes means that the attribute is encrypted with keys from several key classes in sequence. For example, the isDrill attribute in the EmergencyEvent type has to be either encrypted under the *police* and *training*, or under the *fire* and *training* key classes. This prevents anyone receiving emergency-related events in the Active City from finding out whether this is an exercise drill unless they are a training instructor with access to the *training* key class. Unencrypted event attributes are denoted with the empty

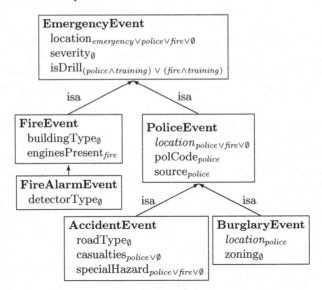

Fig. 8.3. An event type hierarchy with attribute encryption

key class $\emptyset$. In Fig. 8.2 event brokers are annotated with the key classes that they are permitted to use.

Note that the standard subtyping relation between event types must still hold so that a subtype is more specific than its parent type. As a result, key classes can only be removed from inherited event attributes but never added. This is illustrated with the `location` attribute whose access becomes more restrictive as new event types are derived.

Encrypted Attribute Coverage

When an event subscriber submits a content-based subscription for an event type with encrypted attributes, attribute predicates in the subscription must also be encrypted with appropriate key classes for the subscription to match events. The subscriber selects one or more key classes for the encryption of the attribute predicate from all the key classes for which it is authorized. As a consequence, the event model of the publish/subscribe system must be extended to support a coverage relation between event subscriptions and notifications, and among event subscriptions that use attribute encryption. Informally, an encrypted attribute predicate can only be matched by an encrypted event attribute in a notification if it was encrypted with the same key classes. When an attribute predicate should match attributes encrypted under several different key classes, it must be disjunctively encrypted multiple times using these key classes and several copies of the attribute predicate must be included in the subscription. For coverage among subscriptions, an attribute predicate encrypted under particular key classes is covered by another encrypted pred-

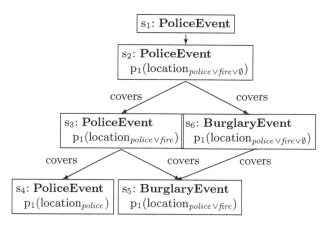

Fig. 8.4. Subscription coverage with attribute encryption

icate if the second predicate covers the first and is encrypted under at least
the key classes of the first predicate.

Definition 8.1 (Encrypted Attribute Coverage). *An encrypted event attribute* a^K *is covered by (or matches) an encrypted attribute predicate* p^L,

$$a^K \sqsubseteq p^L,$$

iff

$$a \sqsubseteq p \wedge K \subseteq L$$

*holds, where K is the set of key classes under which a is conjunctively en-
crypted and L is the set of a conjunction of key classes under which p is
disjunctively encrypted. An encrypted attribute predicate $p_1^{L_1}$ is covered by
another encrypted attribute predicate $p_2^{L_2}$,*

$$p_1^{L_1} \sqsubseteq p_2^{L_2},$$

iff

$$\forall a. \ a \sqsubseteq p_1 \Rightarrow a \sqsubseteq p_2 \wedge L_1 \subseteq L_2,$$

*holds, where a is an event attribute and L_1 and L_2 are sets of conjunctions of
key classes with disjunctive encryption.*

We illustrate this extended coverage relation in Fig. 8.4, which shows six
example subscriptions with regard to the previous event type hierarchy. Subscription s_1 is the most generic because it does not include any attribute
predicates. The attribute predicate in subscription s_3 does not match events
with an unencrypted `location` attribute, and therefore s_3 is covered by s_2.
Subscription s_4 is most specific because the attribute predicate is only encrypted under the *police* key class.

8.1.5 Further Reading

In this section we provide an overview of previous work in the area of security in publish/subscribe systems. Preliminary work on security issues under publish/subscribe semantics can be found in [390]. It identifies the necessity for ensuring the confidentiality of event publications and subscriptions and suggests accountability for billing purposes; however, no mechanisms are provided. In the work by Miklós [260], upper bound filters on advertisements and subscriptions in SIENA are proposed, but the confidentiality of event publications within the publish/subscribe system is not guaranteed.

The *Narada Brokering* project includes a distributed security framework [294, 405] that uses access control lists to control event publishers and subscribers for a topic, limiting the scalability. Cryptographic keys for encrypting publications are centrally managed by a key management center (KMC). An event publisher can choose to use a central topic key from the KMC or the public keys of all event subscribers for encryption, which contradicts decoupled publish/subscribe semantics. Access control can only be provided at the granularity of whole events, and event brokers are implicitly trusted, rather than using different trust levels as supported by our security service.

A publish/subscribe system with scopes (Chap. 6) can be extended to include access control [145]. Scopes, which model visibility in a distributed publish/subscribe system, can be used to split the event system into different trust domains. In effect, this creates multiple distinct overlay networks of event brokers. Secure events only stay within a scope and are therefore never handled by untrusted event clients or brokers. Interactions between trust domains are precisely specified through scope interfaces. Scope interfaces express the access control policies for events crossing trust boundaries. Partitioned overlay networks can use untrusted brokers to create an encrypted tunnel for secure events. This work also proposes an implementation strategy based on aspect-oriented programming [222] to integrate access control with existing publish/subscribe implementations.

8.2 Fault Tolerance

The behavior of a system in the presence of faults is an important property of the system. In Sect. 2.5.2 we described the formal specification of a simple event system. An important goal for such a system was to guarantee *safety* and *liveness* conditions. Recall that a safety condition ensures that nothing bad will happen, whereas liveness stipulates that eventually something good will occur. In other words, Def. 2.5 requires that the system is correct, i.e., exhibits the desired functionality at its interface under all circumstances. To satisfy the specification, all faults occurring in the real world would have to be masked. However, masking all faults is costly if not impossible. Provided that a temporary failure of the system can be accepted, making a system self-stabilizing is an attractive alternative or supplement to fault masking. We will see that self-stabilization comes at cost, namely the weakening of safety conditions. In the following, we describe fault masking and self-stabilization in an event system with an emphasis on the latter.

8.2.1 Fault Masking

Fault masking requires redundancy either in *time* or in *space*. While *time redundancy* repeats actions (e.g., resending a message to cope with message loss), *space redundancy* uses independent copies of the resources that can be affected by faults (e.g., communication channels). Of course, both approaches can also be combined in a single system. However, research about applying fault masking to publish/subscribe systems is still in an early stage.

What are typical scenarios for fault masking in publish/subscribe systems? For example, assume that communication channels fail with in a fail-stop model. Then, we could connect each pair of neighbored brokers with two instead of one communication channel. If one of the communications fails, the brokers can still communicate using the communication channel that is still working. This way, failed communication channels can be masked by the system as long as for any pair of neighbored brokers only one of the two communication channels fails. To be able to mask broker and communication channel failures, we can use two independent broker topologies that do not share physical communication links or computers hosting brokers. In this case, we would have to modify our model such that a client can connect to two remote brokers. We also must take care that no duplicates are delivered and that—if required—the FIFO-producer or causal ordering of messages is ensured. If Byzantine faults can occur, fault masking is much more complicated than in the fail-stop model. This is due to the fact that in the Byzantine model failed links and brokers can behave arbitrarily.

Another possibility for implementing fault masking is to reconfigure the broker network in case of failures such that the failed resources are no longer used in the system. This approach is feasible but not trivial to implement if concurrent faults can occur. While the reconfiguration is in progress, notifications must be buffered at certain brokers. We must ensure that no notifications are lost or duplicated. Extra effort is needed to keep notification ordering guarantees such as publisher-based FIFO or causal ordering, if required [302].

8.2.2 Self-Stabilizing Publish/Subscribe Systems

An alternative (or sensible addition) to fault masking is *self-stabilization*, a concept introduced by Dijkstra [113] in 1974. He defined a system as being self-stabilizing if "regardless of its initial state, it is guaranteed to arrive at a legitimate state in a finite number of steps". In contrast to that, a system which is not self-stabilizing may stay in illegitimate states forever, leading to a permanent failure of the system. Self-stabilization models the ability of a system to recover from arbitrary transient faults within a finite time without any intervention from the outside. If the time between consecutive faults is long enough, the system will start to work correctly again. Transient faults include temporary network link failures resulting in message duplication, loss, corruption, or insertion, arbitrary sequences of process crashes and subsequent recoveries, and arbitrary perturbations of the data structures of any fraction of the processes. The program code running at the nodes and inputs from the outside, however, cannot be corrupted. Dolev [116] gives a comprehensive discussion of self-stabilization.

However, it is, in general, impossible under the fault assumption of self-stabilization to require *any* property that prohibits certain states, i.e., safety properties. For example, the system could deliver a notification n to a client X although X has no active subscription matching n because a fault corrupted the state of the system such that that it "thinks" that X subscribed to n. Therefore, we require that a self-stabilizing publish/subscribe system satisfies the safety property of Def. 2.5 only eventually. This ensures that the system starting from any state will eventually satisfy the actual safety property and continue to do so if no faults occur. The liveness property of Def. 2.5 can be left unchanged. This leads to the following definition of self-stabilizing publish/subscribe systems:

Definition 8.2. *A* self-stabilizing publish/subscribe system *is a publish/subscribe system satisfying the following requirements:*

1. *Eventual Safety Property: Starting from any state, the system eventually satisfies the safety property of Def. 2.5.*
2. *Liveness Property: Starting from any state, the system satisfies the liveness property of Def. 2.5.*

A formal version of this specification can be found in [263].

8.2.3 Self-Stabilizing Content-Based Routing

Under the fault assumption of self-stabilization, the routing configuration can arbitrarily be corrupted by transient faults. Therefore, the applied routing algorithm must ensure (a) that corrupted routing entries are corrected or deleted from the routing table and (b) that missing routing entries are inserted into the routing table.

We assume that each broker stores the information about its neighbors in its ROM. This ensures that this information cannot be corrupted. If it would be stored in RAM or on harddisk, it could also be corrupted by a fault. In this case, we would have to layer self-stabilizing content-based routing on top of a self-stabilizing spanning tree algorithm. Layered composition of self-stabilizing algorithms is a standard technique which is easy to realize when the individual layers have no cyclic state dependencies [116]. In this case, the stabilization time would be bounded by the sum of the stabilization times of the individual layers.

Basic Idea

The basic idea for making content-based routing self-stabilizing is that routing entries are only *leased*. To keep a routing entry, it must be renewed before the *leasing period* π has expired. If a routing entry is not renewed in time, it is removed from the routing table. Interestingly, this approach does not only allow the publish/subscribe system to recover from internal faults but also from certain external faults. For example, if a client crashes, its subscriptions are automatically removed after their leases have expired.

To support leasing of routing table entries, we use a *second chance* algorithm. Routing entries are extended by a *flag* that can only take the two values 1 and 0. Before a routing entry is (re)inserted into the routing table, all existing routing entries whose filter has the same *ID* (as the ID of the filter of the routing entry to be inserted) are removed from the routing table. This is necessary as the IDs of the routing entries can be corrupted, too. We assume that the clock of a broker can only take values between 0 and $\pi - 1$ to ensure that if the clock is corrupted, it can diverge from the correct clock value by at most π. When its clock overruns, a broker deletes all routing entries, whose flags have the value 0 from the routing table and sets the flags of all remaining routing entries to 0 thereafter (new subscriptions have their flags set to 1 initially). Hence, it must be ensured that an entry is renewed once in π to prevent its expiration. On the other hand, it is guaranteed that an entry which is not renewed will be removed from the routing table after at most 2π.

The renewal of routing entries originates at the clients. To maintain its subscriptions without interruption, a client must renew the lease for each of its subscriptions by "resubscribing" to the respective filter once in a *refresh period* ρ. Resubscribing to a filter is done in the same way as subscribing. In general, π must be chosen to be greater than ρ due to varying link delays. The *link delay* δ is the amount of time needed to forward a message over a communication link and to process this message at the receiving broker. In our model, it is considered a fault when δ is not in the range between $\delta_{\min}$ and $\delta_{\max}$. It is important to note that assuming an upper bound for the link delay is a necessary precondition for realizing self-stabilization.

Flooding

The naïve implementation of a self-stabilizing publish/subscribe system is *flooding*: When a broker receives a notification from a local client, the broker forwards the notification to all neighbor brokers. When it receives a notification from a neighbor broker, the notification is forwarded to all other neighbor brokers. Additionally, each processed notification is delivered to all local clients with a matching subscription. Flooding only requires a broker to keep state about the subscriptions of its local clients. Therefore, errors in this state can be corrected locally by forcing clients to renew their subscriptions once in a leasing period. This means that $\rho = \pi$. The main advantage of this scheme is that a coordination among neighboring brokers is not necessary. Hence, no additional network traffic is generated. Additionally, new subscriptions become active immediately. While a corrupted or erroneously inserted subscription survives at most 2π in a routing table and a missing subscription is reinserted after at most π, an erroneously inserted or corrupted notification disappears from the network after at most $d \cdot \delta_{max}$, where d is the *network diameter*, i.e., the length of the longest path a message can take in the broker network. Hence, for flooding, the *stabilization time* Δ, i.e., the time it takes for the system to reach a legitimate state starting from an arbitrary state, equals $\max\{2\pi, d \cdot \delta_{max}\}$.

Simple Routing

The solution for flooding can be extended to simple routing. *Simple routing* treats each subscription independently of other subscriptions. A (un)subscription is inserted into (removed from) the routing table and flooded into the broker network. If a broker receives a (un)subscription from a local client, it is forwarded to all neighbor brokers. If it was received from a neighbor broker, it is forwarded to all other neighbor brokers. Thus, simple routing is idempotent to resubscriptions, and a subscription is redistributed through the broker network when it is renewed by the client. Note that here subscriptions become active only gradually.

A critical issue is that the timing assumptions must allow the clients to renew their leases everywhere in the network before they expire. How large must π be with respect to ρ in this case? To answer this question, consider two brokers B and B' connected by the longest path a message can take in the broker network. This situation is illustrated in Fig. 8.5. Assume a local client X of B leases a routing table entry of B at time t_0 and renews this lease at time $t_1 = t_0 + \rho$. X's lease causes other leases to be granted all along the path to broker B'. Considering the best- and worst-cases of the link delay, the first lease reaches B' at time $a_0 = t_0 + d \cdot \delta_{min}$ in the best case, and the lease renewal reaches B' at time $a_1 = t_1 + d \cdot \delta_{max}$ in the worst-case. If X refreshes its leases after ρ time and if network delays are unfavorable, two lease renewals will arrive at B' within at most $a_1 - a_0$. Hence, $\pi > a_1 - a_0$ must hold to ensure that the entry is renewed in time. Thus, we get $\pi > \rho + d \cdot (\delta_{max} - \delta_{min})$.

The stabilization time Δ depends on the value of π. Since corrupted or erroneously inserted messages can contaminate the network, a delay of $d \cdot \delta_{max}$ must be assumed before their processing is finished. After at most 2π, their effects will be removed everywhere. Overall, the stabilization time sums up to $\Delta = d \cdot (\delta_{max} - \delta_{min}) + 2\pi$. For example, assume that $d = 10$, $\delta_{max} = 25$ ms, and $\delta_{min} = 5$ ms. To guarantee a stabilization time of $\Delta = 30$ s, $\pi = 14.9$ s and thus $\rho = 14.7$ s follows. There is a tradeoff between π and ρ. To have low

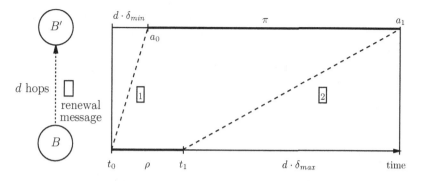

Fig. 8.5. Deriving the minimum leasing time

message overhead, ρ should be as large as possible. However, this implies a large value of π, but π should be as small as possible to facilitate fast recovery.

Advanced Routing Algorithms

The situation is more complicated if advanced content-based routing algorithms such as identity-based, covering-based, or merging-based routing are applied. Contrary to flooding and simple routing these algorithms are—at least the versions presented so far—not idempotent with respect to resubscriptions. However, they can be made idempotent with some minor changes. Note that the maximum stabilization time Δ is not affected by whether an advanced routing algorithm or simple routing is applied because in the worst-case a filter will nevertheless travel all along the longest path in the network.

Consider identity-based routing (for more details we refer to [263]). When a broker B processes a new or canceled subscription F from destination D, it counts the number d of destinations $D' \neq D$ for which a subscription matching the same set of notifications exists in T_B. Depending on the value of d, F is forwarded differently. If $d = 0$, F is forwarded to all neighbors if $D \in L_B$ and to all neighbors except D if $D \in N_B$. If $d = 1$ and $D' \in N_B$, F is forwarded only to D'. If $d = 1$ and $D' \in L_B$ or if $d \geq 2$, F is not forwarded at all. This scheme is not idempotent to resubscriptions because if $d \geq 2$ and one of the identical subscriptions is renewed at B, none of those subscriptions will be forwarded. This can be circumvented if B takes only those subscriptions into account when calculating d whose flag is 1. In this case, in each leasing period that subscription of the identical subscriptions which is renewed first after the broker has run the second chance algorithm is forwarded, ensuring correct forwarding.

Covering-based routing can also be made self-stabilizing. In this case, only routing entries with flag 1 are taken into account when looking for identical subscriptions. However, when looking for subscriptions that really cover a given subscription (i.e., match a real superset of notifications), additionally also those routing entries with flag 0 are considered. This is to avoid sending covered subscriptions unnecessarily to neighbors because they are refreshed before a covering subscription is refreshed. To make merging-based routing self-stabilizing, the refreshing of merged filters must additionally be ensured.

Discussion

The values of π and ρ depend on the delay of the links in the network. So far, we assumed that these values are fixed and equal for every broker in the system. In many scenarios, link delays vary a lot such that it could be advantageous to incorporate this property into the algorithm. We assume that the value of link delay stored at every adjacent broker cannot be corrupted (i.e., it is stored in ROM). The values of π and ρ have then to be calculated individually for every subscription, depending on where the publishers are. Additionally, π and ρ have to be refreshed the same way as described previously for subscriptions. Advertisements that are sent periodically by the publishers could be used for this purpose. Taking this approach, the broker algorithm can take advantage of faster links and stabilize subtrees of the broker topology faster if the links allow for this. The application of leasing is a common way to keep soft states. This technique is used in many protocols and algorithms such as the Routing Information Protocol (RIP, RFC2453) and Directed Diffusion [205].

Simulation

We carried out a discrete event simulation to compare self-stabilizing content-based routing to flooding with respect to their message complexity. Before we discuss the results, we describe the setup of the experiments.

Setup

We consider a broker hierarchy being a completely filled 3-ary tree with five levels. Hence, the hierarchy consists of 121 brokers of which 81 are leaf brokers. Since we use a tree for routing, this implies a total number of 120 communication links. We use hierarchical routing, but similar results can be obtained for peer-to-peer routing, too. With hierarchical routing, subscriptions are only propagated from the broker to which the subscribing client is connected toward the root broker. This suffices because every notification is routed through the root broker. Hence, control messages travel over at most four links. We use identity-based routing and consider 1000 different filter classes (e.g., stocks) to which clients can subscribe.

Subscribers only attach to leaf brokers. Results for scenarios where clients can attach to every broker in the hierarchy can be derived similarly. Instead of dealing with clients directly, we assume independent arrivals of new subscriptions with exponentially distributed interarrival times and an expected time of λ^{-1} between consecutive arrivals. When a new subscription arrives, it is assigned randomly to one of the leaf brokers and one of the filter classes is randomly chosen. The lifetime of individual subscriptions is exponentially distributed with an expected lifetime of μ^{-1}. Each notification is published at a randomly chosen leaf broker. Hence, notifications travel over at most eight links. The corresponding filter class is also chosen randomly. The interarrival times between consecutive publications are exponentially distributed with an expected delay of ω^{-1}. We assume a constant delay in the overlay network of $\delta = 25$ ms, including the communication and the processing delay caused by the receiving broker.

To illustrate the effects of changing the parameters, we considered two possible values for some of the system parameters: For each of the 1000 filter classes, a publication is expected every 1 s (10 s), i.e., $\omega_1 = 1000\,s^{-1}$ ($\omega_2 = 100\,s^{-1}$). The expected subscription lifetime is 600 s (60 s), i.e., $\mu_1 = (600\,s)^{-1}$ ($\mu_2 = (60\,s)^{-1}$). Each client refreshes its subscriptions once in 60 s (600 s), i.e., a refresh period of $\rho_1 = 60$ s ($\rho_2 = 600$ s). Since $d = 8$ in our scenario, the leasing period is $\pi_1 = 60.2$ s ($\pi_2 = 600.2$ s) for ρ_1 (ρ_2). Hence, a subscription will on average be refreshed 10 (100) times before it is canceled by the subscribing client if $\mu = (600\,s)^{-1}$. The resulting stabilization time is $\Delta_1 = 120.6$ s ($\Delta_2 = 1200.6$ s).

We are interested in how the system behaves in equilibrium for different numbers of active subscriptions N. In equilibrium, $dN/dt = 0$ where $dN/dt = \lambda - \mu \cdot N(t)$, implying $N = \lambda/\mu$. Thus, if N and μ is given, λ can be determined. If the system was started with no active subscriptions, we would have to wait until the system approximately reached equilibrium before we begin the measurements. However, in our scenario it is possible to start the system right in the equilibrium. At time 0, we create N subscriptions. For each of these subscriptions, we determine how long it will live, for which filter class it is, and at which leaf broker it is allocated. Since we use an exponential distribution for the lifetime, this approach is feasible because the exponential distribution is memoryless.

Results

The results of our simulation are depicted in Fig. 8.6. Note that the right plot is a magnification of the most interesting part of the left plot. In Fig. 8.6, $b_{s1/2}$ is the notification bandwidth saved if filtering is applied instead of flooding. The figure shows b_{s1} and b_{s2}, which correspond to the publication rate ω_1 and ω_2, respectively. Because b_s linearly depends on ω, a decrease of ω by a factor of 10 leads to a use of one tenth as much notification bandwidth. If there are no subscriptions in the system, $b_{s1} = 116,000\,s^{-1}$ and $b_{s2} = 11,600\,s^{-1}$, respectively. These numbers are $4000\,s^{-1}$ and $400\,s^{-1}$ less than the overall number of notifications published per second. This is because with hierarchical routing, a notification is always propagated to the root broker. The control traffic b_c is caused by subscribing, refreshing, and unsubscribing clients. It only

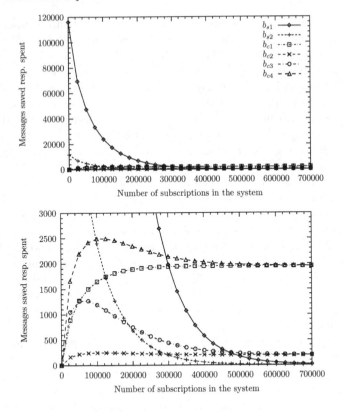

Fig. 8.6. Notification bandwidth saved by doing filtering instead of flooding (b_{s1} : $\omega_1 = 1000\ s^{-1}, b_{s2} : \omega_2 = 100\ s^{-1}$) and control traffic caused by filtering and leasing ($b_{c1}, b_{c4} : \rho_1 = 60\ s,\ b_{c2}, b_{c3} : \rho_2 = 600\ s,\ b_{c1}, b_{c2} : \mu_1 = (600\ s)^{-1},\ b_{c3}, b_{c4} : \mu_2 = (60\ s)^{-1}$). The *lower figure* magnifies the most interesting part of the *upper figure*

arises if filtering is used. The figure shows b_{c1}, b_{c2}, b_{c3}, and b_{c4}, which result from the different combinations of μ and ρ. The value to which b_c converges for large numbers of subscriptions mainly depends on the refresh period ρ. Thus, b_{c1} and b_{c3} converge to $120,000/\rho_1 = 2000s^{-1}$, while b_{c2} and b_{c4} converge to $120,000/\rho_2 = 200s^{-1}$. The evolution of b_c for numbers of subscriptions in the range between 0 and 200,000 is largely influenced by the value of μ. A small μ such as μ_2 leads to a hump (cf. b_{c3} and b_{c4} in Fig. 8.6). Filtering saves bandwidth compared to flooding if b_s exceeds b_c. The points where the curve of the respective variants of b_s and b_c intersect are important: If the number of subscriptions is smaller than at the intersection point, filtering is superior, while for larger numbers flooding is better. For example, the curves of b_{s1} and b_{c1} intersect for about 300,000 subscriptions. Thus, filtering is superior for less than 300,000 subscriptions, while flooding is superior for more than 300,000 subscriptions. Since we consider eight scenarios, we have eight intersection points in Fig. 8.6.

The results gained through the simulation show that applying self-stabilizing filtering makes sense if the average number of subscriptions in the system does not grow beyond a certain point. However, it is important to note that all assumptions taken for the simulation depict worst-case scenarios. For example, the equal distribution of subscriptions to leaf brokers is disadvantageous for filtering. If there was locality in the interests of the clients, filtering would always save a portion of the notification traffic, regardless of how large the number of subscriptions grows [263], and the control traffic would also be smaller. In such scenarios, filtering can be superior to flooding for all numbers of subscriptions. Recently, Jaeger and Mühl [212] have published analytical results that come to the same results as those presented here.

8.2.4 Generic Self-Stabilization Through Periodic Rebuild

In a self-stabilizing system, arbitrary transient faults can occur. The only parts that cannot be corrupted are the program code and the data stored in ROM. In general, we cannot reason about how a routing algorithm (that works correctly in a fault-free system) behaves when it receives corrupted messages or when it is applied to perturbed routing tables. What can merely be assumed is that it will eventually work correctly again when it is restarted from a legitimate initial routing configuration.

In this section, we present a generic wrapper algorithm $\mathcal{A}$ that makes a publish/subscribe system self-stabilizing, regardless of which correct routing algorithm $\mathcal{R}$ it wraps. The only assumptions are that (a) $\mathcal{R}$ has no private state but draws its decision solely on the basis of the respective routing table, that (b) $\mathcal{R}$ terminates after finite time when called, and that (c) each client refreshes its subscriptions once in a refresh period ρ. The wrapper algorithm periodically rebuilds the routing tables starting from an initial routing configuration that is stored in the ROM of each broker. Note that most routing algorithms use an empty initial routing configuration [263]. Our algorithm can be seen as a periodic precautionary distributed reset [16].

Basic Idea

Each broker B maintains two routing tables T_B^0 and T_B^1, which are alternately rebuilt on a periodic basis, and a flag $a_B \in \{0, 1\}$ that determines which of both routing tables is currently rebuilt.[1] However, notification routing always uses both routing tables to determine the target destinations of a notification. A notification is forwarded to a destination if it matches a routing entry for

---

[1] An optimized solution can be implemented with only one table and two flags for every entry indicating to which routing table(s) the entry belongs.

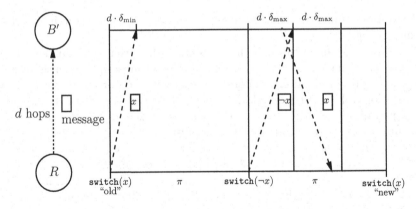

Fig. 8.7. Choosing π such that "old" and "new" update messages do not interleave

this destination in any of the two routing tables. If the routing tables are in a correct state, this does no harm.

Since $\mathcal{A}$ wraps $\mathcal{R}$, every call to $\mathcal{R}$ is intercepted by $\mathcal{A}$. This way, $\mathcal{A}$ determines which routing table the next call of $\mathcal{R}$ operates on in the following way: For every (un)subscription from a local client of B, $T_B^{a_B}$ will be used. If update messages are generated by $\mathcal{R}$ in reaction to the (un)subscription, they will be tagged with a_B. Accordingly, when a broker B' receives an update message tagged with x from a neighbor broker, then $T_{B'}^x$ will be used by $\mathcal{R}$ for this call.

The periodical rebuild is triggered by a modulo clock on the root broker R every π. The rebuild sets $a_R \leftarrow \neg a_R$. Then, it initializes $T_R^{a_R}$ with the initial routing configuration stored in ROM and propagates a switch(a_R) message to all of its neighbors. Similarly, when a broker B' receives a switch(x) message from a neighbor, it sets $a_{B'} \leftarrow x$, initializes $T_{B'}^{a_{B'}}$, and forwards a switch(x) message to all other neighbors. If a (un)subscription is issued twice by a client between two consecutive switch messages without an intervening unsubscription (subscription), this could raise a problem because $\mathcal{R}$ might not tolerate resubscriptions. To avoid this potential problem, a (un)subscription from a local client will be discarded by the wrapper algorithm $\mathcal{A}$ if it is redundant with respect to the contents of the currently active routing table.

Correctness

Before we show the correctness of our scheme, we prove a preparatory lemma.

Lemma 8.1. *In a correct system, if $\pi > 2d \cdot \delta_{max}$, no "old" update messages tagged with x can arrive at any broker after the root broker issued the next "new" switch(x) message.*

Proof. Old update messages tagged with x disappear at most $d \cdot \delta_{max}$ after the last broker has received the switch($\neg x$) message. This means that at most

$2d \cdot \delta_{\max}$ after the root broker has sent the $\texttt{switch}(\neg x)$ message no old update messages tagged with x can arrive. Since π is greater than this value, only new update messages tagged with x can arrive at any broker after the next new $\texttt{switch}(x)$ message is issued by the root broker (Fig. 8.7). $\square$

Theorem 8.1. *When the wrapper algorithm is applied and* $\pi \geq \rho + 2 \cdot d \cdot \delta_{\max}$ *holds, the publish/subscribe system is self-stabilizing and the stabilization time* Δ *is bounded by* $2 \cdot \pi + d \cdot \delta_{\max}$.

Proof. For the correctness, we have to show that (a) the system stays in a correct state if it currently is in a correct state and that (b) the system will eventually enter a correct state if it is currently in an incorrect state.

(a) For the system to stay in a correct state, we have to ensure that (a1) at each broker the rebuild process of the routing table which is currently rebuilt is completed before the next $\texttt{switch}$ message is received, that (a2) the rebuild is based only on new update messages, and that (a3) all new updates messages are received after the respective $\texttt{switch}$ message.

(a1) This means that at each broker the time between two consecutive $\texttt{switch}$ messages must be large enough to ensure that all necessary update messages are received in time. The time difference at which two brokers receive the same $\texttt{switch}$ message cannot be greater than $d \cdot \delta_{\max}$. At all brokers, the clients need at most ρ to reissue all their subscriptions after the broker has received the $\texttt{switch}$ message. The resulting update messages need at most $d \cdot \delta_{\max}$ to travel through the broker network. Therefore, $\pi \geq \rho + 2 \cdot d \cdot \delta_{\max}$ must hold to guarantee that at each broker the rebuild is complete before the next $\texttt{switch}$ message is received.

(a2) By Lemma 8.1 and the fact that $\pi \geq \rho + 2 \cdot d \cdot \delta_{\max}$.

(a3) Due to the FIFO property of the communication channels and the fact that the topology is acyclic, a broker B' can only receive update messages and (un)subscriptions of local clients tagged with x after B' received the corresponding $\texttt{switch}(x)$ message.

(b) Starting from an arbitrary state, every broker receives the next $\texttt{switch}$ message after at most $\pi + d \cdot \delta_{\max}$. This message causes the receiving broker to reinitialize one of its two routing tables. As a result of (a) it is guaranteed that this routing table will be completely rebuilt before the subsequent $\texttt{switch}$ message is received. This second $\texttt{switch}$ message is received by all brokers at most $2\pi + d \cdot \delta_{\max}$ from the beginning. It causes the other routing table to be reinitialized. After all brokers have received and processed the second $\texttt{switch}$ message, the system is guaranteed to be in a correct state again. This is because at all brokers the one routing table is completely rebuilt, while the other is reinitialized. Therefore, the stabilization time Δ is $2 \cdot \pi + d \cdot \delta_{\max}$ in the worst-case (Fig. 8.8). $\square$

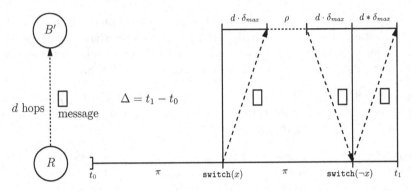

Fig. 8.8. Derivation of the maximum stabilization time

8.2.5 Further Reading

Many self-stabilizing algorithms have been proposed for various kinds of scenarios, while there are only a few contributions that cover publish/subscribe systems. Recently, Shen and Tirthapura [343] presented an alternative approach for self-stabilizing content-based routing. In their approach, all pairs of neighboring brokers periodically exchange sketches of those parts of their routing tables concerning their other neighbors to detect corruption. The sketches that are exchanged are lossy because they are based on bloom filters (which are a generalization of hash functions). However, because of the information loss, it is not guaranteed that an existing corruption is detected deterministically. Hence, the algorithm is not self-stabilizing in the usual sense. Moreover, although generally all data structures can be corrupted arbitrarily, the authors' algorithm computes the bloom filters incrementally. Thus, once a bloom filter is corrupted, it may never be corrected. Furthermore, in their algorithm, clients do not renew their subscriptions. Without this, corrupted routing entries regarding local clients are never corrected. Finally, their algorithm is restricted to simple routing in its current form.

8.3 Congestion Control

Many existing research prototypes of publish/subscribe systems make the assumption that event publication messages are negligible in size and therefore cannot saturate the available network bandwidth or processing power. However, this is not true in practice, and a publish/subscribe system can suffer from congestion leading to a degradation of service to clients. In this section we discuss the connection problem in the context of a publish/subscribe systems. We describe a scalable congestion control mechanism [313] that prevents the occurrence of congestion in a reliable publish/subscribe system. It consists of two algorithms, PDCC and SDCC, that are used in combination to address different aspects of congestion control in the event-based middleware. To motivate the need for congestion control in an event-based middleware, we begin with an overview of the congestion control problem and the requirements for a mechanism to handle congestion. The main part of this section is the description of the two congestion control algorithms as an example of how to perform congestion control in a publish/subscribe system.

8.3.1 The Congestion Problem

We argue that it is necessary to provide congestion control for overlay networks, such as the one established by a distributed publish/subscribe system. *Congestion* occurs when there are not enough resources to sustain the rate at which event publishers send publication messages in an event-based middleware. We distinguish between two kinds of congestion,

1. *network congestion*, where the network bandwidth between event brokers is the limiting resource
2. *event broker congestion*, when the processing of messages at an event broker cannot cope with the data rate

Both kinds of congestion may lead to the loss of messages at event brokers. Message loss is especially undesirable under guaranteed delivery semantics because the resulting retransmission of messages worsens the level of congestion in the system. An event-based middleware suffers from *congestion collapse* when the message loss dominates its operation and prevents event clients from receiving any useful service.

Usually there are two reasons for congestion in an event-based middleware. In many cases, congestion is caused by the underprovisioning of the deployed middleware in terms of network bandwidth or processing power of event brokers so that the middleware cannot handle resource requirements of event dissemination during normal or peak operation. A second, more subtle cause for congestion is the temporary need for more resources as a result of recovery after a failure under guaranteed delivery semantics.

Note that even though connections between event brokers use TCP congestion control, this is not sufficient to prevent congestion in the overlay broker network because of application-level queuing at event brokers. Both network and event broker congestion manifest themselves as the buildup of buffer queues at event brokers. To deal with congestion, current middleware deployments are often vastly overprovisioned, which is a waste of resources. Instead, a congestion control mechanism can address this problem directly.

8.3.2 Requirements

A congestion control mechanism in a publish/subscribe context differs from traditional congestion control found in other networking systems. This is due to the many-to-many communication semantics supported by the publish/ subscribe model and the content-based filtering of messages at application-level event brokers during event dissemination. Not all event subscribers receive the same set of publication messages sent by event publishers, as opposed to the case in application-level multicast, for example. Reliable event dissemination semantics leads to the selective retransmission of publication messages to a subset of recovering event subscribers, which further complicates congestion control. To guide the design of our congestion control mechanism, we formulate six requirements for congestion control in an event-based middleware:

Burstiness. The processing of publication messages at event brokers is bursty because of application-level scheduling and the variable processing cost of content-based filtering of event publications. This means that a congestion condition can arise quickly, requiring early detection by the congestion control mechanism.

Queue Sizes. Due to the burstiness of event routing and the need to cache event streams for retransmission, buffer sizes at event brokers are much higher compared to standard networking components. Buffer overflow only occurs when significant congestion already exists in the system. As a consequence, message loss cannot be used as an indicator for congestion in event-based middleware.

Recovery Control. The congestion control mechanism must ensure that event brokers that are recovering event publications that were previously lost will eventually complete recovery successfully. At the same time, recovering event brokers must be prevented from contributing to congestion. Although negative acknowledgment (NACK) messages are small and themselves cause little congestion, they potentially trigger the retransmission of large event publication messages.

Robustness. It is important that the congestion control mechanism is robust and can protect itself against malicious event clients. A possible design choice is to provide congestion control in the overlay broker network only, ensuring that the publication rate of messages by publisher-hosting brokers can be supported by all interested subscriber-hosting brokers. Flow control between client-hosting brokers and event clients is handled by a separate mechanism that can disconnect malicious clients.

Architecture Independence. The congestion control mechanism should not be tightly coupled to internal implementation details of an event broker. Instead, as a higher-level middleware service, it should support the evolution of the event broker implementation. For example, the detection of congestion should not depend on a particular buffer implementation or queuing discipline used by event brokers.

Fairness. When congestion requires the reduction of publication rates, fair throttling of event publishers must be ensured. The available resources at publisher-hosting brokers should be split equally among all hosted event publishers.

8.3.3 Congestion Control Algorithms

Typically a congestion control mechanism first detects congestion in the system and then adapts system parameters to remove its cause. In this section we describe two such algorithms that provide congestion control for a publish/subscribe system in accordance with the requirements stated in the previous section. The algorithms involve publisher-hosting brokers (PHB) and subscriber-hosting brokers (SHB).

1. A *PHB-driven congestion control algorithm* ensures that publisher-hosting brokers cannot cause congestion because of too high a publication rate. This is achieved by a feedback loop between publishers and subscribers to monitor congestion in the overlay broker network and control the event publication rate at the publishers.
2. An *SHB-driven congestion control algorithm* manages the recovery of subscribers after failure. It limits the rate of NACK messages that cause the retransmission of event publications from publisher-hosting brokers depending on congestion.

These two congestion control algorithms are independent of each other but should be used in conjunction to prevent congestion during both regular operation and recovery. Both algorithms need to distinguish between recovering and nonrecovering event brokers in order to ensure that subscribers can recover successfully. For a simpler presentation of the algorithms, we assume that only event brokers are internal nodes in event dissemination trees with client-hosting brokers constituting the root or leaf nodes. Next we will describe the two algorithms in turn.

PHB-Driven Congestion Control

The *PHB-driven congestion control algorithm* (PDCC) controls the rate at which new publication messages are published by a *publication endpoint* (pubend), such as a set of event publishers. The publication rate is adjusted depending on a *congestion metric*. We use the observed rate of publication messages at subscriber-hosting brokers as our congestion metric, which is similar to the throughput-based metric of TCP Vegas [50]. The rationale behind this is that a decrease in the message rate at a subscriber-hosting broker with an unchanged publication rate at the pubend is an indication of more queuing in the overlay broker network. This queue buildup is considered to be caused by network or event broker congestion in the system. Subscriber-hosting brokers calculate their own congestion metric and notify the publishers upstream whenever they believe that they are suffering from congestion. Congestion indications are aggregated at intermediate brokers so that the pubend is only informed of the worst congestion point. Two types of control messages are used to exchange congestion information between event brokers in an aggregated fashion.

Downstream Congestion Query (DCQ) Messages

The PDCC mechanism is triggered by DCQ messages sent by a publisher-hosting broker down the event dissemination tree to all subscriber-hosting brokers. Since congestion control is performed per tree, a DCQ message carries a tree identifier (`treeID`). A monotonically increasing `sequenceNo` is used for aggregation and the `mPos` field stores the current position in the event stream, which is, for example, the latest assigned event timestamp.

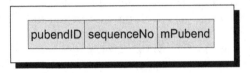

Upstream Congestion Alert (UCA) Messages

UCA messages are sent by subscriber-hosting brokers to inform about congestion. They flow upwards in the event dissemination tree and are aggregated at intermediate brokers so that a publisher-hosting broker only receives a single UCA message in response to a DCQ message. Apart from the tree identifier and the sequence number of the triggering DCQ message, a UCA message contains the minimum throughput rates observed at recovering (`minRecSHBRate`) and nonrecovering (`minNonRecSHBRate`) subscriber-hosting brokers.

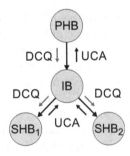

Fig. 8.9. Flow of DCQ and UCA messages

Figure 8.9 summarizes the propagation of DCQ and UCA messages through an overlay broker network in the PHB-driven congestion control algorithm. For the PDCC scheme to be efficient, DCQ and UCA messages must not suffer from congestion and should maintain low delays and loss rates. Next we describe the behavior of the three types of event brokers when processing DCQ and UCA messages in the PDCC algorithm.

Publisher-Hosting Broker (PHB)

A publisher-hosting broker triggers the PDCC mechanism by periodically sending DCQ messages with an incremented sequence number. The interval t_{dcq} at which DCQ messages are dispatched determines the time between UCA responses in a congested system. The higher the rate of responses, the quicker the system adapts to congestion.

When the PHB has not received any UCA messages for a period of time t_{nouca}, it assumes that the system is currently not congested. Therefore, it increases the publication rate if the rate is throttled and the pubend could publish at a higher rate. To increase the publication rate, we use a hybrid scheme with additive and multiplicative increase. The new rate r_{new} is calculated from the old rate r_{old} according to

$$r_{new} = \max \left[\, r_{old} + r_{min}, \, r_{old} + f_{incr} \cdot (r_{old} - r_{decr}) \, \right], \qquad (8.1)$$

where r_{decr} is the publication rate after the last decrease, f_{incr} is a multiplicative increment factor, and r_{min} is the minimum possible increase. The multiplicative use of f_{incr} allows the publication rate to grow faster than a fixed additive increase. However, when the publication rate is already close to the optimal operation point before congestion occurs, it is necessary to limit the increase. This is done by recording the publication rate r_{decr} at which the increase started and using it to restrict the multiplicative increase. This scheme results in the publication rate probing whether the congestion condition has disappeared and, if not, oscillating around the optimal operation point.

When the PHB receives a UCA message, a decision is made about a reduction of the current publication rate. The rate is kept constant if the sequence number in the received UCA message is smaller than the sequence number of the DCQ message that was sent after the last decrease. The reason for this is that the system did not have enough time to adapt to the last change in rate and therefore more time should pass before another adjustment. The rate is also not reduced if the congestion metric in the UCA message is larger than the value in the previous message. This means that the congestion situation in the system is improving, and further reduction of the rate is unnecessary. Otherwise, the publication rate is decreased according to

$$r_{new} = \max \left[\, f_{decr_1} \cdot r_{old}, \, r_{decr} + f_{decr_2} \cdot (r_{old} - r_{decr}) \, \right] \quad \text{iff} \quad r_{decr} \neq r_{old} \quad (8.2)$$

$$r_{\text{new}} = f_{\text{decr}_1} \cdot r_{\text{old}} \quad \text{otherwise,} \tag{8.3}$$

where f_{decr_1} and f_{decr_2} are multiplicative decrement factors. The first term in Eq. (8.2) multiplicatively decreases the rate by a factor f_{decr_1}, whereas the second term reduces the rate relative to the previous decrement r_{decr}. Similar to Eq. (8.1), the second term prevents an aggressive rate reduction when congestion is encountered for the first time after an increase. Since the PDCC mechanism constantly attempts to increase the publication rate in order to achieve a higher throughput, it will eventually cause SHBs to send UCA messages if there is resource shortage in the system, but this should not result in a strong reduction of the publication rate. Taking the maximum of the two decrement values ensures that the publication rate stays close to the optimal operating point. If the congestion situation does not improve after one reduction, the publication rate is reduced again. This time a strong multiplicative decrease according to Eq. (8.3) is performed because the condition $r_{\text{decr}} = r_{\text{old}}$ holds.

Intermediate Broker (IB)

To avoid the problem of feedback implosion [100], aggregation logic for UCA messages at intermediate brokers (IB)must consolidate multiple messages from different SHBs such that the minimum observed rate at any SHB is passed upstream in a UCA message. This enables the publisher-hosting broker to adjust its publication rate to provide for the most congested SHB in the system. Another requirement is that UCA messages that occur for the first time are immediately sent upstream, allowing the publisher-hosting broker to respond as quickly as possible to new congestion in the system.

In Fig. 8.10 the algorithm for processing DCQ and UCA messages at an intermediate broker is given. An IB stores the maximum sequence number `seqNo` and the minimum throughput values for nonrecovering (`minNonRecSHBRate`) and recovering (`minRecSHBRate`) SHBs from the UCA messages that it has processed. After the initialization of these variables (line 1), the function `processDCQ` handles DCQ messages by relaying them down the event dissemination tree in line 6. When a UCA message arrives, the function `processUCAMsg` is called, which first updates the throughput minima (lines 10–11). A new UCA message is only sent upstream if the sequence number of the received message is greater than the maximum sequence number stored at the IB (line 12). This ensures that UCA messages with the same sequence number coming from different SHBs are aggregated before propagation. The first UCA message with a new sequence number immediately triggers a UCA message so that the pubend is quickly informed about new congestion. Subsequent UCA messages from other SHBs that have the same sequence number will be aggregated and contribute toward the throughput minima in the next UCA message. After a UCA message has been sent in line 13, `seqNo` is updated (line 14) and both throughput minima are reset in line 15.

```
1   initialization:
2     seqNo ← 0
3     minNonRecSHBRate ← ∞
4     minRecSHBRate ← ∞
5
6   processDCQ(dcqMsg):
7     sendDownstream(dcqMsg)
8
9   processUCA(ucaMsg):
10    minNonRecSHBRate ←
11      MIN(minNonRecSHBRate, ucaMsg.minNonRecSHBRate)
12    minRecSHBRate ←
13      MIN(minRecSHBRate, ucaMsg.minRecSHBRate)
14    IF ucaMsg.seqNo > seqNo THEN
15      sendUpstream(ucaMsg.seqNo, minNonRecSHBRate,
16        minRecSHBRate)
17      seqNo ← ucaMsg.seqNo
18      minNonRecSHBRate ← ∞
19      minRecSHBRate ← ∞
```

Fig. 8.10. Processing of DCQ and UCA messages at IBs

The example in Fig. 8.11 demonstrates the operation of the aggregation logic at IBs. The topology of six event brokers has two congested event brokers, SHB_1 and SHB_2, and three intermediate brokers $IB_{1,2,3}$ that aggregate UCA messages. Congestion in the system is first detected by SHB_1, and its UCA message with a congestion metric of 0.8 is directly propagated to the PHB. When SHB_2 notices congestion, its UCA message is consolidated at IB_2, which updates its throughput minimum to 0.4. Eventually a UCA message with the congestion value of SHB_2 will propagate up the event dissemination tree in response to a new DCQ message because SHB_2 is more congested than SHB_1.

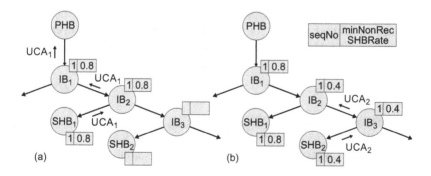

Fig. 8.11. Consolidation of UCA messages at IBs

Subscriber-Hosting Broker (SHB)

The congestion metric used by subscriber-hosting brokers depends on their observed throughput of publication messages and is independent of the actual publication rate of the pubend. An SHB monitors the ratio of PHB and SHB message rate,

$$t = \frac{r_{\text{pubend}}}{r_{\text{SHB}}}, \tag{8.4}$$

and uses this to decide when to send UCA messages with congestion alerts. To allow for burstiness in the throughput due to application-level routing as mentioned previously, t is passed through a standard first-order low-pass filter,

$$\bar{t} = (1 - \alpha)\,\bar{t} + \alpha\,t, \tag{8.5}$$

to obtain a smoothed congestion metric $\bar{t}$ with an empirical value of $\alpha = 0.1$. An SHB has to apply a different strategy for sending UCA messages depending on whether it is recovering event publications or not. We assume that an SHB can determine whether it is a recovering or a nonrecovering event broker. A suitable criterion to detect recovery would be, for example, that the SHB is ignoring new event publications because its event stream is saturated with old events caused by NACK messages.

Nonrecovering SHB. A nonrecovering SHB should receive publication messages at the same rate at which they are sent by the pubend. Therefore, if the smoothed throughput ratio $\bar{t}$ drops below unity by a threshold Δt_{nonrec},

$$\bar{t} < 1 - \Delta t_{\text{nonrec}}, \tag{8.6}$$

the SHB assumes that it has started falling behind in the event stream because of congestion. In rare cases, an SHB could be falling behind slowly because $\bar{t}$ stays below 1 but above $1 - \Delta t_{\text{nonrec}}$ for a long time. Unless there is already significant congestion in the system, this will not cause a queue overflow if buffer sizes are large. An SHB can detect this situation by periodically comparing its current position in its event stream m_{SHB} to the event stream position m_{tree} from the last received DCQ message. If the difference is larger than Δt_s,

$$m_{\text{SHB}} < m_{\text{tree}} + \Delta t_s, \tag{8.7}$$

a UCA message is triggered, even though the congestion metric $\bar{t}$ is above its threshold value.

Recovering SHB. A recovering SHB must receive publication messages at a higher rate than the publication rate, or it will never manage to successfully catch up and recover all lost publication messages. In some applications there is an additional requirement to maintain a minimum recovery rate $1 + \Delta t_{\text{rec}}$

in order to put a bound on recovery time. Thus, a recovering SHB sends a UCA message if

$$\bar{t} < 1 + \Delta t_{\mathrm{rec}}. \tag{8.8}$$

The threshold value Δt_{rec} influences how much of the congested resource will be used for recovery messages as opposed to new publication messages and hence controls the duration of recovery.

SHB-Driven Congestion Control

The *SHB-driven congestion control algorithm (SDCC)* manages the rate at which an SHB requests missed event publications by sending NACK messages upstream to the corresponding PHB. An SHB maintains a *NACK window* to decide which parts of the event stream to request. To control the rate of NACK messages being sent, the NACK window is open and closed additively by the SDCC algorithm depending on the level of congestion in the system. As for the PDCC mechanism, the change in recovery rate throughput is used as a metric for detecting congestion.

At the start of recovery, an SHB uses a small initial NACK window size $nwnd_0$. The NACK window is adjusted during recovery when the recovery rate r_{SHB} changes. The recovery rate r_{SHB} is defined as the ratio between the current NACK window size $nwnd$ and the estimate of the round trip time RTT, which it takes to retrieve a lost event publication from the pubend,

$$r_{\mathrm{SHB}} = \frac{nwnd}{RTT}. \tag{8.9}$$

The NACK window size is changed in a similar fashion to TCP Vegas. When the recovery rate r_{SHB} increases by at least a factor α_{nack}, the NACK window is opened by one additional NACK message per round trip time. When r_{SHB} decreases by at least a factor β_{nack}, the NACK window is reduced by one NACK message,

$$nwnd_{\mathrm{new}} = nwnd_{\mathrm{old}} \pm size_{\mathrm{nack}}. \tag{8.10}$$

This is sufficient to ensure that resent event publications triggered by NACK messages from recovering event brokers do not overload the publish/subscribe system.

8.3.4 Further Reading

A large body of work exists in the area of congestion control in networks, although these solutions do not address the special requirements for congestion control in an publish/subscribe system. In this section we provide a brief overview of applicable work, contrasting it with our approach for congestion control.

Transmission Control Protocol (TCP)

The TCP protocol comes with a point-to-point, end-to-end congestion control algorithm with a congestion window that uses *additive increase, multiplicative decrease (AIMD)* [211]. *Slow start* helps open the congestion window more quickly. Packet loss is the only indicator for congestion in the system, and *fast retransmit* enables the receiver to signal packet loss by ACK repetition to avoid timeouts. TCP Vegas [50] attempts to detect congestion before packet loss occurs by using a throughput-based congestion metric, which is similar to the congestion metric used in the PDCC and SDCC algorithms.

Reliable Multicast

Reliable multicast protocols are similar to reliable publish/subscribe systems due to their one-to-many communication semantics, but typically they have no filtering at intermediate nodes and do not guarantee that all leaves in the multicast tree will eventually catch up with the sender. In general, multicast congestion control schemes can be divided into two categories [407], namely:

1. *sender-based* schemes, in which all receivers support the same message rate
2. *receiver-based* schemes with different message rates by means of transcoded versions of data

Since we can make few assumptions about the content of event publications, a receiver-based approach is not feasible. Congestion control for multicast is often implemented at the transport level relying on router support. It must adhere to existing standards to ensure fairness and compatibility with TCP [149, 179]. Since there are many receivers in the multicast tree, scalable feedback processing of congestion information is important. Unlike *feedback suppression* [107], our approach does not discard information because it consolidates feedback in a scalable way.

The *PGMCC* congestion control protocol [328] forms a feedback loop between the sender and the most congested receiver. The sender chooses this receiver depending on receiver reports in NACK messages. The congestion control protocol for *SRM* [344] is similar except that the feedback agent can give positive and negative feedback, and a receiver locally decides whether to send a congestion notification upstream to compete for becoming the new feedback agent. An approach that does not rely on network support, except minimal congestion feedback in NACK messages, is *LE-SBCC* [376]. Here a cascaded filter model transforms the NACK messages from the multicast tree to appear like unicast NACKs before feeding them into an AIMD module. However, no consolidation of NACK messages can be performed. All these schemes have in common that they use a loss-based congestion metric, which is not a good indicator for congestion in an application-level overlay network.

Multicast Available Bit Rate (ABR) ATM

The ATM Forum Traffic Management Specification [334] includes an available bit rate (ABR) category for traffic though an ATM network. At connection setup, *forward and backward resource management* (FRM/BRM) cells are exchanged between the sender and receiver to create a resource reservation, which is modified at intermediate ATM switches. All involved parties agree on an acceptable cell rate depending on the congestion in the system. In our case, it is difficult to determine an acceptable message rate for an IB since the cost of processing event publications varies depending on size, content, and event subscriptions.

Multicast ABR requires flow control for one-to-many communication. An FRM cell is sent by the source and all receivers in the multicast tree respond with BRM cells, which are consolidated at ATM switches [329]. Different ways of consolidating feedback cells have been proposed [134]. These algorithms have a trade-off between timely response to congestion and the introduction of *consolidation noise* when new BRM cells do not include feedback from all downstream branches. Our consolidation logic at intermediate brokers tries to balance this trade-off by aggregating UCA messages with the same sequence number, but also short-cutting new UCA messages. The scalable flow control protocol in [409] follows a soft synchronization approach, where BRM cells triggered by different FRM cells can be consolidated at a branch point.

Overlay Networks

Congestion control for application-level overlay networks is sparse, mainly because application-level routing is a novel research focus. A hybrid system for application-level reliable multicast in heterogeneous networks that addresses congestion control is *RMX* [76]. It uses a receiver-based scheme with the transcoding of application data. Global flow control in an overlay network can be viewed as a dynamic optimization problem [13], in which a cost-benefit approach helps find an optimal solution.

8.4 Mobility

The emergence of mobile computing has opened up a whole new field of services provided for the benefit of the mobile user. Many such services can exploit the fact that the mobile device is aware of its current location. For example, car navigation systems use knowledge about current and past locations to aid drivers in finding their way through unknown cities. Location information can even be combined with other sources of data, e.g., the weather report, information on traffic jams, or free parking spaces. In such cases, the system can propose routes that avoid places where traffic is high or weather conditions are unpleasant, or can direct the driver to the nearest free parking space. All these are examples for *location-based services*.

A convenient way to construct location-based services is to build them using event infrastructures, such as those provided by publish/subscribe systems. Here, producers and consumers are enabled to exchange information based on message type or content rather than particular destination identifiers or addresses. This *loose coupling* of producers and consumers is the premier advantage of publish/subscribe systems, which facilitates mobile communication. Producers are relieved from managing interested consumers, and vice versa. In the following we study how these advantages can be exploited and what extensions are eligible in the context of mobile services.

We argue that support for mobility should be an issue of the publish/subscribe middleware itself and not be delegated to the application layer. Three kinds of application scenarios have to be supported: (i) existing applications in a static environment, (ii) existing applications in a mobile environment, and (iii) mobility-aware applications. Since publish/subscribe systems and applications have been deployed very successfully, extending existing systems and models is preferred to creating new "mobile" middleware from scratch in order to facilitate the integration of the first two scenarios. As a consequence, the middleware must transparently handle some of the new mobility issues. This allows existing event-based applications to directly interact with and even to be deployed as mobile applications. On the other hand, mobility-aware applications (the third scenario) require the middleware to support a semiautomated handling of location changes. If no such support is available, mobility is actually controlled by the application and not by the movement of the client.

We differentiate among support for two different and orthogonal types of mobility. The first type of mobility is called *physical* mobility, where clients may temporarily disconnect from the publish/subscribe system (due to power-saving requirements or the network characteristics). This means that applications are not necessarily aware of the fact that the client is moving, allowing existing applications to be transferred to mobile environments. The second type of mobility is called *logical mobility*, where clients remain attached to the their broker and have an application-level notion of location, which is described by *location-dependent subscriptions*. As an example, consider a car looking for a free parking space in the street it is currently driving along. In this situation it may subscribe to "New free parking space on Rebeca Drive". However, if Rebeca Drive is a very long street, the same driver will also receive notifications about free parking spaces very far down the road (or behind him), which are impossible to reach in good time. What the user would like to do is to specify a subscription such that he receives all notifications about "vacancies in the vicinity of his current location". We call these subscriptions *location-dependent*.

In this section we analyze and discuss the basic issues involved when adding mobility support to a publish/subscribe infrastructure. We identify and define two orthogonal forms of mobility (physical and logical mobility) and discuss the requirements of a system supporting both types of mobility.

8.4.1 Mobility Issues in Publish/Subscribe Middleware

Mobile clients have many characteristics, among them the need to disconnect from the network for different reasons. Be it for geographical, administrative, or power saving reasons, being connected to the same broker all the time is no longer possible. Hence, we have to take into account that clients will disconnect from their border broker once in a while. The middleware has to deal with moving clients and the possibility that a disconnected client reconnects at the same or a different broker later.

A first step toward mobility is to enhance existing publish/subscribe middleware to allow for roaming clients so that existing applications can be used in mobile environments. This means that the existing interface operations for accessing the middleware and the applications on top are not required to change. More important, the quality of service offered by the middleware must not degrade substantially. The resulting location transparency is necessary to make existing applications mobile, e.g., stock quote monitoring seamlessly transferred from PCs to PDAs.

On the other hand, future applications do not want complete transparency, but rely on *mobility awareness*. More specifically, mobility support should blend out unwanted phenomena, like disconnectedness, and enforce wanted behavior, like the location awareness in location-based services. Consequently, extending the interface of the publish/subscribe middleware to facilitate location awareness is a promising open issue, since most existing work concentrated on the transparency only.

When roaming, clients change (at least some portion of) the context they are operating in, and they might want to react to these changes, e.g., to adapt their subscriptions. However, an appropriate infrastructure support has to relieve the application from having to react "manually" to all changes. The middleware should rather offer an automated adaptation to context changes, i.e., facilitating location dependency. This leads to different notions of mobility and we distinguish:

- *Physical mobility*: A client that is physically mobile disconnects for certain periods of time and has different border brokers along its itinerary through the infrastructure. The main concern of physical mobility is *location transparency*.
- *Logical mobility*: A client that is logically mobile is aware of its location changes. In order to relieve the client from adapting *manually* to new locations, the main concern of logical mobility is *automated* location awareness within the publish/subscribe middleware.

Physical and logical mobility are two orthogonal aspects of mobility. Since the physical layout of a publish/subscribe system does usually not correspond to geographical realities, it seems reasonable to separate the two notions of mobility. In the following, we assume logical mobility to be a refinement of physical mobility in that a client remains connected to the same broker when roaming logically. The two notions have different quality of service requirements and therefore different solutions are developed to match both.

8.4.2 Physical Mobility

Physical mobility is similar to what in the area of mobile computing is called *terminal mobility* or *roaming*. A client accesses the system through a certain number of *access points* (GSM base stations, WLAN access points, or border brokers). When moving physically, the client may get out of reach of one access point and move into the reach of a second access point which are not necessarily overlapping. In general we cannot expect to have seamless access to the broker network but more a sequence of phases of connectedness, e.g., on the daily route between home and office. In this setting we analyze the quality of service requirements from the viewpoint of roaming clients:

- **Interface.** Obviously, the existing interface to the publish/subscribe system must not change as legacy applications are not aware of mobility.
- **Completeness.** Despite intermittent disconnects, the liveness condition of Def. 2.5 must be satisfied, i.e., a finite time after subscribing, the delivery of notifications that are published after this time and match the subscription is guaranteed.
- **Ordering.** In Sect. 2.5.3 FIFO-producer and causal ordering were discussed; they are eligible features in the mobile case, too.
- **Responsiveness.** The delay of relocating a roaming client should be minimal to maximize the responsiveness of the system. This has to be taken into account when designing a relocation protocol.

Possible Solutions

One solution would be to rely on Mobile IP [306] for connecting clients to border brokers, hiding physical mobility in the network layer. The drawback, however, is that the communication is also hidden from the publish/subscribe middleware, which is then not able to draw from any notification delivery localities or routing optimizations, thereby possibly violating the requirement of responsiveness. Such an approach might only be feasible if the physical and logical layout of a given system is completely orthogonal.

A different, naïve solution to implement physical mobility would be to use sequences of *sub-unsub-sub* calls to register a client at a new broker. When a client moves from border broker B_1 to B_2, it simply unsubscribes at B_1 and resubscribes at B_2, without any support in the middleware. But a client may not detect leaving the range of a broker and is in this case not able to unsubscribe at its old location. Even more severely, during its

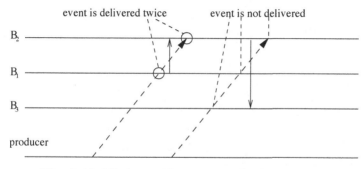

Fig. 8.12. Missing notifications in a flooding scenario

time of disconnectedness, the client might miss several notifications or or get duplicates, even if notifications are flooded in the network and the location change is instantaneous. This problem is depicted in Fig. 8.12. Hence, this solution is not complete and we outline an algorithm in Sect. 8.4.2 that takes into account all requirements stated above. The complete algorithm is detailed by Zeidler and Fiege [408].

Notification Delivery with Roaming Clients

In this section we introduce an algorithm for extending standard brokers (cf. Chap. 4) to cope with mobile clients, maintaining their subscriptions as well as guaranteeing the required quality of service as described in the previous section. Apart from guaranteeing complete notification delivery, our algorithm also ensures that the old border broker will eventually receive an equivalent to an explicit *sign-off* from the client, even if an explicit unsubscribe was not possible.

Our mechanism uses a natural way of distributed caching, which seems in general preferable to a potentially problematic central caching proxy.

Prerequisites

The solution sketched below can be used in every environment that meets the following requirements:

1. Border brokers have to install and maintain a buffer for all notifications that are not yet delivered in order to deal with disconnects.
2. The underlying routing infrastructure uses advertisements. Although not strictly necessary, the relocation effort is reduced substantially in that they guide the search for the old delivery path. Simple routing is assumed as routing strategy for now, and more advanced routing algorithms are discussed later.
3. Border brokers or clients must have some means of detecting the new configuration that a client has entered the range of the broker. Some form of beacon or heartbeat is presupposed; we do not go into the details here.

4. For now, we assume that only subscribers are mobile and that clients acting as producers remain stationary.

Algorithm Outline

We use a stepwise refinement of traditional subscription forwarding, as discusses in Chap. 4, to devise the algorithm:

1. When reconnecting to a broker, subscriptions are automatically reissued so that clients do not need to resubscribe manually.
2. The broker network configuration is updated to accommodate to client relocation rather than handling an independent new (re)subscription from a new location.
3. Notifications forwarded to the old location have to be replayed to the new one in order to bridge disconnectedness.
4. Delivery of new notifications has to be postponed until the replay is finished. In this way, moving does not influence the FIFO-producer order of notifications, fulfilling the ordering requirement.

Consider the scenario of Fig. 8.13(a) with a single consumer. Client C is moving from broker B_6 to broker B_1 (step 1 in the figure). The local broker, which resides on the client, e.g., in the form of libraries, is informed by the new border broker (i.e., B_6) about its relocation, according to the prerequisites. The local broker then reissues the active subscriptions, which were previously forwarded through and recorded in the local broker anyway. By avoiding manual resubscriptions of the client application, the first requirement stated at the beginning of this section is achieved, i.e., the interface to the middleware is not changed.

In the second step, we enable the publish/subscribe middleware to relocate the client. The goal of the relocation process is to update the routing configuration by redirecting the delivery paths currently leading to the old destination of C to the new destination. During this process, reissued subscriptions are propagated as usual, e.g., in the direction of any received advertisement if advertisements are used, through B_2 and B_3 to broker B_4, setting up their routing tables. At B_4 the old and new path from producer P to client C meet (dotted and dashed line, respectively). Broker B_4 is aware of the junction because an entry of the old path of this subscription/ client is already in its routing table.[2] When the routing table in the junction is updated, new published notifications will be delivered to the relocated client. Without assuming any knowledge about the old location of the moving client, the system is able to draw from localities in that only a portion of the delivery path is changed. Changes are limited to the smallest subgraph necessary for diverting routing paths, facilitating the timeliness/efficiency requirement, which is only available with inherent middleware support.

---

[2] Subscriptions can be identified if simple routing is used.

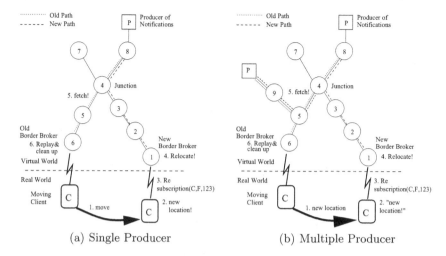

(a) Single Producer   (b) Multiple Producer

Fig. 8.13. Moving client scenarios with one and multiple producers

The third step ensures completeness over phases of disconnectedness during movement. The junction broker B_4 sends a fetch request along the old path to B_6 following the routing table entries for the given subscription. All brokers along this path update their routing tables such that they are pointing into the direction the fetch originates from, i.e., B_4. Border broker B_6 as last recipient replays all buffered notifications. If delivered notifications are annotated with sequence numbers by the border broker, reissued subscriptions can in turn carry the last received number to qualify the replay. Note that replays are forwarded only in the direction of a specific subscription and do not mingle with other clients' data. After replaying, the path from the old broker to the junction broker can be shut down by deleting the subscription's routing table entries as long as advertisement and routing entry point into the same direction; thereby excluding and stopping at the junction. In this way the notifications that passed the junction broker before its update are collected and sent toward the new location, ensuring the required completeness.

The last step finally reorders the notifications so that the sender FIFO condition remains valid after relocation. The new border broker has to block and cache all incoming notifications that are to be delivered to the given client (not impeding communication of other clients) until the replay is finished. As with all buffering, consistency can always only be guaranteed for a predefined, finite amount of time or space.

Figure 8.13(b) shows a scenario with multiple producers. In this case, several junctions exist which all lie on the path from the first junction to the old border broker of the client. For the two producers, the junctions are at brokers B_4 and B_5, respectively.

Extensions

Mobile Producers. So far we have assumed that only consumers can be mobile. When a producer is mobile, the notifications it publishes while it is disconnected from the system are not forwarded but are queued by its local broker. When the producer reconnects to a new border broker, the local broker reissues the advertisements currently active, while still queuing newly published notifications. The forwarding of these advertisements will in turn lead to overlapping subscriptions being forwarded to the new location of the producer. When this process has finished, the queued notifications are forwarded if a matching subscription exists or they are discarded, otherwise. Then, the normal handling of published notifications starts again. Delivery paths that lead to the old location of the producer and which are no longer needed are similarly dropped as described above.

Covering-Based Routing. If covering-based routing instead of simple routing is used, the fetch phase of the algorithm has to be extended. Now, the junction is reached if an entry with a covering subscription $F' \supset F$ is already registered. At this point the delivery path to the new location is correctly built up, but we do not know whether the old location lies in the direction of F' or in the direction of the advertisements. The fetch phase is extended in that fetch requests are sent toward all advertisements and all covering subscriptions; it is a kind of flooding in the overlay network of matching producers and consumers of similar interests. Only one of the fetch requests will not get dropped and reach the old border broker. The replay has to be flooded in the same overlay network if no tunneling mechanisms, internal or external, are used.

Merging-Based Routing. The extension for covering stated above can also cope with a broker network applying merging. Only the number of potential covers increases, and hence the size of those parts of the overlay network that are flooded. Both covering and merging promise to increase routing efficiency, but, on the other hand, aggravate relocation management.

Movement Speed. For simplicity reasons we assume that the client's movement speed is not too fast for the relocation process to terminate before the client moves again, i.e., the process always terminates at the correct broker. However, if resubscriptions of the local broker are annotated with a relocation counter, which is reset after a successful replay, concurrent relocation processes can be identified and controlled in the middleware, avoiding the speed limit.

Cache Management. Even if storage constraints in the border brokers are not of concern, mobile clients may be disconnected for a long period of time in which more missed notifications are cached than the client can handle during replay. The possibly limited resources of mobile clients must be taken into account when designing cache sizes or limiting the replay by semantic filtering [195].

Discussion

The above algorithm shows how relocation and adaptation of the delivery paths is performed in a fully distributed fashion. Many optimizations exist for this algorithm (e.g., [59, 92]). They typically reduce the number of necessary messages, but they also impose further constraints on network layout or require additional information about client movement. The approach presented here is a generalization that is robust and simple. Its central features are:

- **No explicit moveOut.** The algorithm ensures by design that the new broker can identify a relocated client and handle this appropriately. Moreover, the algorithm ensures that the broker at the old location eventually receives an equivalent to a moveOut for proper garbage collection.
- **No central caching proxy.** The algorithm is fully distributed and buffers information wherever necessary, thereby drawing optimally from localities.
- **No information loss.** By buffering information appropriately, the algorithm ensures that no information is lost due to relocation. As with all buffering schemes, this is only true modulo space and/or time constrains.
- **No "out-of-band" communication.** All messages sent related to a relocation process are sent *explicitly* within the broker network and not leaving the paradigm of publish/subscribe. Therefore, we do not need globally unique sequence numbers and can guarantee FIFO-producer ordering as well as not sending duplicates.
- **Optimal use of localities.** The algorithm draws optimally from localities and ensures that only the least necessary subgraph is reconfigured.

8.4.3 Logical Mobility

While physical mobility is a rather technical issue invisible to the application, logical mobility involves location awareness. An example for logical mobility is when clients move around a house or building that is served by only one border broker. In this case, the user might be interested to receive just those notifications that refer to the room in which he is currently located. Note that a client can be both logically and physically mobile at the same time.

A logically mobile client moving from one location to another, e.g., from one room to the other in a company building, will expect a frictionless change of location explicitly without a notable setup time after having changed from its own office to the conference room next door. The adaptation of some location-dependent subscription should take place "instantaneously". Intuitively, we would like to experience the notion of being subscribed to "everything, everywhere, all the time" and increase the reactivity of the system to moving clients.

Location-Dependent Filters

A publish/subscribe system offering *location-dependent filters* has the same interface as a regular publish/subscribe system (i.e., it offers the *pub*, *sub*, *unsub*, *notify* primitives). However, in specifying subscription filters for name/value pairs referring to *"location"*, it supports a new primitive to specify things like "all notifications where the attribute *location* equals my current location". More precisely, we postulate a specific marker *myloc* that can be used in a subscription. The marker stands for a specific set of locations that depend on the current location of the client. For example, a client could issue a subscription for all free parking spaces in the vicinity of his current location as follows: (*service* = "parking"), (*location* ∈ *myloc*), (*car-type* ≥ "compact").

The set of locations associated with the marker is taken from a particular range L of locations. This set is application dependent and can, for instance, contain all the different rooms of a building, all the streets of a town, or all the geographical coordinates given by a GPS system up to a certain granularity. Given a notification with the attribute *location*, the subscription (*location* ∈ *myloc*) will evaluate to true for a particular client at location y iff $x \in myloc(y)$, where $myloc(y)$ is the specific set of locations associated with y. Then, we say that the notification matches the location-dependent filter.

The simplest form of $myloc(y)$ is simply the set $\{y\}$. In this case a notification matches the subscription if $x = y$. But in the car example, the car driver looking for a parking space might want to specify:

$$(location = \text{``at most two blocks away from } myloc\text{''})$$

Then, *myloc* corresponds to all elements of L that satisfy this requirement.

A Tentative but Incomplete Solution for Logical Mobility

While location-dependent filters are not directly supported by current publish/subscribe middleware, one might argue that it is not very difficult to emulate them on top of currently available systems in this case. The idea would be to build a wrapper around an existing system that follows the location changes of the users and transparently unsubscribes to the old location and subscribes to the new one when the user moves. However, depending on the internal routing strategy of the event system, it may lead to unexpected results. The routing strategies deployed in many existing content-based event systems such as SIENA [71], Elvin [341], and REBECA [136] lead to *blackout periods* where no notifications are delivered. The problem is that it usually takes a significant time delay to process a new subscription. After subscribing to a filter, it takes some time t_d until the subscription is propagated to a potential source. Then it takes at least another t_d time until a notification reaches the subscriber. This phenomenon is depicted in Fig. 8.14. (Note that the delay t_d may be different for different notification sources and may change over time.) If the client remains at any new location less than $2t_d$ time, then the subscriber will "starve", i.e., it will receive few or no notifications.

An intuitive but inefficient solution

Another basic solution that can immediately be built using existing technology is again based on flooding. The local broker can then decide to deliver a

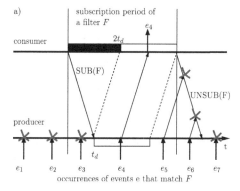

Fig. 8.14. Blackout period after subscribing with simple routing

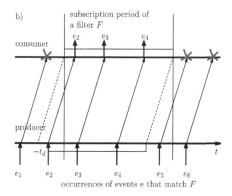

Fig. 8.15. Blackout period with flooding and client-side filtering

notification to a client depending on the client's current location (Fig. 8.15). Obviously, flooding prevents the blackout periods, which were present in the previous solution, but it should be equally clear that flooding is a very expensive routing strategy, especially for large pub/sub systems [267].

Quality of Service of Logical Mobility

Interestingly, while flooding is very expensive and therefore not desirable, it comes very close to the quality of service that we would like to achieve for logical mobility, namely to the notion of being subscribed to "everything, everywhere, all the time". The problem is that it is hard to precisely define the behavior of flooding without reverting to some unpleasantly theoretical constructions of operational semantics.

With logical mobility there is, however, no danger of receiving a notification twice because the consumer remains attached to the same "delivery path". The quality of service we require for logical mobility therefore is simply stated as follows: On change of location from x to y, all notifications should be delivered to the consumer "as if" flooding were used as underlying

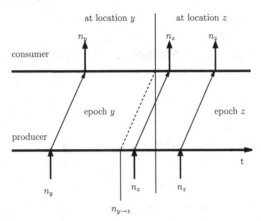

Fig. 8.16. Defining the quality of service for logical mobility using virtual notifications $n_{y \to z}$ that arrives at the consumer just at the time of the location change from y to z

routing strategy. This statement is made a little more concrete in Fig. 8.16, where the sequence of notifications generated by any consumer is divided into epochs that correspond to when the notification actually arrives at the consumer (the epoch borders between locations y and z are drawn as a virtual notification $n_{y \to z}$). We require that all notifications matching the current location-dependent subscription from every such epoch must be delivered. Intuitively, the epochs define the semantics of flooding.

Location-Dependent Filters for Logical Mobility

We now describe the algorithmic solution to the scenario where clients are only logically mobile, i.e., they remain attached to a single border broker.

Main Idea

Consider an arbitrary routing path between a producer and a consumer. This path consists of a sequence of brokers $B_1, B_2, \ldots, B_{k-1}, B_k$, where B_1 is the local broker of the consumer and B_k is the local broker of the producer (Fig. 8.17 shows the setup for $k = 3$). Assume the consumer has issued a location-dependent subscription F. Using the "usual" content-based routing algorithms, the current value $\tilde{F}$ of F, which instantiates the marker variable with the current location, would permeate the network in such a way that the filters along the routing path allow a matching subscription published by the producer to reach the consumer. Formally, the filters $F_1, F_2, \ldots, F_k$ along the links between the brokers should maintain a set-inclusion property (cf. Sect. 4.3.2))

$$F_k \supseteq F_{k-1} \supseteq \ldots \supseteq F_2 \supseteq F_1 \supseteq F_0 = \tilde{F}.$$

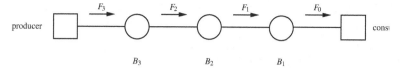

Fig. 8.17. Network setting for the example

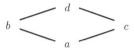

Fig. 8.18. Movement graph defining movement restrictions of a consumer

Obviously, if for any new value $\tilde{F}$ of F a new subscription must flow through the network toward the producers, notifications published in the meantime might go unnoticed. The idea of the proposed scheme is to always have the local broker of the consumer do perfect client-side filtering (i.e., set $F_0 = \tilde{F}$), but to let possible future notifications reach brokers that are nearer to the consumer so that their delay to reach the consumer is lower once the consumer switches to a new location.

Let T denote the set of time values, which for simplicity we will assume to be the set of natural numbers $\mathbb{N}$. Let L denote the set of all consumer locations. Then we define a function $loc : T \to L$ that describes the movement of the consumer over time. For example, for a location set $L = \{a, b, c, d\}$ a possible value of loc is $\{(1, a), (2, b), (3, d), \ldots\}$, meaning that at time 1, the consumer's location is a, at time 2 it is b, and so on.

We assume that loc is subject to some movement restrictions, which in effect define a maximum speed of movement for the consumer. We assume that such a restriction is given by a *movement graph* such as the one depicted in Fig. 8.18. The graph formalizes which locations can be reached from which locations in one movement step of the consumer. One movement step has some application-defined correspondence to one time step.

Given the function loc and a movement graph, it is possible to define a function $ploc : L \times \mathbb{N} \to 2^L$ of possible (future) locations (the notation 2^L denotes the powerset of L, i.e., the set of all subsets of L). The function takes a current location x and a number of consumer steps $q \geq 0$ and returns the set of possible locations, which the consumer could be in starting from x after q steps in the movement graph.

Since a possible move of the consumer always is to remain at the same location, for all locations $x \in L$ and all $q \in \mathbb{N}$ we should require that

$$ploc(x, q) \subseteq ploc(x, q + 1). \tag{8.11}$$

Taking the example values from above, possible values for $ploc$ are as follows:

$$ploc(a, 0) = \{a\} \qquad ploc(a, 1) = \{a, b, c\} \qquad ploc(a, 2) = \{a, b, c, d\}$$

Now, if the consumer is at location a, for example, every broker B_i along the path toward a producer should subscribe for $ploc(a, q)$ for some q, which is an increasing sequence of natural numbers depending on i and the network characteristics. If the time it takes for a broker to process a new subscription is on the order of the time a client remains at one particular location, then the individual filters F_i along the sample network setting in Fig. 8.17 should be set as $F_i = ploc(a, i)$, e.g., $F_0 = ploc(a, 0) = \{a\}$, $F_1 = ploc(a, 1) = \{a, b, c\}$, and so on. This requirement should be maintained throughout location changes by the consumer. For example, whenever a consumer moves from an old location x to a new location y, this will cause B_1 to change the location-dependent part of filter F_0 for client-side filtering from the old to the new location. Broker B_1 updates its routing table appropriately.

In general, broker B_i sends a message with the new location to B_{i+1} instructing it to change F_i from $ploc(x, i)$ to $ploc(y, i)$ and consequently to update the routing table by removing certain locations and adding new locations. Removing and adding new locations corresponds to unsubscribing and subscribing to the corresponding filters. The normal administration messages can be used to do this. Note that Eq. (8.11) guarantees the subset relationship, which should always hold on every path between a producer and a consumer.

Example

As an example, consider the value of *loc* where at time 1 the client is in location a, at time 2 at b, and at time 3 at d in the movement graph depicted in Fig. 8.18. Table 8.1 gives the values of *ploc* for all locations and the first four time instances. For $t = 0$ the value of *ploc* is equal to the current location. For $t = 1$ it returns all locations reachable in one time step in the movement graph, etc.

Table 8.1. Values of $ploc(x, t)$ for the example setting

t	$x = a$	$x = b$	$x = c$	$x = d$
0	$\{a\}$	$\{b\}$	$\{c\}$	$\{d\}$
1	$\{a, b, c\}$	$\{a, b, d\}$	$\{a, c, d\}$	$\{b, c, d\}$
2	$\{a, b, c, d\}$	$\{a, b, c, d\}$	$\{a, b, c, d\}$	$\{a, b, c, d\}$
3	$\{a, b, c, d\}$	$\{a, b, c, d\}$	$\{a, b, c, d\}$	$\{a, b, c, d\}$

Now assume again the setting depicted in Fig. 8.17. The values of Table 8.1 directly determine the filter settings for $F_0, \ldots, F_3$ as shown in Table 8.2. At time $t = 1$ the client moves to location b. This means that F_0 changes from $\{a\}$ to $\{b\}$ and that F_1 must unsubscribe to c and subscribe to d, yielding $F_1 = \{a, b, d\}$. At time $t = 2$ the client moves to d, causing F_0 to change to $\{d\}$ and F_1 to unsubscribe to a and subscribe to c. All other filters remain unchanged.

Table 8.2. Values of filters in example setting

time t	F_3	F_2	F_1	F_0
0	$\{a, b, c, d\}$	$\{a, b, c, d\}$	$\{a, b, c\}$	$\{a\}$
1	$\{a, b, c, d\}$	$\{a, b, c, d\}$	$\{a, b, d\}$	$\{b\}$
2	$\{a, b, c, d\}$	$\{a, b, c, d\}$	$\{b, c, d\}$	$\{d\}$

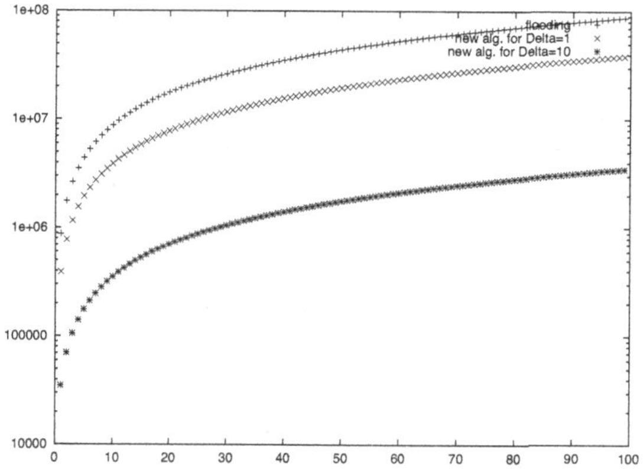

Fig. 8.19. Total number of messages generated for flooding and two scenarios of the new algorithm ($\Delta = 1s$ and $\Delta = 10s$). Note that the y-axis has a logarithmic scale. The x-axis denotes time in seconds

The example nicely shows that the method does some sort of "restricted flooding", i.e., all notifications reach broker B_2, but from there the uncertainty is restricted and so is the flow of notifications forwarded by B_2. In fact, the method described above using the *ploc* function can be regarded as an abstraction of both "trivial" implementations discussed (i.e., both implementations are instantiations of our scheme).

We have informally analyzed the total number of messages (notifications and administrative messages) generated by our new algorithm for an arguably realistic network setting, exactly one consumer and two different speeds of consumer movement: fast movement ($\Delta = 1s$) and slow ($\Delta = 10s$). We compare the results of these calculations with the total number of messages generated by flooding in Fig. 8.19 (see [143] for a detailed description of the system assumptions and the derivation of these numbers). It is interesting to see that although our algorithm generates administrative messages on all network links for every location change of the consumer, the fraction of messages saved is still considerable. We also note that many of the assumptions made in calculating these figures have been very conservative. For example, we assume that there is only one consumer in the network and that notifications are generated by the producers according to a uniform distribution over set of locations. Both assumptions prevent routing strategy optimizations to play to their strengths.

Concluding Mobility

The presented approach to support mobility in publish/subscribe middleware can only be seen as a first start for generic mobility support. We have analyzed the problem of mobility from the viewpoint of the event-based paradigm and have identified two separate flavors of mobility. While physical mobility is tied to the notion of rebinding a client to different brokers and can be implemented transparently, logical mobility refers to a certain form of location awareness offering a client a fine-grained control over notification delivery in the form of location-dependent filters.

Many other interesting problems concerning the combination of mobility and publish/subscribe infrastructures remain. For example, location-dependent filters may be generalized to "dynamic filters" that depend on a function of the local state of the client (not only its current location), like a client interested in receiving notifications for sales that he still can afford.

8.4.4 Further Reading

Further details on the movement algorithms can be found in [141, 142, 408]. Work on middleware for mobile computing usually concentrated on classical synchronous middleware like CORBA. Only recently, position papers have stated that publish/subscribe systems have an enormous potential to better accommodate the needs of large mobile communities [89, 208]. Research in publish/subscribe systems has mainly focused on *static* systems, where clients do not move and the publish/subscribe infrastructure remains relatively stable throughout the system's lifetime, e.g., Elvin [341], Gryphon [197], REBECA [144], and SIENA [71]. If present at all, mobility support is a concern of the application layer. Applications detect the need to change a subscription and have to react explicitly and manually to this detection.

Huang and Garcia-Molina [195, 196] provide a good overview of possible options for supporting mobility in publish/subscribe systems. They describe algorithms for a "new" middleware system tailored and optimized to mobile and ad hoc networks, not so much an extension of an existing system. Cambridge Event Architecture (CEA) [20] and JEDI [92] also address problems of mobility. JEDI uses explicit moveIn and moveOut operations to relocate clients. Hence, mobility is controlled by the application, which is not transparent and even is unrealistic since clients usually only can react *after* having been moved. The mobility extensions of SIENA [59] are very similar. Explicit sign-offs are required and interim notifications stored during disconnectedness are directly forwarded to a new location upon request. Cugola et al. [89] proposes a leader election and group management protocol for dynamic dispatching trees to dynamically adapt the internals of the JEDI event system, their implementation model is based on multicast and it groups identical subscribers. An extension for Elvin allows for disconnectedness using a central caching proxy [368], which is a potential performance bottleneck. Jacobsen [208] presents some very interesting ideas on location-based services and the possible expressiveness of subscription languages. STEAM [257] is an event service designed for wireless ad hoc networks. Subscribers consume only events produced by geographically close-by publishers. It relies on proximity-based group communication.

9

Existing Notification Services

In this chapter we describe some standards (Sect. 9.1), commercial systems (Sect. 9.2), and research prototypes (Sect. 9.3) that are closely related to event-based systems.

9.1 Standards

In this section we describe standards which are related to event-based systems. This includes the CORBA Event Service, the CORBA Notification Service, the Java Message Service (JMS), and the Data Distribution Service (DDS).

9.1.1 CORBA Event and Notification Service

The *Common Object Request Broker Architecture* (CORBA) [283] is a platform- and language-independent object-oriented middleware architecture fa-

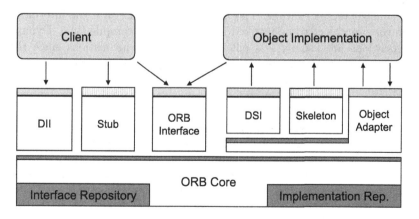

Fig. 9.1. Internal structure of an object request broker (ORB)

cilitating interoperability. CORBA is standardized by the Object Management Group (OMG), and vendors can implement the specification with their products. CORBA is a mature middleware technology that is widely used in financial and telecommunication systems and has inspired many recent middleware initiatives. Some reasons for CORBA's success are its good programming language integration across several mainstream languages, the extensibility of the platform using object services, and its adaptation to heterogeneous distributed systems. The CORBA specification describes the functionality, structure, and the interfaces of the *object request broker (ORB)*. An ORB consists of the following main components (cf. Fig. 9.1):

Object Request Broker (ORB) Core. The ORB core forms the heart of the middleware and handles communication. It resolves object references to locations, performs the marshaling and unmarshaling of method parameters, and sends invocations and results over the network.

Interface Definition Language (IDL). Interfaces of remote objects are defined in IDL, which is purely declarative and is independent of the programming language(s) used for implementations. IDL supports the usual primitive (integers, floats, etc.) and composite data types (e.g., structs). Programming language mappings define how an IDL type is mapped to a type of the programming language that is used.

Static Invocation Interface (SII). An IDL compiler transforms the static interface definitions given in IDL into client-side stub and server-side skeleton source code (in a given programming language). The stubs are called by the client, do the marshaling and unmarshaling of method arguments and method results, and pass the results back to the client. The stubs are also called *Static Invocation Interface (SII)* because their use requires the interface of the called object to be known at compile time. This approach has the advantage that remote method invocations can be statically type-checked by the compiler.

Dynamic Invocation Interface (DII). The DII allows the remote call to be constructed at runtime. This is, for example, useful if the interface of the remote object to be invoked is not known at compile time. Dynamic invocations are usually less efficient than static invocations since they require more code and type-checking must be done at runtime. The server-side complement of the DII is the *Dynamic Service Interface (DSI)*. DII and DSI can together be used for implementing general-purpose gateway, proxy, or browser objects.

Object Adapter (OA). An object adapter is interposed between the ORB and the skeletons. The OA dispatches upcalls received from the ORB to the skeleton of the called object implementation or to the DSI. Other responsibilities of the OA include generation and interpretation of object references, security of interactions, object and implementation activation and deactivation, mapping object references to implementations, and registration of implementations. There can be different types of OAs. The *Portable Object Adapter (POA)* is currently most commonly used.

Interface Repositories. The interface repository contains the IDL definitions of interfaces. The repository can be queried either at compile time or at runtime.

Implementation Repository. The implementation repository contains all implementations of a remote interface at the server side so that remote objects can be located and activated on demand.

Although the need for asynchronicity has been recognized by the OMG, the core design of CORBA is still based on synchronous communication. Before the *Asynchronous Message Invocation (AMI)* was standardized, the only possibility to issue asynchronous two-way calls had been to use deferred synchronous calls, which depend on the tedious dynamic invocation interface. With the static invocation interface, only one-way calls had been possible which provide best-effort method invocations not expecting a return value and thus not requiring blocking. The AMI closes this gap and enables asynchronous two-way calls using the SII. It supports two models, the *polling model* and the *callback model*. In the polling model, the issuer of a call can poll a collocated value-type object to test whether or not the results are now available. In the callback model, the results are delivered to the client by calling a handler method with the results as parameters.

The CORBA platform is extensible by means of object services that address different facets of a distributed computing environment, ranging from transactional support to security. In the next sections, we will take a closer look at the CORBA Event and Notification Services that explicitly deal with anonymous asynchronous communication by providing publish/subscribe functionality.

CORBA **Event Service**

The OMG acknowledged the need for publish/subscribe communication by introducing the CORBA *Event Service* [277] as a CORBA service in 1994. The current version as of 2005 is 1.2 [285]. With the Event Service, communication among suppliers and consumers can be in *push mode*, in which case a supplier pushes data to a consumer, or in *pull mode*, in which case a consumer requests data from a supplier. Instead of communicating directly with each other, consumers and suppliers are decoupled by an *event channel*. This way it is possible to use push and pull communication at both sides.

The Event Service specification supports two models: typed and untyped event communication. With the untyped model, which is most common, events are of the CORBA datatype **any** and can thus contain any IDL datatype. Suppliers can call **push** on the **PushConsumer** interface to deliver data and pull-based consumers can call **pull** on the **PullSupplier** interface to get data (Fig. 9.2). Since consumers and suppliers are decoupled by the event channel, they call these methods not on each other but on the event channel's interface

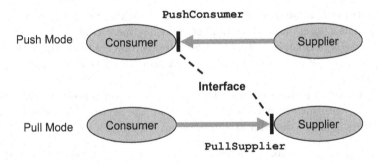

Fig. 9.2. Push mode vs. pull mode (typed event communication)

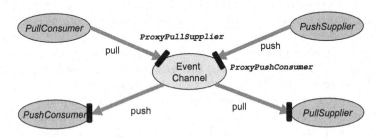

Fig. 9.3. Typed event communication using an event channel

(Fig. 9.3). For typed event communication, which is less common, suppliers and consumers agree on a particular IDL interface and use its methods to exchange information in pull or in push mode.

The Event Service enables CORBA clients to participate in many-to-many communication through an event channel. However, the asynchronous communication is implemented on top of CORBA's synchronous method invocation and thus has the substantial overhead of performing a remote method invocation for every event communication. Moreover, event consumers cannot filter the events they receive from an event channel because no event filtering is supported. In particular, the lack of filtering mechanisms has led to the development of the Notification Service, which can be seen as the successor of the Event Service.

CORBA Notification Service

As the successor of the Event Service, the CORBA *Notification Service* [287] addresses the shortcomings of the Event Service by providing event filtering, quality of service (QoS), and a lightweight form of typed events, called *structured events*. With the Notification Service, suppliers can discover which event types are currently required by all consumers of a channel so that suppliers can produce events on demand, or avoid transmitting events in

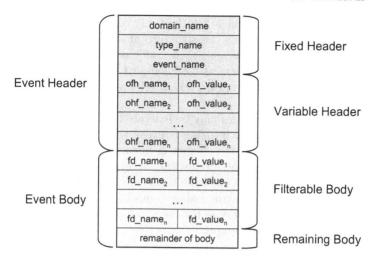

Fig. 9.4. The structure of a structured event (from [287])

which no consumers have interest. Similarly, consumers can discover all event types offered by suppliers so that consumers may subscribe to new event types as they become available. The Notification Service Specification also addresses an optional event type repository that, if present, makes information about the structure of events which may flow through the channel available.

Structured events are divided into a header and a body (Fig. 9.4). The header consists of a fixed and an optional variable header, while the body comprises a filterable and the remaining body. The fixed header contains the domain name, the type, and the unique name of the event. The variable header and the filterable body both contain name/value pairs that hold the data associated with the event. Event consumers can restrict the events that they receive from the event channel by specifying filters over the name/value pairs. The Notification Service Specification requires that an implementation support the *Default Filter Constraint Language*, which is an expressive content-based filtering language that also allows users to filter events based on QoS constraints. Besides the default filter constraint language, an implementation may support any number of additional filter constraint languages.

The Notification Service suffers from the same problems with regard to communication efficiency as the Event Service since both use synchronous two-way calls for event delivery. Moreover, there are still a number of problems inherent to a channel-based solution. Producers and consumers, that is, the application components, have to deal with channels explicitly. They have to select the right ones moving information about the application structure into the components—there is no support for the role of an administrator to arrange channels, producers, and consumers from a system point of view. Using channels also limits system evolution, since the set of channels referenced by applications is static, a problem which is only recently addressed by reflective middleware [96]. Regarding the structure of a system, CORBA channels cannot reflect any hierarchy because their traffic is completely separated. Although event management domains [282] support the federation of multiple channels in arbitrary topologies, they do not offer any filtering of notifications between coupled channels.

Notification Service instances may be federated with the help of ORB domains. However, the necessary bridging between these domains has to be set up manually. The domains are mostly seen as means to model the network and broker infrastructure [318]; they are not targeted at engineering issues of application design. So, in the end one can only assess that the standardized API does not support visibility control and system management sufficiently well, but the CORBA Notification Service may serve as a communication technique to realize a subset of a scope graph.

9.1.2 Jini

The Java programming language is popular for network programming and therefore has some built-in middleware functionality. The *Java Remote Method Invocation* (RMI) specification [367] describes how to synchronously invoke methods of remote objects using request/reply communication between two *Java Virtual Machines (JVMs)* running on separate nodes. The Java RMI compiler generates marshaling code for the proxy object and the server skeleton. Because of the homogeneous environment created by JVMs, in which there is only a single programming language, the burden on the middleware is lower. It is even possible to move executable code between JVMs by using Java's *object serialization* to flatten an object implementation into a byte stream for network transport.

Asynchronous event communication within a single JVM is mainly used in the *abstract window toolkit (AWT)* [358] libraries for graphical user interfaces. The EventListener interface can be implemented by a class to become a callback object for asynchronous events, such as mouse or keyboard events. The Jini framework, described below, extends this to provide event communication between different JVMs. Other variants of asynchronous communication in Java are provided by the messaging infrastructure of JMS (Sect. 9.1.3).

The *Jini* specification [360] enables programmers to create network-centric services by defining common functionality for service descriptions to be announced and discovered. For this, it supports distributed events between JVMs. A RemoteEventListener interface is capable of receiving remote callbacks of instances of the RemoteEvent class. A RemoteEvent object contains a reference to the Java object where the event occurred and an eventID that identifies the type of event. A RemoteEventGenerator accepts registrations from objects and returns instances of the EventRegistration class to keep track of registrations. It then sends RemoteEvent objects to all interested RemoteEventListeners. Event generators and listeners can be decoupled by third-party agents, for example, to filter events, but the implementation is outside the Jini specification and left to the programmer.

JavaSpaces [366] are a part of the Jini framework; they are similar to Linda tuple spaces. With JavaSpaces, tuples can be inserted, read, and removed from a space which stores each tuple from the time it is inserted to the time it is removed. The corresponding operations are `write`, `read`, and `take`. `read` and `take` take a template and block until a tuple that matches the given template is present in the space. `readIfExists` and `takeIfExists` are the nonblocking versions of `read` and `take`; they return instantaneously if a matching tuple is not in the space. To reveal clients from polling for matching tuples, clients can be notified when a matching tuple is inserted into the tuple space via the `notify` operation. However, since there can be multiple listeners notified, it is not guaranteed that a notified client will actually retrieve a matching tuple. There can be multiple spaces that can reside on different hosts. Transactions are also supported. For example, a tuple which is written within a transaction becomes visible outside the transaction only after the transaction committed. More details on transactions can be found in the Jini specification. Fairness and ordering of operations is not addressed by the JavaSpaces specification. *TSpaces* [402] developed by IBM are similar to JavaSpaces.

Summarizing, as is the case for CORBA, event communication in Jini is built on top of synchronous communication (Java RMI), so the same restrictions that limit scalability and efficiency apply.

9.1.3 Java Message Service (JMS)

The *Java Message Service* (JMS) [364] defines a messaging API for Java. Differently from the CORBA Event or Notification Service, JMS can be used without the enterprise object platform, i.e., J2EE [365], of which it is part. JMS clients can choose any vendor-specific implementation of the JMS specification, called a *JMS provider*. JMS comes with two communication modes: point-to-point and publish/subscribe communication. *Point-to-point communication* follows the one-to-one communication abstraction of message queues. Queues are stored and managed at a JMS server that decouples clients from each other. Direct communication between a sender and a receiver without an intermediate server is not supported. In *publish/subscribe communication*, the JMS server manages a number of *topics*. Clients can publish messages to a topic and subscribe to messages from a topic.

JMS provides a topic-based publish/subscribe service with limited content-based filtering support in the form of *message selectors*. A message selector allows a client to specify the messages it is interested in by specifying a filter that operates on the fields of the message header; body fields cannot be evaluated. The selector syntax is based on a subset of the SQL92 [101] conditional expression syntax.

Like structured CORBA events, a JMS message is divided into a message header and body. The header contains various fields, including the destination of the message, its delivery mode, a message identifier, the message priority, a type field, and a timestamp. The delivery mode can be set to PERSISTENT to enforce exactly-once delivery semantics; otherwise best-effort delivery applies. The type of a message is an optional field that can be used by a JMS provider for type-checking the message. Apart from predefined fields, the header can also contain any number of user-supplied fields. The message body is in one of several formats: a StreamMessage, a TextMessage, and a ByteMessage containing the corresponding Java primitive types. A MapMessage is a dictionary of name/value pairs similar to the fields found in the header. Finally, an ObjectMessage uses Java's object serialization feature to transmit entire objects between clients.

Messages can be consumed synchronously or asynchronously, i.e., either pull or push can be used to transfer messages to the respective consumer. There exist two ways of message acknowledgment: messages can either be acknowledged automatically or specifically by the client. Moreover, messages can be persistent or volatile. *Persistent messages* are delivered exactly once to a consumer. They also do not get lost if the provider fails; they usually are logged to stable storage. However, this comes at the cost of a much higher overhead. *Volatile messages* are delivered at most once; they may get lost if the provider fails.

With JMS, subscriptions can either be durable or not. With *durable subscriptions*, notifications are retained while the subscriber is disconnected from the provider until they have been delivered or expired. To the contrary, with a nondurable subscription, notifications that are published while the subscriber is disconnected may get lost.

Sessions can be transactional or nontransactional. *Transactional sessions* allow clients to group the publication and the consumption of several messages into an atomic unit of work. On the producer side, produced messages are retained until commit and if transaction aborts messages are discarded. On the consumer side, all consumed messages are kept until commit and are automatically acknowledged on commit. If the transaction aborts, the messages are redelivered. Hence, messages are actually sent and received when the transaction commits. Since the production and the consumption of the same message cannot be part of the same transaction, only *local* transactions are possible. Another consequence is that transacted sessions cannot be used to implement request/reply interaction. Moreover, point-to-point operations and publish/subscribe operations cannot be mixed inside a single transaction.

Although, at first sight, JMS appears to be a strong contestant for a large-scale middleware, it suffers from several shortfalls: First, the entire model is centralized with respect to JMS servers. As a result, JMS servers are heavyweight middleware components and can become bottlenecks because the JMS specification does not address the routing of JMS messages across multiple servers or the distribution of servers to achieve load balancing. Second, content-based filtering of messages in JMS only considers the message header but not the message body. This seriously reduces the usefulness of message filtering. Finally, JMS is tightly integrated with the Java language. This has the advantage that object instances can be published in a message, but comes with the price of only supporting Java clients, which is not feasible in a large-scale, heterogeneous distributed system.

Another main problem is that aspects that will be important for any JMS implementation are not addressed by the JMS specification. This includes, for example, exception handling, load balancing, fault tolerance, end-to-end security, administration, and message type repositories. For example, the specification leaves open how to define topics or how they are interrelated. Many of these aspects are nevertheless addressed and implemented differently by individual vendors. Hence, applications using these products are incompatible if they use these implementation-specific features.

9.1.4 Data Distribution for Real-Time Systems (DDS)

The Data Distribution Service for Real-Time Systems (DDS) [286, 299] was standardized by the OMG in 2004. DDS follows a "data-centric" approach: it creates the illusion of a *global data space* populated by data objects that applications in distributed nodes can access via read and write operations [298]. Related industrial products, e.g., Splice DDS from Thales (US) [375, 384] and NDDS [324] from Real-Time Innovations (US) are available. The specification describes two layers of interfaces:

- The mandatory *data-centric publish/subscribe (DCPS)* level is targeted toward the efficient delivery of information to interested recipients. It allows for content-based publish/subscribe communication between publishers and subscribers and lays an emphasis on quality of service (QoS).
- The optional higher *data local reconstruction layer (DLRL)* level allows for a simple integration of the service into the application layer. The DLRL automatically reconstructs the state of cached objects locally from updates and allows applications to access objects as if they were local.

Since real-time systems are the application domain of the DDS, special care must be taken to design the interfaces such that real-time requirements can be met by the implementation. The service implementation must be able to preallocate resources reducing dynamic resource allocation to a minimum. For example, copying data should be minimized for efficiency reasons and resource usage should be predictable and bounded. Also due to efficiency reasons, typed events with interfaces are used such that type-safety can be ensured at compile time. Here, typed means that for each datatype, specific classes are generated. Generation tools translate event descriptions into the proper interfaces bridging the gap between typed interfaces and the generic service implementation. The specification pays attention to separate producers from consumers such that they can be implemented independently to facilitate extensibility. QoS is an important issue for the DDS. QoS is supported through several QoS policies that declaratively specify which QoS should be provided instead of how this QoS should be realized. Publishers offers a maximum level for each QoS policy, while subscribers request a minimum level for each QoS policy. For example, a subscriber can request that it wants to receive an update at least once in a given time interval. The next two sections describe the DCPS and the DLRL in more detail.

Data-Centric Publish/Subscribe (DCPS)

The Data-Centric Publish/Subscribe (DCPS) layer is responsible for getting data from publishers to interested subscribers. In the following we describe the main components of DCPS (Fig. 9.5).

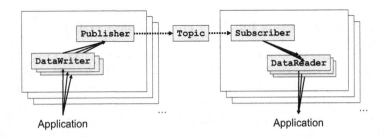

Fig. 9.5. Conceptual overview of data-centric publish/subscribe (DCPS)

A `Publisher` is an object responsible for data distribution. A `DataWriter` is a typed facade that provides access to a publisher. It is bound to exactly one `Topic`, `Publisher`, and application datatype. An application uses a `DataWriter` to communicate with a `Publisher` to the let it know the existence and the value of data objects of a given type. A `Subscriber` is an object responsible for receiving published data. A `DataReader` provides typed access to a subscriber, i.e., to the received data. It is bound to exactly one `Topic`, `Subscriber`, and application datatype. The application associates a `DataReader` to a `Subscriber` to receive the datatype described by the `DataReader`. The QoS experienced by a subscriber is affected by a number of issues. In addition to the `Topic` QoS, the QoS of the `DataWriter`, and the QoS of the `Publisher` affect the QoS on the publisher's side. On the subscriber's side, the QoS is affected by the `Topic` QoS, the `DataReader` QoS, and the `Subscriber` QoS. A `Topic` is conceptually located between publishers and subscribers. It has a name that is unique in the domain and a QoS policy. DCPS differs from other notification services by the fix binding of a `Topic` to a datatype. `ContentFilteredTopic` and `MultiTopic` derive from `Topic`; they can only be used by a `Subscriber`. A `ContentFilteredTopic` provides means for content-based filtering that is similar to the `WHERE` clause of an SQL query. The optional `MultiTopic` class allows users to get data from multiple topics and to combine, filter, and rearrange this data. The data will then be filtered and possibly rearranged using aggregation and projection.

Topic, Publisher, and Subscriber objects are created using the respective create_ operation of DomainParticipant. A DomainParticipant acts as an entry point for an application to the service, serves as a factory for many of the classes, and acts as container for the other objects that make up the service. It represents the local membership of an application in a domain which is a distributed concept, allowing all applications of this domain to communicate with each other.

Datatypes represent information that is sent and received atomically. Instances of a datatype are identified by a *key*. Data with the same key are treated as successive values of the same instance, while data with different keys are treated as referring to different instances. By default, data modifications are disseminated individually, independently, and uncorrelated from other modifications. It is, however, possible that an application requests several modifications to be sent and also received atomically.

To publish data an application first creates a DomainParticipant using DomainParticipantFactory. If the respective Topic does not exist, the applications creates it using the DomainParticipant. Then, the application creates a Publisher using the DomainParticipant and uses the Publisher to create a DataWriter. If the application decides to publish data, it calls the write on the corresponding DataWriter.

To subscribe to data, an application uses a DomainParticipant to find the Topic of interest. Then, it uses the DomainParticipant to create a Subscriber and uses the Subscriber to create a DataReader. To receive data an application can either use a Listener or a WaitSet object. These represent the two basic ways of receiving data and are called *notification-based* and *wait-based*, respectively.

The notification-based interaction style uses listeners. Applications register handlers that are invoked by the middleware to notify the applications about asynchronous events such as the arrival of new data or a QoS violation. From the Listener interface, more specific listeners such as DataReaderListener derive; they add methods depending on the concrete Listener.

The wait-based interaction style uses WaitSet objects that allow an application to wait until one or more of the attached Condition objects are triggered or else until a timeout expires. Condition is subclassed by GuardCondition, StatusCondition, and ReadCondition. A GuardCondition is under the control of an application and can be used by the application to manually wake up the WaitSet. A StatusCondition is attached to any entity; it provides information about the communication status of the respective entity such as the arrival of new data. A ReadCondition allows an application to specify the data samples, in which it is interested. This allows the middleware to enable the condition only when suitable information is available. If data are available, the application can either call read or take on the respective

DataReader. While read allows the data to be read again later, take removes the data from the DataReader.

Data Local Reconstruction Layer (DLRL)

The *Data Local Reconstruction Layer (DLRL)* is an optional layer that may be built on top of DCPS. DLRL allows for a simple integration of the service into the applications by offering an interface on a higher level than DCPS does. DLRL defines an object cache that allows the application to access objects "as if" it were locally available by automatically reconstructing the state of the cached objects from the updates received. To achieve this, object modifications are propagated using DCPS to all parties having a copy of the respective object in their cache and the copies are accordingly updated.

With the DLRL an application can describe DLRL objects with methods, attributes, and relations. Attributes can be either local or shared. As their name suggests, only shared attributes take part in dissemination. To ensure their dissemination, shared attributes are attached to DCPS entities. A DLRL object has at least one shared attribute. DLRL objects can be manipulated using the native language constructs, which in turn triggers changes to the corresponding DCPS entities in the background. Single inheritance of DLRL objects is supported and different kinds of associations can be used to relate DLRL objects to each other. The associations can be used to navigate among the DLRL objects. With the DLRL, the application model is given in OMG IDL. In addition, for example, the mapping from application types to topics and which attributes should be shared are defined using XML. The required classes are then generated automatically by an IDL compiler.

To achieve the dissemination of object modifications, the DLRL specification defines several mappings between the DCPS and the DLRL layer:

1. The *structural mapping* defines the relation between DLRL objects and DCPS data. It is very similar to an object to relation mapping known from database management. Each DLRL object is mapped to a DCPS data sample. Topics correspond to database tables, and data samples correspond to tuples.
2. The *operational mapping* defines the relation between DLRL objects and DCPS entities (e.g., Topic). For example, each DLRL class is mapped to several DCPS topics. The use of the DCPS entities is totally transparent to the application using DLRL.
3. The *functional mapping* defines the relation between DLRL functions (mainly access to the DLRL objects) and the DCPS functions.

Several classes are used by an application to access DLRL objects at runtime. A Cache contains a set of objects that are locally available and that are managed consistently. Its contents are updated transparently when updates arrive. A Cache is created using a CacheFactory. At creation time its mode is set to read-only, write-only, or read/write. A Cache comprises one or more

CacheAccess objects that isolate a set of objects in a given access mode. A CacheAccess allows users to globally manipulate DLRL objects in isolation.

9.1.5 WS Eventing and WS Notification

Most early Web services were based on synchronous request/reply interaction. After Web services had been on the market for some years, the need for asynchronous push capabilities was recognized. These capabilities are needed for services such as stock quoting services if they should not be based on resource-intensive polling. Pushing information to a service requires that the service can be contacted using a communication endpoint. *Web Services Addressing (WS-Addressing)* introduces service endpoint references for Web services. These endpoints can be passed as message parameters, for example, to register a subscription, and to subsequently deliver messages to the registered service. The different parts of WS-Addressing are currently being standardized by the World Wide Web Consortium (W3C). On top of WS-Addressing, *Web Services Eventing (WS-Eventing)* [201, 387] resides. It lets a Web service, called *event sink*, register at another Web service, called *event source*, such that the former can receive notification messages from the latter. A subscription is only valid until an expiration time, which is passed by the event source to the event sink as part of the subscription reply message. The event sink can request the notifications to be filtered by an *event filter*, which is a Boolean expression that is by default given as an XML XPath expression.

Web Services Notification (WSN) [200, 386] is an alternative to WS-Eventing. WSN is currently being standardized by the OASIS (Organization for the Advancement of Structured Information Standards). It consists of *Web Services Base Notification (WS-BaseNotification)* [274], *Web Services Brokered Notification (WS-BrokeredNotification)* [275], and *Web Services Topics (WS-Topics)* [276]. WSN also builds upon WS-Addressing. WS-BaseNotification defines the NotificationConsumer interface and the NotificationProducer interface used for *direct notification*, and specifies messages and message exchanges to be implemented by services that wish to act in these roles along with operational requirements expected of them. Consumers register their subscriptions directly at the producers. Content-based filtering is supported by *selector expressions*. A producer sends a notification directly to the consumers that registered a matching subscription. WS-BrokeredNotification defines interfaces, messages, and message exchanges needed for *brokered notification*, which uses notification brokers as intermediaries to decouple producers from consumers. WS-Topics define the concepts centered around topic-based publish/subscribe such as topics, topic spaces, topic trees, and topic expressions.

9.1.6 The High-Level Architecture (HLA)

The *High-Level Architecture(HLA)* [98] originated at the U.S. Department of Defense in 1996 and was later standardized by the IEEE (Standard 1516) and by the OMG. The corresponding standard of the OMG is described in the *Distributed Simulation Systems (DSS)* specification [284]. The HLA is mainly used to deploy distributed simulations. It provides the specification of a common technical architecture for use across all classes of simulations in the US Department of Defense serving as a structural basis for simulation inter-operability. With the HLA a simulation is carried out by a set of *federates*. Each federate manages a set of *objects* (e.g., tanks), which move in a *routing space*. Inside the HLA, *Data Distribution Management (DDM)* services support the routing of data among federates during the course of a federation execution. Especially, DDM allows for content-based subscriptions based on object attributes. However, content-based filtering is usually done on the client side. Federates express their interest to receive updates by subscribing to all updates that occur in a rectangular *region* of the routing space. Besides these *subscription regions*, there are *update regions*. Regions may change, for example, when an object moves in the routing space.

For distributing the updates, region-based and grid-based approaches are used [48]. With the *region-based* approach usually one multicast group is used for every update region and the subscribing federates join those groups that overlap with their subscription regions. With the *grid-based* approach, the routing space is divided into *cells* and for each cell a multicast group is used. Publishing federates publish updates to those multicast groups the update belongs to and subscribing federates join all groups that overlap with their subscription regions.

9.2 Commercial Systems

We discuss IBM WebSphere MQ in Sect. 9.2.1, TIBCO Rendezvous in Sect. 9.2.2, and Oracle Advanced Queuing in Sect. 9.2.3. Instead of describing all features of these commercial systems, we put an emphasis on those features which are related to publish/subscribe.

9.2.1 IBM WebSphere MQ

IBM *WebSphere MQ (MQ)* [198, 199] (formerly known as IBM MQSeries) is a messaging platform that is part of IBM's WebSphere suite. MQ is a powerful middleware, whose strength lies in the simple integration of legacy applications through loosely coupled queues. A particular strength of WebSphere MQ is its availability for many platforms including Windows, Linux, Solaris, and many others. Its main focus is on point-to-point messaging using queues, especially request/reply on communication. A *queue manager* is a process that manages a set of queues and offers the queuing services to applications via an API. Several programming language bindings of the API to send and receive messages to and from queues exist. WebSphere MQ comes with advanced messaging features, such as transactions, clustered queue managers for load balancing and availability, and built-in security mechanisms. Additionally, a queue manager provides functions to administrators so that they can create new queues, alter the properties of existing queues, and control the operation of the queue manager. For a program to use the services of a queue manager, it must establish a connection to that queue manager.

WebSphere MQ Publish/Subscribe (MQPS)

WebSphere MQ Publish/Subscribe (MQPS) allows MQ applications to communicate using publish/subscribe communication. MQPS was originally a supplement for MQSeries but was later incorporated into WebSphere MQ. It offers topic-based publish/subscribe communication; no content-based subscriptions are supported. In topic-based subscriptions, two wildcards can be used: while a ? can be replaced by any single character, an * can be replaced by any sequence of characters. It is suggested to use the / to organize the topics into a hierarchy. The publisher specifies the topic of a publication when it publishes the information, and the subscriber specifies the topics on which it wants to receive publications. The routing of messages from producers to subscribers is carried out by a *broker* that uses standard MQ functionality to achieve this. Hence, an application using MQPS can use all the features available to existing MQ applications. Publishers can optionally register their intention to publish information on a certain topic at the broker. Publishers and subscribers do not have to be on the same machine as a broker. They can reside anywhere in the network, provided there is a route from their queue manager to the broker.

Related topics can be grouped together to form a *stream*. Streams separate the information flow of the grouped topics from topics in other streams. At each broker that supports a stream, there is a queue with name of the stream. There is a default stream. Streams can also be used to restrict the types of publication a broker has to deal with. This can, for example, be used for load balancing. Access control is also done based on streams.

Brokers can be connected to each other to form a hierarchy. Subscriptions flow to all nodes in the network that support the respective stream. A broker consolidates all the subscriptions that are registered with it, whether from applications directly or from other brokers. In turn, it registers subscriptions for these topics with its neighbors, unless a subscription already exists. Hence, forwarding of identical subscriptions is avoided. When an application publishes information, the receiving broker forwards it (possibly through one or more other brokers) to any applications that have valid subscriptions for it, including applications registered at other brokers supporting this stream.

MQPS allows publications to be retained such that they can be delivered to subsequent subscribers. This way, new subscribers can gather information without having to wait until it (or an updated version) is published again.

WebSphere Business Integration Event Broker (BIEB)

The *WebSphere Business Integration Event Broker (BIEB)* is a complement to WebSphere MQ. BIEB provides high-performance nonpersistent publish/subscribe functionality to clients that can then use content-based subscriptions in addition to topic-based subscriptions. Brokers can be connected to form a hierarchy. Brokers can also be grouped together to form fully connected *collectives*; in this case, the collectives are then connected to form a hierarchy. Brokers can also be *cloned* to improve the availability of the publish/subscribe system. Subscriptions are propagated through the broker network. However, only the topic filter is propagated and not the content filter. Hence, a broker might receive publications in which none of its subscribers is interested. Additionally, it is possible to use IP multicast to distribute subscriptions and publications in LANs.

Message *flows* can be defined that describe operations to be performed on an incoming message, and the sequence in which they are carried out. A flow consists of a number of *flow nodes*, each of which corresponds to a processing step. The *flow connections*, which connect flow nodes, define which processing steps are carried out, in which order, and under which conditions. A flow node can also contain a *subflow* which allows message flows to be composed. Message flows run in a container called *message flow project* which are deployed at a broker. *Subscription points* can be used to make information associated with a particular topic available in a number of different formats. For example, stock prices might be published with a default currency of dollars, but might be required by subscribers expressed in other currencies. Subscription nodes are implicitly connected to *publication nodes* of message flows.

9.2.2 TIBCO Rendezvous

TIBCO Software (US) is a major player in the publish/subscribe middleware market. Its publish/subscribe middleware product TIBCO Rendezvous has been available for many years and has been applied by major customers, especially in the area of financial services. For example, the NASDAQ has implemented its trading floor using TIB Rendezvous. According to TIBCO, the trading floor infrastructure handles 1.8 billion real-time messages per day and 25 thousand trades per second. The current version of TIBCO Rendezvous, as of January 2006, is Version 7.4 [381]. TIBCO Rendezvous is available for many platforms including Linux, Windows, Solaris, and FreeBSD, and APIs are available for many programming languages including Java, C, C++, and Perl 5. TIBCO Rendezvous originally was called TIBCO's Information Bus (TIB) and was renamed later. It is based on ideas presented by Oki et al. [289], who proposed a distributed implementation of a subject-based publish/subscribe system called the *Information Bus*.

TIBCO Rendezvous uses patented subject-based addressing [345]. Subscriptions select subjects from a subject hierarchy. A single subject is selected by its dotted name (e.g., stocks.technology.fooInc), where the parts of the name that are separated by dots are called *elements*. An application can use *wildcards* to select more than one subject. The wildcard * can be replaced by any element, while the wildcard > can be replaced by any dot-separated sequence of elements. Hence, `stocks.technology.*` matches `stocks.technology.foo` but not `stocks.technology.software.bar`, while `stocks.technology.>` matches both. The mapping from subjects to underlying transport protocols, in particular to specific IP multicast addresses, has to be done manually, and it is statically encoded in every producer and consumer. Although the inherent communication efficiency of IP multicast is appealing, it comes at the cost of a rather static configuration, which not only complicates maintenance, but also restricts configurability and integration, and thus the range of possible application domains [382].

A program, which wants to participate in a distributed system in which hosts communicate by the means of TIBCO Rendezvous, uses a TIBCO Rendezvous API library matching the used platform and programming language. In such a system each participating host runs a *rendezvous daemon* (`rvd`), which runs as a separate process. Each message published by a program is handed out to the local daemon via the API library and is then multicast to all daemons in this network. Programs attempt to connect to a local daemon. If a local daemon process is not yet running, the program starts one automatically and connects to it. The daemons hide many details from the programs such as data transport, packet ordering, receipt acknowledgment, and retransmission requests.

With TIBCO Rendezvous *messages* are the entities that travel among programs. A message comprises *data fields*, a subject indicating its destination, and an optional reply-to subject. Each field contains one data item which can be identified by either by its name or by its numerical identifier. Programs do not have to know the wire format of messages; conversions to and from the wire format are transparent to the application. The wire format contains, besides the data itself, also metainformation about the data contained such that the data is "self-describing" in the sense that the receiver is able to interpret and use the data properly.

To register interest in a set of *event occurrences*, a program creates an *event object* whose parameters specify that set. The programmer can specify in which event queue an occurred event is inserted and which callback function is invoked when the event is dispatched. Dispatching can be done in several ways. Queues can be prioritized and grouped to have a fine-grained control of dispatching. Discarding policies can be chosen that specify which event (e.g., the first in a queue) is discarded when the queue size exceeds a given limit. Besides *message events*, which signal the arrival of a message, *timer events* and *I/O events* are supported. The *event driver* recognizes the occurrence of events and places them in the appropriate event queues for dispatch. To receive messages, programs create *listener events*, which specify that messages which match a subject name (that may contain wildcards) are of interest, define callback functions to process the inbound messages, and dispatch events in a loop. A *transport* defines the delivery scope of messages. While *network transports* deliver messages across a network, *intraprocess transports* deliver messages only between program threads within a single process. The creation of a transport takes a `service` parameter. Messages do not travel among transports having different `service` parameters. Together with all listener events bound to a transport, a transport defines the actual set of receivers of a published message.

TIBCO Rendezvous supports two levels of message reliability. With *reliable delivery* the middleware tries to do its best to ensure that a message reaches all participants. However, certain faults, such as daemon crashes, can lead to applications not getting all messages they would have gotten without this fault. The advantage of this scheme is its good performance. With *certified delivery*, the delivery of messages is guaranteed. Messages additionally carry the sender's name, a subject-independent message ID, and an expiration time. This information is used by daemons and routers to request retransmissions of missing messages and to discard expired messages. Despite retransmissions, the order in which messages are delivered satisfies a FIFO-sender policy. To ensure that messages can be delivered even in case of daemon crashes and subsequent restarts, messages are stored persistently. However, reliability comes at a cost: certified delivery greatly degrades the performance of the system.

Independent networks of TIBCO Rendezvous instances can be connected with *information routers*. They forward messages between distinct networks so that subscribers can transparently listen for subject names and receive messages from other networks. Administrators managing the routers have control over the subject names (and associated messages) that are relayed and flow in or out of a network. These routers offer a basic means of structuring.

TIBCO Rendezvous has proven to be scalable to large-scale systems. However, if the subject-based filtering is not expressive enough, extra filtering of events is left to the subscribers. In these cases, scalability can become a problem and the network might be overwhelmed by too many event broadcasts. A JMS implementation is also available from TIBCO (cf. Sect. 9.1.3).

9.2.3 Oracle Streams Advanced Queuing

Oracle Streams Advanced Queuing (AQ) was the first database-integrated messaging system in the industry. This approach is contrary to products such as TIBCO Rendezvous (cf. Sect. 9.2.2), which are not bundled with a database. With the release of Oracle 10, AQ was renamed to *Oracle Streams Advanced Queuing*. AQ offers a JMS implementation (cf. Sect. 9.1.3) called *Oracle JMS*, which is compliant to JMS 1.1 and a proprietary API for queues. Oracle recommends using the standardized JMS API instead of the proprietary AQ API, if Java is used as programming language. As a result of the database integration of AQ, all the functionality offered by the Oracle 10 database can be applied to messaging. This includes query support, indexing, transactions, triggers, consistency constraints, logging, replication, authentication, access control, backup, recovery, data export, and data import.

The basic abstraction of AQ, which decouples producers of messages from consumers of messages, are *queues*. Due to the tight database integration of AQ, queues are normal database tables and messages are normal rows in database tables. Hence, messages can be accessed (i.e., queried) using standard SQL. SQL can be used to access the message properties and the payload. Message histories are available and indexes can be used to optimize access.

Messages can be *enqueued* into or *dequeued* from a queue. Multiple producers can enqueue messages into a queue, and multiple consumers can dequeue messages from a queue. AQ distinguishes among *single-consumer* and *multiconsumer* queues. While single-consumer queues are used for point-to-point messaging, multiconsumer queues can be used for different kinds of point-to-multipoint messaging, including publish/subscribe communication. To allow multiple consumers to dequeue the *same* message from a queue, AQ supports *message recipients* and *queue subscriber*. If a message should be consumed by multiple consumers, it remains in the queue until it is consumed by all its intended consumers. While message recipients are specified by the producer of a message, applications or other queues must subscribe to a queue to become a queue subscriber. Subscriptions can be *rule based*. In this case, not all messages that are enqueued can be dequeued by a queue subscriber, but only those that match the subscription, which is specified in a syntax similar to a WHERE clause of SQL. A subscriber can specify a callback that is invoked to notify it asynchronously about the availability of a new matching message.

There are a number of enqueue and dequeue options available, such as an earliest dequeue time for a message and a message expiration time. Messages are not necessarily dequeued in the order in which they are enqueued. Messages can be grouped to form a set that can only be consumed by one consumer at a time. This feature can, for example, be used to transfer a huge payload by a set of messages. Messages can be retained for a given period after consumption. In a message history also the enqueue time and the dequeue time of a message is saved. Retained messages can be related to each other and applications can track sequences of related messages and produce event journals automatically.

Messages can be propagated based their content from a queue to other queues residing either in the same database or in remote databases. This enables applications to communicate that are not connected to the same queue or to the same database. With message propagation, messages can be fanned out to a large number of recipients without requiring them all to dequeue messages from a single queue. This is known as compositing or funneling messages. Messages can also be propagated using HTTP or HTTPS. AQ allows for message format transformations which are represented by SQL functions. Messages can be transformed during enqueue or during dequeue.

An alternative to *persistent messaging* is *buffered messaging*, which provides a much faster queuing implementation. Buffered messaging is useful for applications not requiring the reliability and transaction support of persistent messaging. It is faster because it stores messages in main memory and only writes messages to disk if the main memory is too small to hold all current messages. Buffered messaging uses the same API as persistent messaging.

In summary, Oracle Streams Advanced Queuing is a feature-rich messaging system that supports different communication styles including publish/-subscribe. Because of its tight database coupling it exhibits many interesting features that other systems do not expose. However, this comes at the cost of a rather heavyweight implementation.

9.3 Research Prototypes

Many research prototypes have emerged since the second half of the 1990s. The pioneers of this area were the Gryphon (Sect. 9.3.1), the SIENA (Sect. 9.3.2), the JEDI (Sect. 9.3.3), the READY (Sect. 9.3.8), and the Elvin (Sect. 9.3.7) event notification services and the Cambridge Event Architecture (CEA) (Sect. 9.3.6). From the newer approaches we present REBECA in Sect 9.3.4 and HERMES in Sect. 9.3.5. Each of the systems we discuss in the following has its own focus (e.g., routing or matching) and differs from the others in some way. With the above selection of systems we try to cover most of the area. Of course, there are many other research prototypes that are not discussed in this book.

9.3.1 Gryphon

The *Gryphon* project at IBM Research [203] led to the development of an industrial-strength, reliable, content-based event broker that is now part of IBM's WebSphere suite as the IBM WebSphere MQ Event Broker [202]. It is a mature publish/subscribe middleware implementation with a JMS interface that provides a redundant, topic- and content-based multibroker publish/subscribe service. The Gryphon event broker has been successfully deployed for large-scale information dissemination at global sports events, such as the Olympic Games. Opyrchal et al. have also investigated how IP multicast can be used to improve the efficiency of event distribution [291] (cf. Sect. 4.6.7). Gryphon includes an efficient event matching engine [6], a scalable routing algorithm, and security features.

Gryphon is based on an information flow model for messaging [28, 354]. An *information flow graph (IFG)* specifies the exchange of information between information producers and consumers. Information flows can be altered by (1) filtering, (2) stateless transformations, and (3) stateful transformations (aggregation). A logical IFG is mapped onto a physical event broker topology. Figure 9.6 shows an example of a Gryphon deployment. Nodes in the IFG are partitioned into a collection of *virtual brokers* PHB, $IB_{1,2}$, and SHB_{1-4}, which are then mapped onto clusters of physical event brokers called *cells*. Similarly,

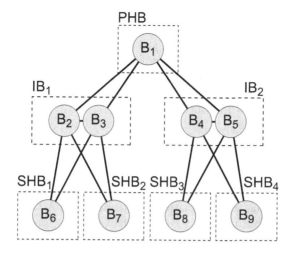

Fig. 9.6. A Gryphon network with virtual event brokers

edges connecting nodes in the IFG are *virtual links* that map onto *link bundles*, containing multiple redundant connections between event brokers for reliability and load balancing.

An event broker that has publishing clients connected to it is called a *publisher-hosting broker (PHB)*. It contains *publisher endpoints* (or *pubends*), which represent a collection of publishers that enter information into the IFG. Correspondingly, a *subscriber-hosting broker (SHB)* consumes information through one or more *subscriber endpoints* (or *subends*) from the IFG according to its subscriptions. An event broker that is neither publisher-hosting nor subscriber-hosting is an *intermediate broker* (IB). The topology mapping is statically defined at deployment time, although more recent work [410] includes dynamic topology changes due to failure and evolution. Several extensions are implemented as part of the Gryphon event broker; these are discussed in the following.

Guaranteed Delivery

A *guaranteed delivery service* [39] provides exactly-once delivery of events, as required for JMS persistent events. The propagation of information (*knowledge*) from pubends to subends is modeled with a *knowledge graph*. Lost knowledge due to message loss causes *curiosity* to propagate up the knowledge graph and trigger the retransmission of events. Curiosity is implemented as *negative acknowledgment (NACK)* messages sent by SHBs. A subscriber that remains connected to the system is guaranteed to receive a gapless ordered filtered subsequence of the *event stream* published at a pubend. A more detailed description of guaranteed delivery and how it can be extended to address congestion in an event-based middleware is given in Sect. 8.3.

Durable Subscriptions

The *durable subscription service* [40] guarantees exactly-once delivery despite periods of disconnection of event subscribers from the system. This means that the event stream is buffered while a subscriber is not available and replayed upon reconnection. As for the guaranteed delivery service, an event log is kept at PHBs and cached at intermediate brokers.

Relational Subscriptions

The final extension is the *relational subscription service* [214]. Its goal is to implement the stateful transformations supported by Gryphon's IFG model, combining messaging with a relational data model. Relational subscriptions can be seen as a continuous query over event streams, providing event subscribers with the expressiveness of a relational language. This relates to the requirement for composite event detection in an event-based middleware, which is discussed in Chap. 7.

The Gryphon event broker includes many of the features that a distributed systems' programmer expects from an event-based middleware. However, the overlay network of event brokers is static, as it is defined in configuration files at deployment time. This makes it difficult for the middleware to adapt to changing network conditions. Failure within a cell of event brokers can be tolerated, but major changes to the IFG cannot be compensated for. Although composite event detection is provided by relational subscriptions, a relational data model for messaging might be too heavy-weight for many applications.

9.3.2 SIENA

One of the first implementations of a distributed content-based publish/subscribe system was the *scalable internet event notification architecture* (SIENA) [65, 71]. SIENA is a multibroker event notification service that targets at Internet-scale deployment. Brokers are called *servers* in SIENA. As usual, event publishers and subscribers connect to a server in the logical overlay network. Events published by publishers are then routed through the overlay network of servers depending on the subscriptions submitted by subscribers.

SIENA uses covering-based routing in its hierarchical and its peer-to-peer variants. Other routing algorithms are not supported. The algorithms used by SIENA are similar to those presented in Sects. 4.5.4 and 4.6.2. In case the peer-to-peer variant is applied, advertisements are supported. The algorithms build upon a *partially ordered set (POSET)*, which allows brokers to keep track of the covering relations among filters. More precisely, the transitive reflexive reduction of the partial order induced by the covering relation is stored. Each server manages a POSET that is accordingly updated when a subscription or unsubscription is processed by the server.

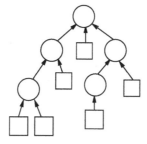

Fig. 9.7. A hierarchical topology in SIENA

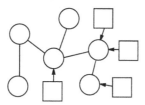

Fig. 9.8. An acyclic peer-to-peer topology in SIENA

The POSET can also be used for matching [71] by traversing it, for example in depth-first order, starting from the root filters, i.e., from those filters which cover all other filters. If a visited filter does not match, then no child filter can match the notification. Carzaniga et al. [67, 70] also presented an alternative matching algorithm that is based on the counting algorithm (cf. Sect. 3.2.2) and that is similar to those presented by Mühl [262].

In SIENA a notification consists of a set of typed attributes. Subscriptions and advertisements are conjunctions of attribute filters, which are simple predicates (e.g., comparisons) over the event attributes. If there is only one attribute filter per attribute, a notification matches a subscription (an advertisement) if it satisfies all attribute filters. However, the interpretation is different for subscriptions and advertisements if there is more than one attribute filter for an attribute. For a subscription, a notification has to match *all* of these attribute filters, while for an advertisement, a notification has to match *at least one* of these attribute filters. Hence, the models of subscriptions and advertisements differ. This fact complicates computing overlapping and covering among filters for more complex data types.

SIENA considers three different types of topologies: hierarchical (Fig. 9.7), acyclic peer-to-peer (Fig. 9.8), and generic peer-to peer (Fig. 9.9). In contrast to an acyclic topology, a generic peer-to-peer topology is not restricted to be a tree. Here, peer-to-peer only means that there is no master/slave relation among servers as there is for hierarchical topologies. In a hierarchical topology, hierarchical covering-based routing is used. In this case, the protocol that

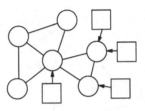

Fig. 9.9. A generic peer-to-peer topology in SIENA

clients use to interact with the respective server they are connected to is the same that a server uses to interact with its master server. Hence, there is an unidirectional flow of subscriptions from servers to their parent servers. In an acyclic or a generic peer-to-peer topology, peer-to-peer covering-based routing is applied. In this case, for the communication among servers a different protocol is used that allows for a bidirectional flow of subscriptions and advertisements. While in an acyclic peer-to-peer topology one common tree is used for filter and notification propagation, in a generic peer-to-peer topology for each producer the minimum spanning tree is used that connects this server with all others servers. A filter is then only forwarded by a server B if it comes from those neighbor servers being on the shortest path from the originating server to B.

There exists no precise specification of the semantics of notification delivery, and the informally described semantics has several peculiarities. A notification should only be delivered to a client if the client had a matching subscription at the time the notification was published, and notifications may be delivered after cancellation of the respective subscriptions. A client that unsubscribes to a filter implicitly unsubscribes to all filters that are covered by the former filter, too. This approach burdens the client with keeping track of covering relations among the issued subscriptions. Hence, it makes clients depend on the applied routing algorithm. The benefit of this approach is that it simplifies routing because (un)subscriptions from neighbors and local clients can then be treated in the same way.

SIENA lacks support for type-checking of events. The complete freedom given to publishers to advertise and publish any event makes it harder to catch type-mismatch errors during system development. SIENA also addressed security issues [390]. Even though the idea of *event patterns* is introduced as a higher-level service, little detail is given on detection and temporal issues. Only the detection of sequences of events is discussed. The topology of the overlay network of event servers is static and must be specified at deployment time. The efficiency of the content-based routing will therefore depend on the quality of the overlay network topology.

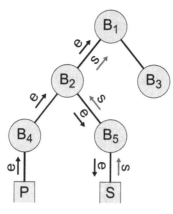

Fig. 9.10. Hierarchical event routing in JEDI

9.3.3 JEDI

The *Java Event-Based Distributed Infrastructure* (JEDI) [92] is a Java-based implementation of a distributed content-based publish/subscribe system from the Politecnico di Milano, Italy. Events in JEDI are *tuples* having a name and a list of values called *event parameters*. Subscriptions are specified as *templates* (cf. Sect 3.1.1). A JEDI system consists of *active objects*, which publish or subscribe to events, and *event dispatchers*, which route events. Event dispatchers are organized in a tree structure, and routing is performed according to hierarchical covering-based routing. Subscriptions propagate upwards in the tree, and state about them is maintained at the event dispatchers. Events also propagate upwards but follow downward branches whenever they encounter a matching subscription, as shown in Fig. 9.10. Since hierarchical routing is applied, advertisements are not used to restrict the propagation of subscriptions.

Support for Mobile Clients

The system has been extended to support mobile computing [89]. Event dispatchers support `moveOut` and `moveIn` operations that enable subscribers to disconnect and reconnect at a different dispatcher in the network. There is no single event dissemination tree for all subscriptions, but instead a tree is built dynamically as a *core-based tree* [24]. The core, called a *group leader*, has to make a global broadcast to announce its presence. A new event dispatcher, wanting to become part of the dissemination tree, directly contacts the group leader. The group leader then delegates the request to an appropriate event dispatcher in the dissemination tree, which becomes the parent of the new node. As a downside, this algorithm requires that every event dispatcher must have knowledge of all group leaders in the system.

Dynamic Reconfigurations

An approach for dynamically reconfiguring the dissemination tree is proposed by Cugola et al. [93, 308]. They focus on the reconfigurations that substitute one link by another one (Fig. 9.11). Instead of intentionally reconfigurations

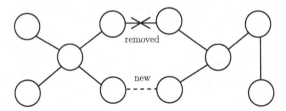

Fig. 9.11. Substituting one link with another link

(e.g., triggered by an administrator), their approach also works for reconfigurations caused by link faults. Regarding routing algorithms, they only consider simple and identity-based routing; however, they state that their algorithms could be generalized to covering-based routing. The use of advertisements is not discussed.

First, the authors describe more precisely than previous work the *strawman approach*. With this approach, both endpoints of the removed link behave as if they had received an unsubscription for each of the subscriptions of the other that are currently active. The endpoints of the added link exchange all to establish the delivery of all notifications needed at the other side. The processes of tearing down the old link and establishing the new link are carried out concurrently. As the authors explain, this has the consequence that notifications might get lost, duplicated, or reordered (violating FIFO-producer or causal ordering). The approach is also inefficient with respect to the filter forwarding overhead because subscriptions might be canceled that are shortly later reinserted, and vice versa. The strawman approach also leads to correct routing tables if multiple links are exchanged concurrently.

After discussing the strawman approach, the authors also propose a solution that (is according to their simulation results) more efficient than the strawman approach but which exhibits the same deficiencies with respect to notification loss, duplication, and reordering. With this solution, the new link is established a bounded delay (i.e., a timeout is used) before the old link is removed, i.e., subscription propagation starts earlier than unsubscription propagation. However, choosing a sensible value for the timeout seems difficult. To avoid the propagation of subscriptions that would otherwise be removed a short time later, subscriptions located at an endpoint of a removed link are removed from the routing tables of the respective brokers instantaneously and only their propagation is delayed.

More recently, the authors presented a more advanced approach to deal with reconfigurations [94] based on *reconfiguration paths* which identify the minimal portion of the system affected by a fault. This approach is better suited for controlled administration than for dealing with faults.

9.3.4 REBECA

The REBECA notification service [136] implements the publish/subscribe interface and conforms to the definition of simple event systems (cf. Sect. 2.1). Its basic architecture is a representative example of a distributed notification service, which is comparable to that of other services such as SIENA, JEDI. However, REBECA is different from other services:

Formal Specification. REBECA is based on a formal specification that defines the intended behavior of the notification unambiguously.

Extensible Data and Filter Model. The default data model of REBECA is the name/value pair model. However, the set of datatypes and constraints that can be used is not fixed but extensible.

Extensible Routing Framework. REBECA is designed to support various routing algorithms [263, 267]. Peer-to-peer and hierarchical variants of the algorithms as well as advertisements can be used.

Visibility Control. With REBECA it is possible to control the visibility of notifications [139, 144] by using the scopes.

Architecture

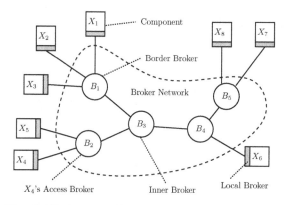

Fig. 9.12. An exemplary router network of REBECA

The constituents of the system are the components (i.e., producers and consumers)and the notification service (Fig. 9.12). The notification service

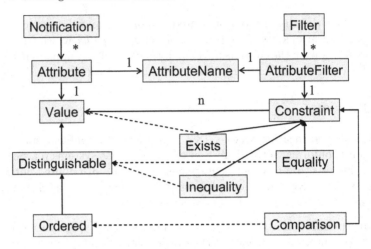

Fig. 9.13. The filtering framework of REBECA

consists of a number of brokers that form an overlay network in the underlying physical network. Brokers are processes that run on physical nodes. The communication topology of the overlay is an acyclic graph. Edges are communication links that are mapped to TCP/IP connections. As an alternative, IP multicast can be used. Obviously, an acyclic topology is can become a bottleneck, but extensions exploiting redundancy are available to tackle problems of scalability and single points of failure [86, 311, 374].

REBECA distinguishes three types of brokers: local, border, and inner brokers. *Local brokers* provide access to the middleware by offering the publish/ subscribe interface to the components. Usually, they are part of the communication library loaded into application components; they are not represented in the graph, but only used for implementation issues. A local broker is connected to one border broker. *Border brokers* form the boundary of the distributed communication middleware and maintain connections to local brokers, i.e., the clients of the service. *Inner brokers* are connected to other inner or border brokers; they do not maintain connections to clients.

Local brokers put the first message containing a newly published notification into the network. Border and inner brokers forward the messages to neighbor brokers according to filter-based routing tables and respective routing strategies. At the end, the messages are sent to the local brokers of the consumers, and from there the notifications are delivered to the application components.

Extensible Data and Filter Model

In the default data model of REBECA, a notification consists of a set of attributes that are name/value pairs. Attribute values can be of different types, including the usual primitive types such as integers, strings, Booleans, and floats but also composite types such as points or rectangles. It is possible to add new datatypes to the filtering framework (Fig. 9.13) easily. New data types should support the operations which are needed by the applied routing algorithms such that routing optimizations become possible. The set of constraints that can be imposed on attributes contains the usual operator such as equality, inequality, and comparisons. It can be extended by new constraints. For more details regarding the data and filter model of REBECA please refer to Chap. 3.

Extensible Routing Framework

REBECA is based on a flexible routing framework which allows new routing algorithms to be added easily. If a new algorithm is added, it can be used for subscriptions and for advertisement propagation. It can also be combined with other routing algorithms in the sense that, for example, the new algorithm is used for subscription forwarding and a previously existing algorithm is used for advertisement forwarding. In contrast to, for example, SIENA, the publish/subscribe interface used by components is independent of the applied routing algorithm. Thus, applications need not to be changed if a new routing algorithm is applied.

Currently, REBECA supports flooding, simple, identify-based, covering-based, and merging-based routing (cf. Sect. 4.5). The implementation of the routing algorithms closely follows the pseudocode we have presented and so we can place high confidence on the correctness of the implementation. The following combinations of routing algorithms are possible: If only subscriptions are used, any of the four filter-based routing algorithms can be applied. If advertisements are used, for subscription forwarding and for advertisement forwarding one of the filter-based routing algorithms can be used, resulting in ten possible combinations. The use of advertisements can greatly enhance the efficiency of the system if certain kinds of notifications can only be produced in certain parts of the broker network. In this case, the size of the subscription routing tables and the filter forwarding overhead is reduced. In the hierarchical setting, again any of the four filter-based routing algorithms can be used. Together with flooding, this results in altogether 19 different combinations of routing algorithms. Flooding can only be combined with a filter-based routing algorithm in a hybrid routing scheme. In this case, in a subtopology notifications are flooded and filters are only forwarded to the root broker of this subtopology. For more details regarding the routing framework please refer to Chap. 4.

Visibility Control

In large-scale publish/subscribe systems, the ability to control the visibility of notifications is a crucial feature. If a notification should not be visible in some part of the system, then it is also not necessary to distribute the notification into this part. The visibility of events can be controlled with scopes that facilitate information hiding. Together with input and output interfaces this points the way toward event-based components. Event mapping can be used to transform notifications from one representation to another, which is a necessity in heterogeneous systems. For more details regarding scopes please refer to Chap. 6.

Available Prototypes

Two prototypes have emerged and are available: a Java-based prototype and a prototype based on Microsoft's .NET platform. We are implementing a bridge between the two prototypes to make them interoperable. Other developers in the REBECA project are currently implementing the scoping concept [138, 140, 144] that allows the visibility of notifications to be constrained using a scope graph. Histories supporting caching of past notifications [81] and that support client and broker mobility [141, 142, 408] as well as P2P-based routing [374] are also part of current implementation and research efforts.

9.3.5 Hermes

Another research prototype is HERMES [310], a distributed, event-based middleware platform. HERMES is aimed at a generic class of large-scale data dissemination applications, such as Internet-wide news distribution and a sensor-rich, active building. It follows a type- and attribute-based publish/subscribe model that places particular emphasis on programming language integration by supporting type-checking of event data and event type inheritance.

To handle dynamic, large-scale environments, HERMES uses peer-to-peer techniques for autonomic management of its overlay network of event brokers and for scalable event dissemination. It is based on an implementation of a peer-to-peer routing layer to create a self-managed overlay network of event brokers for routing events. Its content-based routing algorithm is scalable because it does not require global state to be established at all event brokers. Its routing algorithms use rendezvous nodes, as explained in Sec. 4.6.3, to reduce routing state in the system, and include fault tolerance features for repairing event dissemination trees. HERMES is also resilient against failure through the automatic adaptation of the overlay broker network and the routing state at event brokers. An emphasis is put on the middleware aspects of HERMES so that its typed events support a tight integration with an application programming language.

A primary feature of the HERMES event-based middleware is scalability. HERMES includes two content-based routing algorithms to disseminate events from event publishers to subscribers. The *type-based routing algorithm* only supports subscriptions depending on the event type of event publications. It is comparable to a topic-based publish/subscribe service but differs by

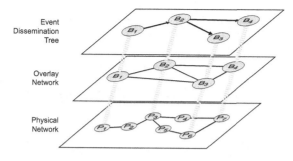

Fig. 9.14. Layered networks in HERMES

observing inheritance relationships between event types. The second algorithm is *type- and attribute-based routing*, which extends type-based routing with content-based filtering on event attributes in publications. In both algorithms, event-type specific advertisements are sent by publisher-hosting brokers to set up routing state. Advertisements are not broadcast to all event brokers, but instead event brokers can act as special rendezvous nodes that guarantee that event subscriptions and advertisements join in the network in order to form valid event dissemination trees.

System Model

Both routing algorithms use a distributed hash table to set up state for event dissemination trees. The distributed hash table functionality is implemented by a peer-to-peer routing substrate, called PAN, formed by the event brokers in HERMES. PAN is an extended implementation of the Pastry routing algorithm. The advantage of such peer-to-peer overlay networks are threefold: first, the overlay network can react to failure by changing its topology and thus adding fault tolerance to HERMES. Second, the peer-to-peer routing substrate that manages the overlay network is responsible for handling membership of event brokers in a HERMES deployment. Third, the discovery of rendezvous nodes, which must be well-known in the network, is simplified by the standard properties of the distributed hash table.

The three layers of networks in HERMES are illustrated in Fig. 9.14. The bottom layer is the physical network with routers and links that HERMES is deployed in. The middle layer constitutes the peer-to-peer overlay network that offers a distributed hash table abstraction. The top layer consists of multiple event dissemination trees that are constructed by HERMES to realize the event-based middleware service. When a message is routed using the peer-to-peer overlay network, a callback to the upper layer is performed at every hop, which allows the event broker to process the message by altering it or its own state.

In addition to scalable event dissemination, HERMES supports event typing, the creation of event type hierarchies through inheritance, and generic, su-

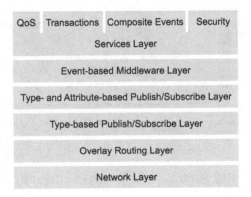

Fig. 9.15. Overview of the HERMES architecture

pertype event subscriptions. This enhances its integration with current object-oriented programming languages such as Java or C++.

Architecture

As shown in Fig. 9.15, the architecture of HERMES has six layers. Each layer builds on top of the functionality provided by the layer underneath and exports a clearly defined interface to the layer above. Apart from that, the layers are independent of each other. A layered architecture for a communications system has the advantage that each layer can have its implementation easily replaced by a different implementation if necessary. For example, if a more efficient implementation of a distributed hash table becomes available, HERMES can benefit from this without major modification. Since HERMES is implemented by the event brokers, its layered structure is also reflected in the implementation of an event broker. Next, we describe the role of each layer, starting with the lowest one.

Network Layer. The lowest layer is the network layer that represents the unicast communication service of the underlying physical network. This assumes that HERMES is deployed in a network with full unicast connectivity between nodes, such as the Internet. No other network-level services, such as group communication primitives, are necessary.

Overlay Routing Layer. This layer implements an application-level routing algorithm that provides the abstraction of a distributed hash table. A peer-to-peer implementation of this layer is chosen for reasons of scalability and robustness. It takes application-level nodes, which are HERMES event brokers, and creates routing state in order to hash keys to nodes. It also handles the addition, removal, and failure of nodes in the overlay network. The topology of the overlay routing layer is optimized with respect to a proximity metric of the underlying physical network.

Type-Based Publish/Subscribe Layer. This layer exports a primitive type-based publish/subscribe service on top of the distributed hash table established by the previous layer. Type-based routing supports subscriptions according to an event type and observes the inheritance relationships between event types. Event dissemination trees are then created with the help of rendezvous nodes in the system. Trees are also repaired by retransmitting messages after state at event brokers has been lost.

Type- and Attribute-Based Publish/Subscribe Layer. This layer extends the type-based service with content-based filtering on event attributes. The same rendezvous node mechanism is used for the construction of event dissemination trees. However, the trees are annotated with filtering expressions derived from the type- and attribute-based subscriptions. These filtering expressions are placed at strategic locations in the network, usually as close to event producers as possible in order to discard unnecessary events as early as possible.

Event-Based Middleware Layer. At this layer, event-based middleware functionality is added to the content-based publish/subscribe system of the previous layers. Typing information is maintained by the rendezvous nodes so that event publications and subscriptions can be type-checked automatically by HERMES. The event-based middleware layer also extends the API used by event clients to invoke HERMES.

Services Layer. The services layer is a set of pluggable extensions to the event-based middleware layer. It allows the HERMES middleware to provide a wide range of higher-level middleware services. For example, different guarantees of publication and subscription semantics can be supported by a QoS module at the services layer. Another service may deal with composite event detection or transaction support. Services may violate the strict layering of the architecture and obtain direct access to lower layers if this is necessary for their functionality.

9.3.6 Cambridge Event Architecture (CEA)

The *Cambridge Event Architecture* (CEA) [18, 20] was created in the early 1990s to address the emerging need for asynchronous communication in multimedia and sensor-rich applications. It introduced the *publish–register–notify* paradigm for building distributed applications. This design paradigm allows the simple extension of synchronous request/reply middleware, such as CORBA, with asynchronous publish/subscribe communication. Middleware clients that become *event sources* (publishers) or *event sinks* (subscribers) are standard middleware objects.

The interaction between an event source and sink is illustrated in Fig. 9.16. First, an event source has to advertise the events that it produces, for example, in a name service. In addition to regular methods in its synchronous interface, an event source has a special `register` method so that event sinks can subscribe (*register*) to events produced by this source. Finally, the event source

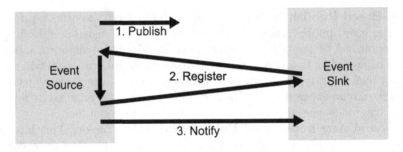

Fig. 9.16. The publish–register–notify paradigm in the CEA

performs an asynchronous callback to the event sink's `notify` method (*notify*) according to a previous subscription. Note that event filtering happens at the event sources, thus reducing communication overhead. The drawback of this is that the implementation of an event source becomes more complex since it has to handle event filtering. Another drawback is that the transmission of notifications to multiple consumers are independent unicast communications.

Direct communication between event sources and sinks causes a tight coupling between clients. To address this, the CEA includes *event mediators*, which can decouple event sources from sinks by implementing both the source and sink interfaces, acting as a buffer between them. Chaining of event mediators is supported, but general content-based routing, as done by other distributed publish/subscribe systems, is not part of the architecture. More recent work [192] investigates the federation of separate CEA event domains using contracts that are enforced by special mediators acting as gateways between domains. A Java implementation of the CEA, *Herald* [346], supports storage of events.

The design goal of the CEA is to seamlessly integrate publish/subscribe with standard middleware technology. Therefore, events are strongly typed objects of a particular event class and are statically type-checked at compile time. Initially, subscriptions were template-based for equality matching only, but they were then extended with a predicate-based language withname/value pairs. These subscriptions are type-checked dynamically at runtime. Furthermore, the CEA provides a service for complex subscriptions based on composite event patterns [189]. This is an important requirement for an event-based middleware. We presented our approach for detecting composite events in Chap. 7.

COBEA

The CEA was implemented on top of CORBA in the CORBA-*based event architecture* (COBEA) [244]. Events are passed between event sources and sinks as parameters in CORBA method calls. Event clients can by typed or untyped: a typed client encodes the structure of an event type in an IDL `struct` datatype, whereas an untyped client uses the generic `any` datatype. Type-checking for typed clients is done by the IDL compiler. The subscription language consists of a conjunction of predicates over the attributes defined in the event type.

ODL-COBEA

The use of CORBA IDL to express event types is cumbersome since its original purpose is the specification of interfaces for remote method calls. In [309], COBEA is extended with an event type compiler that transforms event type definitions in the *Object Definition Language* (ODL) [72] into appropriate CORBA IDL interfaces. ODL is a schema language defined by the Object Data Management Group (ODMG). With ODL, objects can be described language-independently for storage in an object-oriented database. The advantage of using ODL for event definitions is that it provides support for persistent events because it unifies the mechanisms for transmission and storage of events [19].

An example of an ODL-defined event type, as it would be used in the Active Office application scenario, is given in Fig. 9.17. Event types consist of a set of typed attributes and form an ODL inheritance hierarchy, in which all types are derived from the `BaseEvent` ancestor class. The `BaseEvent` type has attributes that all event types inherit, namely a unique `id` field, a `priority` field, a `source` field with the name of the event source that generated this event, and a `timestamp`. ODL-COBEA is aware of inheritance relationships between event types and supports *supertype subscriptions*. When an event subscriber subscribes to an event type, it will also receive any published events that are of a subtype of the type specified in the subscription. This means that an event subscriber that subscribes to the `BaseEvent` type will consequently receive all events published at a given event source.

The CEA and in particular the ODL-COBEA implementation recognize the importance of type-checking for events in a publish/subscribe system. The object-oriented approach for defining event types cleanly integrates with current object-oriented programming languages and middleware architectures. Static type-checking, as done by an event type compiler, does not introduce a runtime cost, but it tightly couples event sinks to sources.

The main disadvantage of the CEA is the lack of content-based event routing between event mediators. This limits the scalability of the architecture as it forces a subscriber to know the publisher (or mediator) that offers

```
1  class LocationEvent extends BaseEvent {
2      attribute short id;
3      attribute string location;
4      attribute long lastSighting;
5  };
```

Fig. 9.17. An ODL definition of event types in ODL-COBEA

a particular event type. In addition, it makes the implementation of event sources challenging because they are required to perform event filtering depending on subscriptions. Several distributed content-based publish/subscribe systems were proposed after the CEA to address these problems.

9.3.7 Elvin

Elvin [341] is a notification service for application integration and distributed systems monitoring developed by the Distributed Systems Technology Centre in Australia. It features a security framework, internationalization, and pluggable transport protocols, and has been extended to provide content-based routing of events [340]. Events are name/value pairs with a predicate-based subscription language. An interesting feature of Elvin is a *source quenching mechanism*, where event publishers can request information from event brokers about the subscribers currently interested in their events. This enables publishers to stop publishing events when there are no subscriptions, reducing computation and communication overheads.

Clients for a wide range of programming languages are available, which led to the implementation of many notification applications. Applications, such as a ticker-tape, were evaluated as means for collaboration in a pervasive office environment [148]. Other work investigates event correlation and support for disconnected operation in mobile applications [368].

9.3.8 READY

The READY event notification service [184] introduced *event zones* to partition components based on logical, administrative, or geographical boundaries and to delimit the visibility of events. Boundary brokers connect zones and control the communication between them, and may enforce security policies on connected clients. Although similar to scoping, zones resemble more the domain idea of CORBA as it mainly addresses control on the physical routing network; the engineering aspect is lacking. For instance, in READY a component belongs to exactly one zone so that there is only a two-level hierarchy. The system is structured only based on one specific point of view, prohibiting composition and mixing of aspects [188]. Heterogeneity issues are only mentioned in READY: boundary brokers could apply transformations on crossing notifications. Following the idea of CORBA domains, brokers operate here on a rather coarse and static granularity, whereas event mappings (Sect. 6.4) allow for syntactic and semantic mappings in the formal model and at every layer of abstraction in a scoped system.

9.3.9 Narada Brokering

The *Narada Brokering* project [293] aims to provide a unified messaging environment for grid computing, which integrates grid services, JMS, and JXTA.

It is JMS compliant (Sect. 9.1.3), but also supports a distributed network of brokers as opposed to the centralized client/server solution advocated by JMS. The JXTA specification [180] is used for peer-to-peer interactions between clients and brokers.

Events can be XML messages that are matched against XPath [398] subscriptions by an XML matching engine. The network of brokers is hierarchical, built recursively out of clusters of brokers. Every broker has complete knowledge of the topology, so that events can be routed on shortest paths following the broker hierarchy. In general, there is the additional overhead of keeping event brokers organized hierarchically, which can be costly. Dynamic changes of the topology are propagated to all affected brokers.

10

Outlook

Events are of increasing importance in modern distributed systems. Growing interconnectivity, continuous evolution, and real-time adaptation demand a loose coupling of communicating parties and asynchronous communication that traditional approaches such as request/reply cannot provide. The event-based computing paradigm offers the required behavior and flexibility. Surveying the current state of distributed applications, event-based techniques can already be found in a wide-range of domains, including telecommunication systems, network management, mobile and ad hoc networking, application integration, control systems, and user interfaces. They have proven to help in constructing loosely coupled systems in a pragmatic way. With a thorough understanding of the principles of event-based computing, which we attempted to further with this book, we feel that the applicability of event-based techniques will expand so that an increasing number of engineers can exploit the inherent benefits of this communication paradigm.

In this book we gave an overview of the broad area of event-based systems. We introduced the basic concepts of publishing and subscribing for notifications along with theoretical foundations necessary to specify the behavior of event-based systems. We also presented different content-based models and matching algorithms that can be used to implement event-based systems. Many of these algorithms can be extended to the distributed case in order to provide truly scalable notification services for Internet-wide applications. To show the breadth of event-based systems, we also investigated several current research directions: *scoping* as a software engineering technique for structuring event-based systems, *composite events* as a way of increasing the expressiveness of subscriptions, and a range of advanced research topics such as security models, fault tolerance, and congestion control, that are likely to play a major role in future event-based systems.

Open Topics

Event-based systems have come a long way from their beginnings as active database triggers to current sophisticated Internet-wide notification services. Nevertheless, there is much scope for future work in many areas before event-based systems become a dominating paradigm for the engineering of large-scale distributed applications. This section provides a subjective overview of topics that we believe are important for future investigation. This will help remove the remaining stumbling blocks that hinder a larger deployment of event-based systems.

Algorithms

Although algorithms are already a major focus of current work in event-based systems, many questions remain unanswered. As distributed systems embrace Web service techniques and XML becomes the lingua franca of the Internet, routing and matching algorithms for publish/subscribe systems must acknowledge this. Initial work on semistructured data models for notification systems (Sect. 3.1.3) and on efficient matching and routing algorithms for XML (Sect. 3.2.5) already exists but is bound to receive more attention in the near future. Specifications for asynchronous event communication in Web services such as the WS Eventing and Notification efforts (Sect. 9.1.5) will bring the benefits of event-based techniques to Web services engineers.

Future content-based routing algorithms will be affected by the desire of applications to increase the amount of expressiveness when specifying subscriptions. As a result, composite event detection will become an important way to support a more fine-grained expression of subscriber interests in many applications and will become an integral part of the basic architecture of a large-scale notification service.

Another requirement for routing and matching algorithms is the support for QoS features. QoS covers a broad range of issues from bandwidth and real-time constraints to reliability and transactional processing. Many real-world applications expect hard guarantees on notifications that must be respected by the event-based system. This can only be achieved if routing and matching algorithms have been designed with QoS features from the beginning. The majority of current algorithms only operate on a best-effort basis.

A natural step in the evolution of overlay networks is that their functionality will eventually be included in the networking hardware itself. In recent years, Internet routers have already shown limited filtering capabilities to carry out efficient and intelligent multicast routing [88]. As advances in router hardware reach widespread availability, content-based routing techniques can be supported natively in hardware and replace address-based point-to-point routing as the dominant communication paradigm.

From a methodological point of view, the design of routing algorithms should be guided by real-world workloads that publish/subscribe applications experience. With larger deployment of such application, sets of work-load traces should become available that researchers and engineers can use to assess the efficiency and performance of their routing algorithms and implementations. Similar to standardized benchmark suites that evaluate processor or database performance, notification services require widely accepted benchmark standards that help users choose between different event-based systems.

Data Management and Processing

The goal of notification systems is to disseminate events to interested clients without changing the data itself. Services for composite event detection perform processing on event data to discovery patterns that are of interest to clients. As the next step in this direction, generic stream-processing systems carry out arbitrary data transformations on stream of events. This richer processing abstraction enables applications to push the business logic into the middleware infrastructure. Such a unification of data dissemination and processing blurs the boundary between the network and the application. As a result, in-network resources can be exploited for data processing. The network is used more efficiently because data is only transported in its most aggregated form to the application. Long-term event flows can be set up in the form of continuous database queries, taking the burden of event processing away from the distributed application.

A related but orthogonal issue is the life cycle management of event data. Notifications have a temporal component and are tied to current events but often applications also want to access historical events. Ideally, a notification system should provide access to future events (in the form of subscriptions), current events (in the form of notifications), and past events (in the form of historical queries). For example, caches [81] and histories [263] can be installed to make past notifications available to applications. Such an approach removes the boundary between a notification service and a classical database system.

Software Engineering

To achieve a larger adoption of event-based techniques in industry, it is important to provide the software engineering toolbox necessary to implement such applications. However, existing work on event-based systems has focused so far on scalability issues in terms of communication efficiency and size, and problems of software engineering and management were often neglected. Scalable distributed application design using event-based techniques cannot be exploited in practice if the adopted design methodologies are incompatible or cumbersome to use with such an approach. For example, the loose coupling between components in a publish/subscribe system is often not expressible in traditional design frameworks. Many existing asynchronous design frameworks are illustrated with target application scenarios that have a rather simple structure, e.g., one-way information dissemination such as stock monitoring. Instead, we need design frameworks that naturally match the properties of complex event-based systems.

Our discussion of scoping in Chap. 6 has demonstrated the power of new design abstractions that asynchronous event-based communication can bring to the table. Visibility in event-based systems, from which scoping is derived, is at the heart of many problems, and scoping provides a way to address these problems. It separates coordination from computation and allows for model-driven development of distributed event-based systems. Existing design methodologies need to embrace these techniques and expose their flexibility in a sensible and controlled manner to the application designer.

Programming Paradigms and Tools

Most popular programming languages are firmly rooted in a synchronous request/reply approach and require the system designer to jump through hoops when building a truly asynchronous distributed system. For example, event-driven programming can simplify the implementation of distributed algorithms by reducing the potential for race conditions and deadlocks. It also facilitates debugging because at each node in the system it creates a linear execution trace of events that can be inspected to track problems. However, modern programming languages such as Java make it difficult to write entirely nonblocking code as most of their network functionality has been written with multithreading in mind. Mainstream languages could also aid the implementation of large-scale distributed applications by integrating asynchronous publish/subscribe primitives as first class citizens in the language. Unfortunately, most languages require the system designer to conceptualize their designs at the level of direct point-to-point connections between the components of an application. A higher-level, event-based view could help mask unnecessary complexity. Finally, current programming languages do not handle asynchronous failure notification, which is the common case in Internet-scale distributed applications.

A related concern to software engineering is the availability of tool support when implementing and debugging event-based systems. In particular, development environments must support the implementation of event-based designs. For example, a recent masters thesis [268] successfully used Eclipse [377] as an integration platform for design and management plug-ins. This is a promising starting point for more sophisticated tools to facilitate the design, programming, deployment, and management of event-based systems. Design methodologies such as the CORBA Method Driven Architecture (MDA) [155] can be used to define an abstract model with loosely coupled components, which is then transformed into skeleton code for the implementation. The event-based interaction specified by the model can then be implemented with one of several routing algorithms depending on application requirements and expected workloads. Likewise, there are many ways to implement scoping as prescribed in a model. To improve efficiency, a middleware architecture could pick the most appropriate implementation dynamically at runtime.

Security

Security is a major concern in Internet-wide applications with potentially millions of users under different administrative domains. As system designers gain more experience with the deployment and management of large-scale event-based applications, the security requirements of such systems should be explored in more detail, affecting the design of suitable access control models. The loose coupling of components in event-based systems makes it harder to establish traditional authentication and trust relationships. Therefore, new access control models are necessary to support the specifics of event-based communication. In Sect. 8.1, we have given the flavor of one such secure publish/subscribe model based on role-based access control. As an alternative approach, since any form of trust corresponds to some correlation of the participants, scoping suggests itself as a mechanism to incorporate security policies and implementation [145].

Adaptability

The management complexity of large-scale distributed systems has led to efforts in the direction of self-organizing and autonomic computing [219]. Distributed event-based systems can take advantage of these techniques for overlay management. Currently content-based routing algorithms using peer-to-peer substrates exist, in which the overlay network adapts to mask failure and improve routing performance [312]. As a next step, a more dynamic overlay broker network could start and shut down brokers on demand in response to event flows in the system. For example, an event broker that only has a single downstream child can be removed from the forwarding path because it is redundant. Similarly, the event flows along certain parts of the overlay topology could control the connectivity among brokers to increase the efficiency of covering or merging relations between the subscriptions hosted in the system. We believe that such dynamic adaptation techniques will hide the complexity of running an efficient large-scale notification from system administrators.

Dependability

Dependability and fault tolerance are other topics in event-based systems that have not received the attention that they deserve. Any deployed Internet-scale notification service must handle network and node failure. However, dependability can only be achieved if it is part of the entire system architecture. In Sect. 8.2, we described how self-stabilizing routing algorithms can help build dependable event-based systems. Other techniques, such as the routing of events via multiple paths, can further improve the availability of notification services in practice.

Another important technique to achieve robust system behavior is transactions. Transactions enable applications to roll-back system state to recover after a failure condition has been encountered. To support transactions, suitable transactional semantics needs to be defined for event-based systems, and only initial work exists in this area [236]. A challenge is that the implementation of distributed transactions can have high overhead, thus limiting the feasibility of transactions for performance-aware applications. Also, the decoupled nature of subscribers and publishers makes it hard to follow the progress of a transaction in the system and to collect sufficient state for a potential roll-back.

Theory

Distributed notification systems are harder to deploy in practice because of the lack of models for the provisioning of such systems. Addressing this issue requires the development of formal techniques to model the behavior of real-world event-based systems. Formal approaches enable a system designer to reason about the correct execution of complex distributed applications that use asynchronous event-based communication. Research into theoretical computer science has resulted in a number of promising asynchronous calculi such as Π-calculus [261] and join calculus [150] that are directly applicable to event-based systems. This theoretical work should enable the design and implementation of event-based systems that stand on firmer theoretical grounds.

References

[1] D. Abadi, D. Carney, U. Cetintemel, M. Cherniack, C. Convey, S. Lee, M. Stonebraker, N. Tatbul, and S. Zdonik. Aurora: A new model and architecture for data stream management. *VLDB*, 12(2):120–139, August 2003.

[2] M. Abadi and L. Lamport. Composing specifications. *ACM Transactions on Programming Languages and Systems*, 15(1):73–132, January 1993.

[3] G. D. Abowd, R. Allen, and D. Garlan. Using style to understand descriptions of software architectures. *ACM Software Engineering Notes*, 18(5):9–20, 1993.

[4] D. Abrahams and A. Gurtovoy. *C++ Template Metaprogramming: Concepts, Tools, and Techniques from Boost and Beyond.* Addison-Wesley Professional, 2004.

[5] M. Addlesee, R. Curwen, S. Hodges, J. Newman, P. Steggles, A. Ward, and A. Hopper. Implementing a sentient computing system. *IEEE Computer Magazine*, 34(8):50–56, August 2001.

[6] M. Aguilera, R. Strom, D. Sturman, M. Astley, and T. Chandra. Matching events in a content-based subscription system. In *Proceedings of the 18th ACM Symposium on Principles of Distributed Computing (PODC 1999)*, pages 53–61, 1999.

[7] S. Ahuja, N. Carriero, and D. Gelernter. Linda and friends. *Computer*, 19(8):26–34, August 1986.

[8] Akamai Technologies, Inc. Content and application delivery. Online information: http://www.akamai.com/en/html/services/content/application/delivery.html, 2003.

[9] I. F. Akyildiz, W. Su, Y. Sankarasubramaniam, and E. Cayirci. Wireless sensor networks: A survey. *Computer Networks*, 38(4):393–422, 2002.

[10] B. Alpern and F. B. Schneider. Defining liveness. *Information Processing Letters*, 21:181–185, 1985.

[11] P. A. Alsberg and J. D. Day. A principle for resilient sharing of distributed resources. In *International Conference on Software Engineering (ICSE'76)*, pages 562–570, October 1976. IEEE Computer Society.

[12] M. Altinel and M. J. Franklin. Efficient filtering of XML documents for selective dissemination of information. In *The VLDB Journal*, pages 53–64, 2000.

[13] Y. Amir, B. Awerbuch, C. Danilov, and J. Stanton. Global flow control for wide area overlay networks: A cost-benefit approach. In *Proceedings of OPENARCH'02*, pages 155–166, June 2002.

[14] A. Arasu, S. Babu, and J. Widom. The CQL continuous query language: Semantic foundations and query execution. Technical report, Stanford University, 2003.

[15] F. Arbab and C. Talcott, editors. *5th International Conference on Coordination Models and Languages (COORDINATION 2002)*, volume 2315 of *LNCS*, 2002. Springer.

[16] A. Arora and M. Gouda. Distributed reset. *IEEE Transactions on Computers*, 43(9):1026–1038, September 1994.

[17] S. Babu and J. Widom. Continuous queries over data streams. *SIGMOD Record*, 30(3):109–120, 2001. ISSN 0163-5808.

[18] J. Bacon, J. Bates, R. Hayton, and K. Moody. Using events to build distributed applications. In *IEEE SDNE Services in Distributed and Networked Environments*, pages 148–155, June 1995.

[19] J. Bacon, A. Hombrecher, C. Ma, K. Moody, and W. Yao. Event storage and federation using ODMG. In *Proceedings of the 9th International Workshop on Persistent Object Systems (POS9)*, pages 265–281, September 2000.

[20] J. Bacon, K. Moody, J. Bates, R. Hayton, C. Ma, A. McNeil, O. Seidel, and M. Spiteri. Generic support for distributed applications. *IEEE Computer*, 33(3):68–76, 2000.

[21] J. Bacon, L. Fiege, R. Guerraoui, H.-A. Jacobsen, and G. Mühl, editors. *1st Intl. Workshop on Distributed Event-Based Systems (DEBS'02)*, 2002. IEEE. ISBN 0-7695-1588-6. Published as part of the ICDCS '02 Workshop Proceedings.

[22] J. Bacon, K. Moody, and W. Yao. A model of OASIS role-based access control and its support for active security. *ACM Transactions on Information and System Security (TISSEC)*, 5(4):492–540, November 2002.

[23] M. Balazinska, H. Balakrishnan, S. Madden, and M. Stonebraker. The design of the Borealis stream processing engine. In *Proceedings of the 2005 ACM SIGMOD international conference on Management of data*, pages 13–24. ACM, January 2005.

[24] T. Ballardie, P. Francis, and J. Crowcroft. Core based trees (CBT). In *Proceedings of ACM SIGCOMM'93*, pages 85–95, September 1993. ISBN 0-89791-619-0.

[25] G. Banavar, editor. *Advanced Topic Workshop Middleware for Mobile Computing (Middleware 2001)*, November 2001.

[26] G. Banavar, T. Chandra, B. Mukherjee, J. Nagarajarao, R. E. Strom, and D. C. Sturman. An efficient multicast protocol for content-based publish-subscribe systems. In *Proceedings of the 19th IEEE International Conference on Distributed Computing Systems*, pages 262–272, 1999.

[27] G. Banavar, T. D. Chandra, R. E. Strom, and D. C. Sturman. A case for message oriented middleware. In P. Jayanti, editor, *13th International Symposium on Distributed Computing (DISC'99)*, volume 1693 of *LNCS*, pages 1–17. Springer, 1999.

[28] G. Banavar, M. Kaplan, K. Shaw, R. Strom, D. Sturman, and W. Tao. Information flow based event distribution middleware. In W. Sun, S. Chanson, D. Tygar, and P. Dasgupta, editors, *ICDCS Workshop on Electronic Commerce and Web-based Applications/Middleware*, pages 114–121, 1999.

[29] E. Baralis, S. Ceri, and S. Paraboschi. Modularization techniques for active rules design. *ACM Transactions on Database Systems (TODS)*, 21(1):1–29, 1996.

[30] P. Barham, B. Dragovic, K. Fraser, S. Hand, T. Harris, A. Ho, R. Neugebauer, I. Pratt, and A. Warfield. Xen and the art of virtualization. In *Proceedings of the 19th ACM Symposium on Operating Systems Principles (SOSP'03)*, pages 164–177, October 2003.

[31] D. J. Barrett, L. A. Clarke, P. L. Tarr, and A. E. Wise. A framework for event-based software integration. *ACM Transactions on Software Engineering and Methodology*, 5(4):378–421, October 1996.

[32] J. Bates, J. Bacon, K. Moody, and M. Spiteri. Using events for the scalable federation of heterogeneous components. In P. Guedes and J. Bacon, editors, *Proceedings of the 8th ACM SIGOPS European Workshop: Support for Composing Distributed Applications*, pages 58–65, September 1998.

[33] P. C. Bates. Debugging heterogeneous distributed systems using event-based models of behavior. *ACM Transactions on Computer Systems*, 13(1):1–31, February 1995.

[34] A. Belokosztolszki, D. M. Eyers, P. R. Pietzuch, J. Bacon, and K. Moody. Role-based access control for publish/subscribe middleware architectures. In Jacobsen [210], pages 1–8.

[35] J. A. Bergstra, A. Ponse, and S. A. Smolka, editors. *Handbook of Process Algebra*. North-Holland, 2001.

[36] M. Bernardo and F. Franzè. Exogenous and endogenous extensions of architectural types. In Arbab and Talcott [15], pages 40–55.

[37] P. A. Bernstein. Transaction processing monitors. *Communications of the ACM*, 33(11):75–86, Nov. 1990.

[38] B. Betts and C. Heinrich. *Adapt or Die: Transforming Your Supply Chain into an Adaptive Business Network*. John Wiley, 2003.

[39] S. Bhola, R. Strom, S. Bagchi, Y. Zhao, and J. Auerbach. Exactly-once delivery in a content-based publish-subscribe system. In Fabre and Jahanian [131], pages 7–16.

[40] S. Bhola, Y. Zhao, and J. Auerbach. Scalably supporting durable subscriptions in a publish/subscribe system. In *Proceedings of the International Conference on Dependable Systems and Networks (DSN'03)*, pages 57–66, June 2003.

[41] K. Birman. The surprising power of epidemic communication. In A. Schiper, A. Shvartsman, H. Weatherspoon, and B. Zhao, editors, *International Workshop on Future Directions in Distributed Computing (FuDiCo 2002)*, volume 2584 of *LNCS*, pages 97–102, 2002. Springer.

[42] K. P. Birman. The process group approach to reliable distributed computing. *Communications of the ACM*, 36(12):37–53, December 1993.

[43] K. P. Birman and T. A. Joseph. Reliable communication in the presence of failures. *ACM Transactions on Computer Systems (TOCS)*, 5(1):47–76, 1987.

[44] A. Birrell and B. Nelson. Implementing remote procedure calls. *ACM Transactions on Computer Systems*, 2(1):39–59, February 1984.

[45] C. Bockisch, M. Haupt, M. Mezini, and K. Ostermann. Virtual machine support for dynamic join points. In *Proceedings of the 3rd International Conference on Aspect-Oriented Software Development (AOSD'04)*, pages 83–92, 2004. ACM Press.

[46] C. Bornhövd and A. P. Buchmann. A prototype for metadata-based integration of Internet sources. In M. Jarke and A. Oberweis, editors, *11th International Conference on Advanced Information Systems Engineering (CAiSE'99)*, volume 1626 of *LNCS*, pages 439–445, 1999. Springer.

[47] C. Bornhövd, M. Cilia, C. Liebig, and A. P. Buchmann. An infrastructure for meta-auctions. In *Second International Workshop on Advance Issues of E-Commerce and Web-Based Information Systems (WECWIS'00)*, pages 21–30, June 2000.

[48] A. Boukerche and C. Dzermajko. Dynamic grid-based vs. region-based data distribution management strategies in multi-resolution large-scale distributed systems. In *Proceedings of the 18th International Parallel and Distributed Processing Symposium*, pages 243–248. IEEE, April 2004. doi: 10.1109/IPDPS.2004.1303296.

[49] A. T. Bouloutas, S. Calo, and A. Finkel. Alarm correlation and fault identification in communication networks. *IEEE Transactions on Communications*, 42(2/3/4):523–533, February 1994.

[50] L. S. Brakmo, S. W. O'Malley, and L. L. Peterson. TCP vegas: New techniques for congestion detection and avoidance. In *Proceedings of ACM SIGCOMM'94*, pages 24–35, August 1994. ACM.

[51] G. Bricconi, E. D. Nitto, A. Fuggetta, and E. Tracanella. Analyzing the behavior of event dispatching systems through simulation. In *Proceedings of the 7th International Conference on High Performance Comput-*

ing, volume 1970 of *Lecture Notes In Computer Science*, pages 131–140. Springer, 2000.

[52] G. Bricconi, E. D. Nitto, and E. Tracanella. Issues in analyzing the behavior of event dispatching systems. In *Proceedings of the 10th International Workshop on Software Specification and Design (IWSSD-10)*, pages 95–103. IEEE Computer Society, 2000.

[53] M. Broy and E.-R. Olderog. Trace-oriented models of concurrency. In Bergstra et al. [35], chapter 2.

[54] A. Buchmann, C. Bornhövd, M. Cilia, L. Fiege, F. Gärtner, C. Liebig, M. Meixner, and G. Mühl. Dream: Distributed reliable event-based application management. In M. Levene and A. Poulovassilis, editors, *Web Dynamics—Adapting to Change in Content, Size, Topology and Use*, pages 319–349. Springer, 2004. ISBN 3-540-40676-X.

[55] P. Buneman. Semistructured data. In *Proceedings of the 16th ACM SIGACT SIGMOD SIGART Symposium on Principles of Database Systems (PODS'97)*, pages 117–121, 1997.

[56] F. Buschmann and K. Henney. A distributed computing pattern language. In *Seventh European Conference on Pattern Languages of Programs (EuroPLoP 2002)*, 2002.

[57] F. Buschmann, R. Meunier, H. Rohnert, P. Sommerlad, and M. Stal. *Pattern-Oriented Software Architecture: A System of Patterns*. Wiley, 1996.

[58] A. Campailla, S. Chaki, E. Clarke, S. Jha, and H. Veith. Efficient filtering in publish-subscribe systems using binary decision diagrams. In *Proceedings of the 19th Conference on Software Engineering*, pages 443–452, May 2001. IEEE Computer Society.

[59] M. Caporuscio, P. Inverardi, and P. Pelliccione. Formal analysis of clients mobility in the Siena publish/subscribe middleware. Technical report, Department of Computer Science, University of L'Aquila, October 2002.

[60] L. Capra, W. Emmerich, and C. Mascolo. Middleware for mobile computing (a survey). Research Note RN/30/01, University College London, July 2001.

[61] B. Carbunar, M. Valente, and J. Vitek. CoreLime: A coordination model for mobile agents. In *International Workshop on Concurrency and Coordination (ConCoord 2001)*, 2001.

[62] L. Cardelli and A. D. Gordon. Mobile ambients. In M. Nivat, editor, *Proceedings of Foundations of Software Science and Computation Structures (FoSSaCS)*, volume 1378 of *LNCS*, pages 140–155, 1998. Springer.

[63] D. Carney, U. Çetintemel, M. Cherniack, C. Convey, S. Lee, G. Seidman, M. Stonebraker, N. Tatbul, and S. B. Zdonik. Monitoring streams — a new class of data management applications. In *VLDB*, pages 215–226, 2002.

[64] N. Carriero and D. Gelernter. Linda in context. *Communication of the ACM*, 32(4):444–458, April 1989.

[65] A. Carzaniga. *Architectures for an Event Notification Service Scalable to Wide-area Networks*. PhD thesis, Politecnico di Milano, Milan, Italy, December 1998.

[66] A. Carzaniga and P. Fenkam, editors. *3rd Intl. Workshop on Distributed Event-Based Systems (DEBS'04)*, May 2004. IEE.

[67] A. Carzaniga and A. L. Wolf. Forwarding in a content-based network. In A. Feldmann, M. Zitterbart, J. Crowcroft, and D. Wetherall, editors, *Proceedings of the 2003 Conference on Applications, Technologies, Architectures, and Protocols for Computer Communications (SIGCOMM'03)*, pages 163–174, 2003. ACM.

[68] A. Carzaniga, E. Di Nitto, D. S. Rosenblum, and A. L. Wolf. Issues in supporting event-based architectural styles. In *Proceedings of the Third International Workshop on Software Architecture (ISAW '98)*, pages 17–20, 1998.

[69] A. Carzaniga, D. R. Rosenblum, and A. L. Wolf. Challenges for distributed event services: Scalability vs. expressiveness. In W. Emmerich and V. Gruhn, editors, *ICSE '99 Workshop on Engineering Distributed Objects (EDO '99)*, May 1999.

[70] A. Carzaniga, J. Deng, and A. L. Wolf. Fast forwarding for content-based networking. Technical Report CU-CS-922-0, Department of Computer Science, University of Colorado, Boulder, Colorado, November 2001.

[71] A. Carzaniga, D. S. Rosenblum, and A. L. Wolf. Design and evaluation of a wide-area event notification service. *ACM Transactions on Computer Systems*, 19(3):332–383, 2001.

[72] R. G. G. Cattell, D. Barry, D. Bartels, M. Berler, J. Eastman, S. Gamerman, D. Jordan, A. Springer, H. Strickland, and D. Wade. *The Object Database Standard: ODMG 2.0*. Morgan Kaufmann, San Francisco, CA, USA, 1997.

[73] A. Celik, A. Datta, and S. Narasimhan. Supporting subscription oriented information commerce in a push-based environment. *IEEE Transactions on Systems, Man and Cybernetics*, 30(4):433–445, July 2000.

[74] S. Chakravarthy and D. Mishra. Snoop: An expressive event specification language for active databases. Technical Report UF-CIS-TR-93-007, Department of Computer and Information Sciences, University of Florida, Gainesville, FL, March 1993.

[75] S. Chandrasekaran, O. Cooper, A. Deshpande, et al. TelegraphCQ: Continuous Dataflow Processing for an Uncertain World. In *Proc. of the 1st Biennial Conf. on Innovative Data Systems Research (CIDR'03)*, January 2003.

[76] Y. Chawathe, S. McCanne, and E. A. Brewer. RMX: Reliable Multicast for Heterogeneous Networks. In *Proceedings of INFOCOM'00*, pages 795–804, March 2000.

[77] J. Chen, D. J. DeWitt, F. Tian, and Y. Wang. NiagaraCQ: A scalable continuous query system for internet databases. In *Proceedings of the*

2000 ACM SIGMOD International Conference on Management of Data, pages 379–390. SIGMOD, 2000.

[78] P. Ciancarini. Coordination models and languages as software integrators. *ACM Computing Surveys (CSUR)*, 28(2):300–302, 1996.

[79] M. Cilia. *An Active Functionality Service for Open Distributed Heterogeneous Environments*. PhD thesis, TU Darmstadt, Darmstadt, Germany, 2002.

[80] M. Cilia, C. Bornhövd, and A. P. Buchmann. Moving active functionality from centralized to open distributed heterogeneous environments. In C. Batini, F. Giunchiglia, P. Giorgini, and M. Mecella, editors, *Proceedings of the 6th International Conference on Cooperative Information Systems (CoopIS '01)*, volume 2172 of *LNCS*, pages 195–210, 2001. Springer.

[81] M. Cilia, L. Fiege, C. Haul, A. Zeidler, and A. Buchmann. Looking into the past: Enhancing mobile publish/subscribe middleware. In Jacobsen [210]. doi: 10.1145/966618.966631.

[82] M. Cilia, M. Haupt, M. Mezini, and A. P. Buchmann. The convergence of AOP and active databases: Towards reactive middleware. In F. Pfenning and Y. Smaragdakis, editors, *Proceedings of the International Conference on Generative Programming and Component Engineering (GPEC'03)*, volume 2830 of *LNCS*, pages 169–188, 2003. Springer.

[83] M. Cilia, M. Antollini, C. Bornhoevd, and A. Buchmann. Dealing with heterogeneous data in pub/sub systems: The concept-based approach. In Carzaniga and Fenkam [66].

[84] M. Colan. *InfoBus 1.2 Specification*. Lotus, 1999.

[85] P. Costa and D. Frey. Publish-subscribe tree maintenance over a DHT. In Dingel and Strom [114], pages 414–420.

[86] P. Costa, M. Migliavacca, G. P. Picco, and G. Cugola. Introducing reliability in content-based publish-subscribe through epidemic algorithms. In Jacobsen [210].

[87] A. Crespo, O. Buyukkokten, and H. Garcia-Molina. Efficient query subscription processing in a multicast environment. In *Proceedings of the 16th International Conference on Data Engineering (ICDE)*, page 83, 2000.

[88] J. Crowcroft, J. Bacon, P. Pietzuch, G. Coulouris, and H. Naguib. Channel islands in a reflective ocean: Large-scale event distribution in heterogeneous networks. *IEEE Communications Magazine*, 40(9):112–115, September 2002.

[89] G. Cugola and E. Di Nitto. Using a publish/subscribe middleware to support mobile computing. In Banavar [25].

[90] G. Cugola and H.-A. Jacobsen. Using publish/subscribe middleware for mobile systems. *ACM SIGMOBILE Mobile Computing and Communications Review*, 6(4):25–33, 2002.

[91] G. Cugola, E. Di Nitto, and A. Fuggetta. Exploiting an event-based infrastructure to develop complex distributed systems. In *Proceedings*

of the 1998 International Conference on Software Engineering, pages 261–270. IEEE Computer Society, 1998.

[92] G. Cugola, E. Di Nitto, and A. Fuggetta. The JEDI event-based infrastructure and its application to the development of the OPSS WFMS. *IEEE Transactions on Software Engineering*, 27(9):827–850, 2001.

[93] G. Cugola, G. P. Picco, and A. L. Murphy. Towards dynamic reconfiguration of distributed publish-subscribe middleware. In W. Emmerich, A. Coen-Porisini, and A. van der Hoek, editors, *3rd International Workshop on Software Engineering and Middleware (SEM 2002)*, volume 2596 of *Lecture Notes in Computer Science*, pages 187–202. Springer, 2002.

[94] G. Cugola, D. Frey, A. L. Murphy, and G. P. Picco. Minimizing the reconfiguration overhead in content-based publish-subscribe. In H. M. Haddad, A. Omicini, R. L. Wainwright, and L. M. Liebrock, editors, *Proceedings of the 2004 ACM Symposium on Applied Computing (SAC'04)*, pages 1134–1140, 2004. ACM.

[95] D. E. Culler and W. Hong. Special issue: Wireless sensor networks — introduction. *Communications of the ACM*, 47(6):30–33, 2004.

[96] E. Curry, D. Chambers, and G. Lyons. Reflective channel hierarchies. In *The 2nd Workshop on Reflective and Adaptive Middleware, Middleware 2003*, 2003.

[97] E. Curry, D. Chambers, and G. Lyons. Extending message-oriented middleware using interception. In Carzaniga and Fenkam [66].

[98] J. S. Dahmann, R. Fujimoto, and R. M. Weatherly. The Department of Defense High Level Architecture. In S. Andradóttir, K. J. Healy, D. H. Withers, and B. L. Nelson, editors, *Proceedings of 29th Winter Simulation Conference*, pages 142–149, 1997.

[99] C. H. Damm, P. T. Eugster, and R. Guerraoui. Linguistic support for distributed programming abstractions. In T. H. Lai and K. Okada, editors, *Proceedings of the 24th International Conference on Distributed Computing Systems (ICDCS'04)*, pages 244–251, March 2004. IEEE Computer Society Press.

[100] P. B. Danzig. *Optimally Selecting the Parameters of Adaptive Backoff Algorithms for Computer Networks and Multiprocessors*. PhD thesis, University of California, Berkeley, CA, 1989.

[101] C. Date. *An Introduction to Database Systems*. Addison-Wesley, 8th edition, 2003.

[102] C. T. Davies, Jr. Data processing spheres of control. *IBM Systems Journal*, 17(2):179–198, 1978.

[103] U. Dayal, B. T. Blaustein, A. P. Buchmann, U. S. Chakravarthy, M. Hsu, R. Ledin, D. R. McCarthy, A. Rosenthal, S. K. Sarin, M. J. Carey, and R. J. Miron Livny. The HiPAC project: Combining active databases and timing constraints. *SIGMOD Record*, 17(1):51–70, 1988.

[104] U. Dayal, A. Buchmann, and D. McCarthy. Rules are objects too: A knowledge model for an active, object-oriented database system. In

Proceedings of the 2nd International Workshop on Object-Oriented Database Systems, volume 334 of *LNCS*, pages 129–143. Springer, 1988.

[105] U. Dayal, A. P. Buchmann, and S. Chakravarthy. The HiPAC Project. *Active Database Systems: Triggers and Rules For Advanced Database Processing*, pages 177–206, 1996.

[106] S. E. Deering and D. R. Cheriton. Multicast routing in datagram internetworks and extended LANs. *ACM Transactions on Computer Systems*, 8(2):85–110, May 1990.

[107] D. DeLucia and K. Obraczka. Multicast feedback suppression using representatives. In *Proceedings of INFOCOM'97*, pages 463–470, April 1997.

[108] A. Demers, D. Greene, C. Hauser, W. Irish, and J. Larson. Epidemic algorithms for replicated database maintenance. In *Proceedings of the Sixth Annual ACM Symposium on Principles of Distributed Computing*, pages 1–12. ACM, 1987.

[109] P. Deolasee, A. Katkar, A. Panchbudhe, K. Ramamritham, and P. Shenoy. Adaptive push-pull: Dissemination of dynamic web data. In *10th International World Wide Web Conference*, pages 265–274, May 2001. ACM.

[110] Y. Diao and M. J. Franklin. High-performance XML filtering: An overview of YFilter. *IEEE Data Engineering Bulletin*, March 2003.

[111] Y. Diao, M. Altinel, M. J. Franklin, H. Zhang, and P. Fischer. Path sharing and predicate evaluation for high-performance XML filtering. *ACM Transactions on Database Systems*, 28(4):467–516, 2003.

[112] Y. Diao, S. Rizvi, and M. J. Franklin. Towards an Internet-scale XML dissemination service. In M. A. Nascimento, M. T. Özsu, D. Kossmann, R. J. Miller, J. A. Blakeley, and K. B. Schiefer, editors, *Proc. of VLDB'04*, pages 612–623, 2004. Morgan Kaufmann.

[113] E. W. Dijkstra. Self-stabilizing systems in spite of distributed control. *Communications of the ACM*, 17(11):643–644, 1974. ISSN 0001-0782.

[114] J. Dingel and R. Strom, editors. *4th Intl. Workshop on Distributed Event-Based Systems (DEBS'05)*, June 2005. IEEE.

[115] J. Dingel, D. Garlan, S. Jha, and D. Notkin. Reasoning about implicit invocation. In *Proceedings of of the 6th International Symposium on the Foundations of Software Engineering (FSE-6)*, pages 209–221, November 1998. ACM.

[116] S. Dolev. *Self-Stabilization*. MIT Press, Cambridge, MA, 2000.

[117] S. Duarte, J. L. Martins, H. J. Domingos, and N. Preguiça. A case study on event dissemination in an active overlay network environment. In Jacobsen [210].

[118] J. Eder and E. Panagos. Towards distributed workflow process management. In C. Bussler, P. Grefen, H. Ludwig, and M.-C. Shan, editors, *Proceedings of the Workshop on Cross-Organisational Workflow Management and Coordination*, 1999.

[119] T. Elrad, R. E. Filman, and A. Bader. Aspect-oriented programming: Introduction. *Communications of the ACM*, 44(10):29–32, 2001. Special Issue on Aspect-Oriented Programming.

[120] P. Eugster, R. Guerraoui, and J. Sventek. Type-based publish/subscribe. Technical Report DSC ID:200029, EPFL Lausanne, Lausanne, Switzerland, 2000.

[121] P. Eugster, R. Guerraoui, and C. Damm. Linguistic support for large-scale distributed programming. In *Proceedings of the Intl. Conference on Object-Oriented Programming Systems, Languages and Applications (OOPSLA)*, pages 131–146. ACM, 2001.

[122] P. Eugster, S. Handurukande, R. Guerraoui, A.-M. Kermarrec, and P. Kouznetsov. Lightweight probabilistic broadcast. In *Proceedings of The International Conference on Dependable Systems and Networks (DSN 2001)*, pages 443–452, July 2001. IEEE Computer Society.

[123] P. Eugster, R. Guerraoui, S. Handurukande, P. Kouznetsov, and A.-M. Kermarrec. Lightweight probabilistic broadcast. *ACM Transactions on Computer Systems*, 21(4):341–374, 2003.

[124] P. T. Eugster and R. Guerraoui. Content-based publish/subscribe with structural reflection. In *In 6th USENIX Conference on Object-Oriented Technologies and Systems (COOTS'01)*, pages 131–146. USENIX, 2001.

[125] P. T. Eugster and R. Guerraoui. Probabilistic multicast. In Fabre and Jahanian [131].

[126] P. T. Eugster, R. Guerraoui, and J. Sventek. Distributed asynchronous collections: Abstractions for publish/subscribe interaction. In E. Bertino, editor, *European Conference on Object-Oriented Programming (ECOOP 2000)*, volume 1850 of *LNCS*, pages 252–276, 2000.

[127] P. T. Eugster, R. Guerraoui, and C. H. Damm. On objects and events. In L. Northrop and J. Vlissides, editors, *Proceedings of the OOPSLA '01 Conference on Object Oriented Programming Systems Languages and Applications*, pages 254–269, 2001. ACM.

[128] P. T. Eugster, R. Guerraoui, A.-M. Kermarrec, and L. Massoulieacute. Epidemic information dissemination in distributed systems. *IEEE Computer*, 37(5):60–67, May 2004.

[129] E. Evans. *Domain-Driven Design*. Addison-Wesley Professional, 2003.

[130] H. Evans and P. Dickman. DRASTIC: A run-time architecture for evolving, distributed, persistent systems. In M. Akşit and S. Matsuoka, editors, *European Conference for Object-Oriented Programming (ECOOP '97)*, volume 1241 of *LNCS*, pages 243–275, 1997. Springer.

[131] J.-C. Fabre and F. Jahanian, editors. *International Conference on Dependable Systems and Networks (DSN'02)*, 2002. IEEE.

[132] F. Fabret, F. Llirbat, J. Pereira, and D. Shasha. Efficient matching for content-based publish/subscribe systems. Technical report, INRIA, 2000.

[133] F. Fabret, A. Jacobsen, F. Llirbat, J. Pereira, K. Ross, and D. Shasha. Filtering algorithms and implementation for very fast publish/subscribe.

In T. Sellis and S. Mehrotra, editors, *Proceedings of the 20th Intl. Conference on Management of Data (SIGMOD 2001)*, pages 115–126, 2001.

[134] S. Fahmy, R. Jain, R. Goyal, B. Vandalore, S. Kalyanaraman, S. Kota, and P. Samudraand. Feedback Consolidation Algorithms for ABR Point-to-Multipoint Connections in ATM Networks. In *Proceedings of IEEE INFOCOM'98*, volume 3, pages 1004–1013, March 1998.

[135] L. Fiege. *Visibility in Event-Based Systems*. PhD thesis, Technical University of Darmstadt, Darmstadt, Germany, 2005.

[136] L. Fiege and G. Mühl. Rebeca Event-Based Electronic Commerce Architecture, 2000. http://event-based.org/rebeca.

[137] L. Fiege, G. Mühl, and A. Buchmann. An architectural framework for electronic commerce applications. In *Informatik 2001: Annual Conference of the German Computer Society*, 2001.

[138] L. Fiege, M. Mezini, G. Mühl, and A. P. Buchmann. Engineering event-based systems with scopes. In B. Magnusson, editor, *Proceedings of the European Conference on Object-Oriented Programming (ECOOP)*, volume 2374 of *LNCS*, pages 309–333, June 2002. Springer.

[139] L. Fiege, M. Mezini, G. Mühl, and A. P. Buchmann. Visibility as central abstraction in event-based systems. In A. Beugnard, S. Sadou, L. Duchien, and E. Jul, editors, *Concrete Communication Abstractions of the Next 701 Distributed Object Systems (ECOOP 2002 Workshop)*, volume 2548 of *LNCS*, 2002. Springer.

[140] L. Fiege, G. Mühl, and F. C. Gärtner. A modular approach to build structured event-based systems. In *Proceedings of the 2002 ACM Symposium on Applied Computing (SAC'02)*, pages 385–392, 2002. ACM.

[141] L. Fiege, F. C. Gärtner, S. B. Handurukande, and A. Zeidler. Dealing with uncertainty in mobile publish/subscribe middleware. In *1st International Workshop on Middleware for Pervasive and Ad-Hoc Computing (MPAC 03)*, pages 60–67, 2003. PUC-Rio.

[142] L. Fiege, F. C. Gärtner, O. Kasten, and A. Zeidler. Supporting mobility in content-based publish/subscribe middleware. In M. Endler and D. C. Schmidt, editors, *ACM/IFIP/USENIX International Middleware Conference (Middleware 2003)*, volume 2672 of *LNCS*, pages 103–122, 2003. Springer.

[143] L. Fiege, F. C. Gärtner, O. Kasten, and A. Zeidler. Supporting mobility in content-based publish/subscribe middleware. Technical Report IC/2003/11, Swiss Federal Institute of Technology (EPFL), School of Computer and Communication Sciences, Lausanne, Switzerland, March 2003.

[144] L. Fiege, G. Mühl, and F. C. Gärtner. Modular event-based systems. *The Knowledge Engineering Review*, 17(4):359–388, 2003.

[145] L. Fiege, A. Zeidler, A. Buchmann, R. Kilian-Kehr, and G. Mühl. Security aspects in publish/subscribe systems. In Carzaniga and Fenkam [66].

[146] L. Fiege, M. Cilia, and A. B. Gero Mühl. Publish/subscribe grows up: Support for management, visibility control & heterogeneity. *IEEE Internet Computing: Special Issue — Asynchronous Middleware and Services*, 10(1):48–55, January 2006.

[147] R. E. Filman, T. Elrad, S. Clarke, and M. Aksit, editors. *Aspect-Oriented Software Development*. Addison-Wesly, 2005.

[148] G. Fitzpatrick, S. Kaplan, T. Mansfield, A. David, and B. Segall. Supporting public availability and accessibility with elvin: Experiences and reflections. *Computer Supported Cooperative Work*, 11(3):447–474, 2002. ISSN 0925-9724.

[149] S. Floyd and K. Fall. Promoting the use of end-to-end congestion control in the internet. *IEEE/ACM Transactions on Networking*, 7(4):458–472, 1999.

[150] C. Fournet and G. Gonthier. The reflexive CHAM and the Join-Calculus. In *POPL '96: Proceedings of the 23rd ACM SIGPLAN-SIGACT Symposium on Principles of Programming Languages*, pages 372–385, 1996. ACM.

[151] M. Fowler. *Patterns of Enterprise Application Architecture*. Addison-Wesley, 2003.

[152] M. Fowler. Closure. http://www.martinfowler.com/bliki/Closure.html, September 2004.

[153] M. Fowler. Inversion of control containers and the dependency injection pattern. http://martinfowler.com/articles/injection.html, January 2004.

[154] M. Fowler. *UML Distilled*. Addison-Wesley, 2004.

[155] D. S. Frankel. *Model Driven Architecture*. Wiley, 2003.

[156] M. Franklin, S. Jeffery, S. Krishnamurthy, F. Reiss, S. Rizvi, E. Wu, O. Cooper, A. Edakkunni, and W. Hong. Design considerations for high fan-in systems: The HiFi approach. In *Proc. of CIDR'05*, January 2005.

[157] M. J. Franklin and S. B. Zdonik. A framework for scalable dissemination-based systems. In A. M. Berman, M. Loomis, and T. Bloom, editors, *Proceedings of the 12th ACM Conference on Object-Oriented Programming Systems, Languages, and Applications (OOPSLA '97)*, pages 94–105, Oct. 5–9, 1997.

[158] M. J. Franklin and S. B. Zdonik. "Data In Your Face": Push Technology in Perspective. In L. M. Haas and A. Tiwary, editors, *Proceedings ACM SIGMOD International Conference on Management of Data (SIGMOD'98)*, pages 516–519, 1998. ACM.

[159] L. Fuchs. Area: A cross-application notification service for groupware. In S. Bødker, M. Kyng, and K. Schmidt, editors, *The 6th European Conference on Computer Supported Cooperative Work (ECSCW 1999)*, pages 61–80, 1999. Kluwer Academic.

[160] A. Fuggetta, G. P. Picco, and G. Vigna. Understanding code mobility. *IEEE Transactions on Software Engineering*, 24(5):342–361, 1998.

[161] E. Gamma, R. Helm, R. Johnson, and J. Vlissides. *Design Patterns*. Addison-Wesley, Reading, MA, USA, 1995.

[162] J. García, J. Borrell, M. A. Jaeger, and G. Mühl. An alert communication infrastructure for a decentralized attack prevention framework. In *Proceedings of the IEEE International Carnahan Conference on Security Technology (ICCST)*, pages 234–237, October 2005. IEEE. ISBN 0-7803-9245-0.

[163] J. García, M. A. Jaeger, G. Mühl, and J. Borrell. Decoupling components of an attack prevention system using publish/subscribe. In *Proceedings of the 2005 IFIP conference on Intelligence in Communication Systems*, pages 87–98, October 2005. Springer.

[164] D. Garlan and D. Notkin. Formalizing design spaces: Implicit invocation mechanisms. In S. Prehn and W. J. H. Toetenel, editors, *VDM '91: Formal Software Development Methods*, volume 551 of *LNCS*, pages 31–44, 1991. Springer.

[165] D. Garlan and M. Shaw. An introduction to software architecture. In V. Ambriola and G. Tortora, editors, *Advances in Software Engineering and Knowledge Engineering*, volume 1, pages 1–40. World Scientific, 1993.

[166] D. Garlan, G. E. Kaiser, and D. Notkin. Using tool abstraction to compose systems. *IEEE Computer*, 25(6):30–38, June 1992.

[167] D. Garlan, R. Allen, and J. Ockerbloom. Architectural mismatch: Why reuse is so hard. *IEEE Software*, 12(6):17–26, November 1995.

[168] F. C. Gärtner. Fundamentals of fault-tolerant distributed computing in asynchronous environments. *ACM Computing Surveys*, 31(1):1–26, March 1999.

[169] F. C. Gärtner. *Formale Grundlagen der Fehlertoleranz in verteilten Systemen*. PhD thesis, TU Darmstadt, Darmstadt, Germany, 2001.

[170] S. Gatziu and K. R. Dittrich. Detecting composite events in active database systems using petri nets. In *Proceedings of the 4th International Workshop on Research Issues in Data Engineering: Active Database Systems (RIDE-AIDS'94)*, pages 2–9, February 1994.

[171] S. Gatziu, A. Koschel, G. von Bültzingsloewen, and H. Fritschi. Unbundling active functionality. *SIGMOD Record*, 27(1):35–40, Mar. 1998.

[172] D. Gawlick and S. Mishra. Information sharing with the Oracle database. In Jacobsen [210].

[173] N. H. Gehani, H. V. Jagadish, and O. Shmueli. Event specification in an active object-oriented database. In *Proceedings of ACM International Conference on Management of Data (SIGMOD'92)*, pages 81–90, June 1992.

[174] D. Gelernter. Generative communication in Linda. *ACM Transactions on Programming Languages and Systems*, 7(1):80–112, January 1985.

[175] M. R. Genesereth and S. P. Ketchpel. Software agents. *Communications of the ACM*, 37(7):48–53, July 1994.

[176] D. Georgakopoulos, M. F. Hornick, and A. P. Sheth. An overview of workflow management: From process modeling to workflow automation infrastructure. *Distributed and Parallel Databases*, 3(2):119–153, April 1995.

[177] A. Geppert and D. Tombros. Event-based distributed workflow execution with EVE. In N. Davies, K. Raymond, and J. Seitz, editors, *Middleware '98*. Springer, 1998.

[178] A. Goldberg and D. Robson. *Smalltalk 80: The Language and its Implementation*. Addison-Wesley, 1983.

[179] S. J. Golestani and K. K. Sabnani. Fundamental observations on multicast congestion control in the internet. In *Proceedings of INFOCOM'99*, pages 990–1000, March 1999.

[180] L. Gong. Project JXTA: A Technical Overview. Whitepaper, Sun Microsystems, October 2002. http://www.jxta.org.

[181] J. Gough and G. Smith. Efficient recognition of events in distributed systems. In *Proceedings of 18th Australasian Computer Science Conference (ACSC)*, February 1995.

[182] J. Gray and A. Reuter. *Transaction Processing: Concepts and Techniques*. Morgan Kaufmann, 1993.

[183] T. J. Green, G. Miklau, M. Onizuka, and D. Suciu. Processing XML streams with deterministic automata. In *ICDT '03: Proceedings of the 9th International Conference on Database Theory*, pages 173–189, 2002. Springer. ISBN 3-540-00323-1.

[184] R. Gruber, B. Krishnamurthy, and E. Panagos. The architecture of the READY event notification service. In P. Dasgupta, editor, *Proceedings of the 19th IEEE International Conference on Distributed Computing Systems, Middleware Workshop*, pages 108–113, May 1999. IEEE.

[185] R. Gruber, B. Krishnamurthy, and E. Panagos. READY: A high performance event notification service. In *Proceedings of the 16th International Conference on Data Engineering*, pages 668–669. IEEE Computer Society, 2000.

[186] S. Handurukande, P. T. Eugster, P. Felber, and R. Guerraoui. Event systems: How to have ones cake and eat it too. In Bacon et al. [21]. ISBN 0-7695-1588-6. Published as part of the ICDCS '02 Workshop Proceedings.

[187] E. N. Hanson, M. Chaabouni, C.-H. Kim, and Y.-W. Wang. A predicate matching algorithm for database rule systems. In *19th ACM SIGMOD Conference on the Management of Data (SIGMOD)*, pages 271–280, May 1990.

[188] W. Harrison and H. Ossher. Subject-oriented programming (A critique of pure objects). In A. Paepcke, editor, *Proceedings of the 8th ACM Conference on Object-Oriented Programming Systems, Languages, and Applications (OOPSLA '93)*, pages 411–428, 1993.

[189] R. Hayton. *OASIS: An Open Architecture for Secure Interworking Services*. PhD thesis, University of Cambridge Computer Laboratory, Cambridge, United Kingdom, June 1996. Technical Report No. 399.

[190] C. Heinlein. Workflow and process synchronization with interaction expressions and graphs. In A. Reuter, D. Lomet, A. Buchmann, and D. Georgakopoulos, editors, *Proc. of the 17th International Conference on Data Engeneering (ICDE)*, pages 243–252, 2001. IEEE Computer Society.

[191] G. Hohpe and B. Woolf. *Enterprise Integration Patterns: Designing, Building, and Deploying Messaging Solutions*. Addison-Wesley, 2003.

[192] A. B. Hombrecher. *Reconciling Event Taxonomies Across Administrative Domains*. PhD thesis, University of Cambridge Computer Laboratory, Cambridge, United Kingdom, June 2002.

[193] J. E. Hopcroft, R. Motwani, and J. D. Ullman. *Introduction to Automata Theory, Languages, and Computation*. Addison-Wesley, 2001.

[194] J. Huang, A. Black, J. Walpole, and C. Pu. Infopipes — an abstraction for information flow. In *Proceedings of the ECOOP Workshop on The Next 700 Distributed Object Systems*, June 2001.

[195] Y. Huang and H. Garcia-Molina. Publish/subscribe in a mobile environment. In S. Banerjee, editor, *2nd ACM International Workshop on Data Engineering for Wireless and Mobile Access (MobiDE'01)*, pages 27–34, 2001.

[196] Y. Huang and H. Garcia-Molina. Publish/subscribe tree construction in wireless ad-hoc networks. In M.-S. Chen, P. Chrysanthis, M. Sloman, and A. Zaslavsky, editors, *4th International Conference on Mobile Data Management (MDM 2003)*, volume 2574 of *LNCS*, pages 122–140, 2003. Springer.

[197] IBM. Gryphon: Publish/subscribe over public networks. Technical report, IBM T.J. Watson Research Center, 2001.

[198] IBM. *WebSphere MQ: Application Programming Guide Version 6.0*, May 2005.

[199] IBM. *WebSphere MQ: Publish/Subscribe User's Guide Version 6.0*, May 2005.

[200] IBM, Akamai Technologies, Computer Associates, Fujitsu Laboratories of Europe, Globus, Hewlett-Packard, SAP AG, Sonic Software, and TIBCO Software. Publish-subscribe notification for Web services, March 2004. http://www.ibm.com/developerworks/library/specification/ws-pubsub.

[201] IBM, BEA Systems, Microsoft, Computer Associates, SUN Microsystems, and TIBCO Software. Web Services Eventing (WS-Eventing) Specification, August 2004. http://www.ibm.com/developerworks/webservices/library/specification/ws-eventing/.

[202] IBM Corporation. IBM WebSphere MQ Event Broker, May 2002. http://www.ibm.com/software/integration/mqfamily/eventbroker.

[203] IBM TJ Watson Research Center. Gryphon: Publish/Subscribe over Public Networks. December 2001. http://researchweb.watson.ibm.com/gryphon/Gryphon

[204] C. Intanagonwiwat, R. Govindan, and D. Estrin. Directed diffusion: A scalable and robust communication paradigm for sensor networks. In *Proceedings of the Sixth Annual International Conference on Mobile Computing and Networking (MobiCom'00)*, pages 56–67, 2000.

[205] C. Intanagonwiwat, R. Govindan, D. Estrin, J. Heidemann, and F. Silva. Directed diffusion for wireless sensor networking. *IEEE/ACM Transactions on Networking (TON)*, 11(1):2–16, 2003.

[206] ISO/IEC. Open distributed processing–reference model. International Standard ISO/IEC IS 10746, May 1995.

[207] ITU-T. ITU-T X.509. Recommendation, ITU-T International Telecommunication Union, Geneva, Switzerland, 2000.

[208] H.-A. Jacobsen. Middleware services for selective and location-based information dissemination in mobile wireless networks. In Banavar [25].

[209] H.-A. Jacobsen. Middleware architecture design based on aspects, the open implementation metaphor and modularity. In A. Rashid and L. Blair, editors, *Workshop on Aspect-Oriented Programming and Separation of Concerns*, August 2001.

[210] H.-A. Jacobsen, editor. *2nd Intl. Workshop on Distributed Event-Based Systems (DEBS'03)*, June 2003. ACM.

[211] V. Jacobson and M. J. Karels. Congestion avoidance and control. In *Proceedings of ACM SIGCOMM'88*, pages 314–332, August 1988.

[212] M. A. Jaeger and G. Mühl. Stochastic analysis and comparison of self-stabilizing routing algorithms for publish/subscribe systems. In *The 13th IEEE/ACM International Symposium on Modeling, Analysis and Simulation of Computer and Telecommunication Systems (MASCOTS 2005)*, pages 471–479, September 2005. IEEE.

[213] K. Jenkins, K. Hopkins, and K. Birman. A gossip protocol for subgroup multicast. In M. Raynal and L. Rodrigues, editors, *International Workshop on Applied Reliable Group Communication (WARGC 2001)*, 2001. IEEE.

[214] Y. Jin and R. Strom. Relational subscription middleware for internet-scale publish-subscribe. In Jacobsen [210].

[215] T. Joseph. A messaging-based architecture for enterprise application integration. In *Proceedings of the 15th International Conference on Data Engineering (ICDE'99)*, pages 62–63, 1999.

[216] M. Kahani and H. W. P. Beadle. Decentralised approaches for network management. *ACM SIGCOMM Computer Communication Review*, 27 (3):36–47, 1997.

[217] G. Kappel, S. Rausch-Schott, and W. Retschitzegger. Coordination in workflow management systems—a rule-based approach. In W. Conen and G. Neumann, editors, *Coordination Technology for Collaborative*

Applications (ASIAN 1996 Workshop), volume 1364 of *LNCS*, pages 99–120. Springer, 1998.

[218] A. M. Keller and J. Basu. A predicate-based caching scheme for client-server database architectures. *VLDB Journal*, 5(1):35–47, 1996.

[219] J. O. Kephart and D. M. Chess. The vision of autonomic computing. *IEEE Computer Magazine*, pages 41–50, January 2000.

[220] G. Kiczales. Beyond the black box: Open implementation. *IEEE Software*, 13(1):8–11, January 1996.

[221] G. Kiczales, J. des Rivieres, and D. G. Bobrow. *The Art of the Meta-Object Protocol*. MIT Press, Cambridge, MA, USA, 1991.

[222] G. Kiczales, J. Lamping, A. Mendhekar, C. Maeda, C. Lopes, J.-M. Loingtier, and J. Irwin. Aspect-oriented programming. In M. Akşit and S. Matsuoka, editors, *ECOOP'97—Object-Oriented Programming*, volume 1241 of *LNCS*, pages 220–242. Springer, 1997.

[223] F. Kon, F. Costa, G. Blair, and R. H. Campbell. The case for reflective middleware. *Communications of the ACM*, 45(6):33–38, 2002.

[224] H. Kopetz. Event-triggered versus time-triggered real-time systems. In *Proceedings of the International Workshop on Operating Systems of the 90s and Beyond*, volume 563 of *LNCS*, pages 87–101. Springer, 1991.

[225] E. Kotsovinos, B. Dragovic, S. Hand, and P. R. Pietzuch. Xenotrust: Event-based distributed trust management. In *Proceedings of Trust and Privacy in Digital Business (TrustBus'03). In conjunction with the 14th International Conference on Database and Expert Systems Applications (DEXA'03)*, September 2003.

[226] R. Laddad. *AspectJ in Action*. Manning, 2003.

[227] L. Lamport. Time, clocks, and the ordering of events in a distributed system. *Communications of the ACM*, 21(7):558–565, July 1978.

[228] L. Lamport. Proving the correctness of multiprocess programs. *IEEE Transactions on Software Engineering*, 3(2):125–143, March 1977.

[229] L. Lamport. What good is temporal logic? In R. E. A. Mason, editor, *Proceedings of the IFIP Congress on Information Processing*, pages 657–667, 1983. North-Holland.

[230] L. Lamport and N. Lynch. Distributed computing: Models and methods. In J. van Leeuwen, editor, *Handbook of Theoretical Computer Science, Volume B: Formal Models and Semantics*, pages 1157–1199. Elsevier, 1990.

[231] F. Lange, R. Kröger, and M. Gergeleit. JEWEL: Design and implementation of a distributed measurement system. *IEEE Transactions on Parallel and Distributed Systems*, 3(6):657–671, 1992.

[232] O. Lassila and R. R. Swick. Resource description framework (RDF) model and syntax specification. W3C Recommendation, Feb. 1999. http://www.w3.org/TR/REC-rdf-syntax.

[233] J. Le Boudec. The Asynchronous Transfer Mode: a tutorial. *Computer Networks and ISDN Systems*, 24:279–309, 1992.

[234] G. T. Leavens and M. Sitaraman, editors. *Foundations of Component-Based Systems*. Cambridge University Press, 2000.

[235] J. Liberty. *Programming C#*. O'Reilly, 3rd edition, 2003.

[236] C. Liebig and S. Tai. Advanced transactions. In *Proceedings of the 2nd International Workshop on Engineering Distributed Objects (EDO'00)*, volume 1999 of *Lecture Notes in Computer Science*, pages 188–193, November 2000. Springer.

[237] C. Liebig and S. Tai. Middleware mediated transactions. In G. Blair, D. Schmidt, and M. Takizawa, editors, *3rd Intl. Symposium on Distributed Objects and Applications (DOA'01)*, September 2001. IEEE Computer Society.

[238] C. Liebig, M. Cilia, and A. Buchmann. Event composition in time-dependent distributed systems. In *Proceedings of the 4th Intl. Conference on Cooperative Information Systems (CoopIS '99)*. IEEE Computer Society, September 1999.

[239] B. Liskov and R. Scheifler. Guardians and actions: Linguistics support for robust, distributed systems. *ACM Transactions on Programming Languages and Systems*, 5(3):381–404, 1983.

[240] H. Liu and H.-A. Jacobsen. A-ToPSS — a publish/subscribe system supporting approximate matching. In *Procedings of the 28th VLDB Conference*, 2002. http://www.vldb.org/conf/2004/DEMP8.PDF.

[241] L. Liu, C. Pu, W. Tang, and W. Han. Conquer: A continual query system for update monitoring in the WWW. *International Journal of Computer Systems, Science and Engineering, Special issue on Web semantics*, 14(2):99–112, 1999.

[242] D. Luckham. *The Power of Events*. Addison-Wesley, 2002.

[243] D. C. Luckham. Rapide: A language and toolset for simulation of distributed systems by partial ordering of events. In *DIMACS Partial Order Methods Workshop IV*. Princeton University, July 1996.

[244] C. Ma and J. Bacon. COBEA: A CORBA-based event architecture. In J. Sventek, editor, *Proceedings of the 4th Conference on Object-Oriented Technologies and Systems (COOTS-98)*, pages 117–132, 1998. USENIX Association.

[245] P. Maes. Concepts and experiments in computational reflection. In N. Meyrowitz, editor, *Proceedings of the 2nd ACM Conference on Object-Oriented Programming Systems, Languages and Applications (OOPSLA '87)*, pages 147–155, October 1987. ACM. ISBN 0-89791-247-0.

[246] T. W. Malone and K. Crowston. The interdisciplinary study of coordination. *ACM Computing Surveys*, 26(1):87–119, 1994.

[247] T. W. Malone and R. J. Laubacher. The dawn of the E-Lance economy. *Harvard Business Review*, pages 145–152, September 1998.

[248] Z. Manna and A. Pnueli. *The Temporal Logic of Reactive and Concurrent Systems*. Springer, 1992.

[249] M. Mansouri-Samani. *Monitoring of Distributed Systems*. PhD thesis, Imperial College, London, UK, 1995.

[250] M. Mansouri-Samani and M. Sloman. Gem: A generalised event monitoring language for distributed systems. *IEE/IOP/BCS Distributed Systems Engineering Journal*, 4(2):96–108, June 1997.

[251] M. Mansouri-Samani and M. Sloman. Gem — a generalised event monitoring language for distributed systems. In *Joint International Conference on Open Distributed Processing (ICODP) and Distributed Platforms (ICDP) '97*, 1997.

[252] R. C. Martin. The Dependency Inversion Principle. *C++ Report*, 8(6): 61–66, June 1996.

[253] D. Mason and D. Woit. Problems with software reliability composition. In *Proceedings of 1998 International Symposium on Software Reliability Engineering (ISSRE'98 Fast Abstracts)*, 1998. http://www.chillarege.com/fastabstracts/issre98/98408.html.

[254] F. Mattern. The vision and technical foundations of ubiquitous computing. *Upgrade*, II(5), 2001. Special issue on Ubiquitous Computing.

[255] N. Maxemchuk and D. Shur. An internet multicast system for the stock market. *ACM Transactions on Computer Systems*, 19(3):384–412, 2001.

[256] N. Medvidovic and R. N. Taylor. A framework for classifying and comparing architecture description languages. In M. Jazayeri and H. Schauer, editors, *ESEC/FSE '97*, volume 1301 of *Lecture Notes in Computer Science*, pages 60–76. Springer, 1997.

[257] R. Meier and V. Cahill. Steam: Event-based middleware for wireless ad hoc networks. In Bacon et al. [21]. ISBN 0-7695-1588-6. Published as part of the ICDCS '02 Workshop Proceedings.

[258] R. Meier, M.-O. Killijian, R. Cunningham, and V. Cahill. Towards proximity group communication. In Banavar [25].

[259] D. Meyer. RFC 2365: Administratively scoped IP multicast. http://www.ietf.org/rfc/rfc2365.txt, July 1998. Status: Best Current Practice.

[260] Z. Miklós. Towards an access control mechanism for wide-area publish/-subscribe systems. In Bacon et al. [21], pages 516–521. ISBN 0-7695-1588-6. Published as part of the ICDCS '02 Workshop Proceedings.

[261] R. Milner. *Communicating and Mobile Systems: The Pi Calculus*. Cambridge University Press, May 1999.

[262] G. Mühl. Generic constraints for content-based publish/subscribe systems. In C. Batini, F. Giunchiglia, P. Giorgini, and M. Mecella, editors, *Proceedings of the 6th International Conference on Cooperative Information Systems (CoopIS '01)*, volume 2172 of *LNCS*, pages 211–225, 2001. Springer.

[263] G. Mühl. *Large-Scale Content-Based Publish/Subscribe Systems*. PhD thesis, Darmstadt University of Technology, Darmstadt, Germany, 2002. http://elib.tu-darmstadt.de/diss/000274/.

[264] G. Mühl and L. Fiege. Supporting covering and merging in content-based publish/subscribe systems: Beyond name/value pairs. *IEEE Distributed Systems Online (DSOnline)*, 2(7), 2001.

[265] G. Mühl, L. Fiege, and A. P. Buchmann. Evaluation of cooperation models for electronic business. In *Information Systems for E-Commerce, Conference of German Society for Computer Science*, pages 81–94, November 2000. ISBN 3-85487-194-5.

[266] G. Mühl, L. Fiege, and A. P. Buchmann. Filter similarities in content-based publish/subscribe systems. In H. Schmeck, T. Ungerer, and L. Wolf, editors, *International Conference on Architecture of Computing Systems (ARCS)*, volume 2299 of *Lecture Notes in Computer Science*, pages 224–238, 2002. Springer.

[267] G. Mühl, L. Fiege, F. C. Gärtner, and A. P. Buchmann. Evaluating advanced routing algorithms for content-based publish/subscribe systems. In A. Boukerche, S. K. Das, and S. Majumdar, editors, *The Tenth IEEE/ACM International Symposium on Modeling, Analysis and Simulation of Computer and Telecommunication Systems (MASCOTS 2002)*, pages 167–176, October 2002. IEEE.

[268] M. Mühleisen. Programming and administration of publish-subscribe systems (in German). Master's thesis, Technische Universität Darmstadt, 2005.

[269] S. Mullender, editor. *Distributed Systems*. Addison-Wesley, 2nd edition, 1993.

[270] B. C. Neuman and T. Ts'o. Kerberos: An authentication service for computer networks. *IEEE Communications Magazine*, 32(9):33–38, September 1994.

[271] B. Nguyen, S. Abiteboul, G. Cobena, and M. Preda. Monitoring XML data on the Web. *SIGMOD Record*, 30(2):437–448, 2001.

[272] D. Notkin, D. Garlan, W. G. Griswold, and K. Sullivan. Adding implicit invocation to languages: Three approaches. In *Proceedings of the JSSST International Symposium on Object Technologies for Advanced Software*, volume 742 of *Lecture Notes in Computer Science*, pages 489–510. Springer, November 1993.

[273] *FIXML — A Markup Language for the Financial Information eXchange (FIX) protocol.* Oasis, July 2001. http://www.oasis-open.org/cover/fixml.html.

[274] OASIS. Web Services Base Notification (WS-BaseNotification), July 2005.

[275] OASIS. Web Services Brokered Notification (WS-BrokeredNotification), July 2005.

[276] OASIS. Web Services Web Services Topics (WSTopics), July 2005.

[277] Object Management Group (OMG). CORBA event service specification. OMG Document formal/94-01-01, 1994.

[278] Object Management Group (OMG). Corba components, 1999. OMG document orbos/99-07-01.

[279] Object Management Group (OMG). CORBA notification service. OMG Document telecom/99-07-01, 1999.

[280] Object Management Group (OMG). CORBA event service specification, version 1.0. OMG Document formal/2000-06-15, 2000.

[281] Object Management Group (OMG). CORBA transaction service v1.1. OMG Document formal/00-06-28, 2000.

[282] Object Management Group (OMG). Management of event domains. Version 1.0, Formal Specification, 2001. OMG document formal/01-06-03.

[283] Object Management Group (OMG). The common object request broker: Architecture and specification, version 3.0. OMG document formal/02-06-33, July 2002.

[284] Object Management Group (OMG). Distributed simulation systems specification, version 2.0. OMG Document formal/02-11-01, 2002.

[285] Object Management Group (OMG). CORBA event service specification, version 1.2. OMG Document formal/2004-10-02, 2004.

[286] Object Management Group (OMG). Data distribution service for real-time systems. OMG Document formal/04-12-02, 2004.

[287] Object Management Group (OMG). CORBA notification service, version 1.1. OMG Document formal/2004-10-11, 2004.

[288] Object Management Group (OMG). UML superstructure specification, v2.0. OMG document formal/05-07-04, 2005.

[289] B. Oki, M. Pfluegl, A. Siegel, and D. Skeen. The information bus—an architecture for extensible distributed systems. In B. Liskov, editor, *Proceedings of the 14th Symposium on Operating Systems Principles*, pages 58–68, December 1993. ACM.

[290] L. Opyrchal. *Content-Based Publish/Subscribe Systems: Scalability and Security*. PhD thesis, University of Michigan, Ann Arbor, MI, USA, 2004.

[291] L. Opyrchal, M. Astley, J. Auerbach, G. Banavar, R. Strom, and D. Sturman. Exploiting IP multicast in content-based publish-subscribe systems. In J. Sventek and G. Coulson, editors, *IFIP/ACM International Conference on Distributed Systems Platforms (Middleware 2000)*, volume 1795 of *LNCS*, pages 185–207. Springer, 2000.

[292] Oracle, Inc. Introduction to Oracle Advanced Queuing (AQ). Application Developer's Guide, July 2001.

[293] S. Pallickara and G. Fox. Naradabrokering: A middleware framework and architecture for enabling durable peer-to-peer grids. In M. Endler and D. Schmidt, editors, *Proceedings of the 4th International Conference on Middleware (Middleware'03)*, volume 2672 of *LNCS*, pages 41–61, June 2003.

[294] S. Pallickara, M. Pierce, G. Fox, Y. Yan, and Y. Huang. A Security Framework for Distributed Brokering Systems. http://www.naradabrokering.org/papers/NB-SecurityFramework.pdf, 2003.

[295] G. A. Papadopoulos and F. Arbab. Coordination models and languages. In M. Zelkowitz, editor, *The Engineering of Large Systems*, volume 46 of *Advances in Computers*. Academic, August 1998.

[296] G. A. Papadopoulos and F. Arbab. Modelling activities in information systems using the coordination language manifold. In K. M. George and G. B. Lamong, editors, *Proceedings of the ACM Symposium on Applied Computing (SAC '98)*, pages 185–193, 1998. ACM.

[297] G. A. Papadopoulos and F. Arbab. Configuration and dynamic reconfiguration of components using the coordination paradigm. *Future Generation Computer Systems*, 17(8):1023–1038, June 2001.

[298] G. Pardo-Castellote. OMG data distribution service: Real-time publish/subscribe becomes a standard. *RTC Magazine*, jan 2005.

[299] G. Pardo-Castellote. OMG data distribution service: Architectural overview. In Wu [401], pages 200–206.

[300] D. L. Parnas. On the criteria to be used in decomposing systems into modules. *Communications of the ACM*, 15(12):1053–1058, December 1972.

[301] C. Partridge, T. Mendez, and W. Milliken. RFC 1546: host anycasting service, November 1993. Status: Informational, http://www.ietf.org/rfc/rfc1546.txt.

[302] H. Parzyjegla. Ein adaptives brokernetz für publish/subscribe systeme. Master's thesis, Technische Universität Berlin, Berlin, Germany, October 2005.

[303] H. Parzyjegla, G. Mühl, and M. A. Jaeger. Reconfiguring publish/subscribe overlay topologies. In *5th Intl. Workshop on Distributed Event-based Systems (DEBS'06)*, July 2006. IEEE Press.

[304] N. W. Paton and O. Diaz. Active Database Systems. *ACM Computing Surveys*, 31(1):63–103, 1999.

[305] J. Pereira, F. Fabret, F. Llirbat, and D. Shasha. Efficient matching for Web-based publish/subscribe systems. In O. Etzion and P. Scheuermann, editors, *Proc. of the Int. Conf. on Cooperative Information Systems (CoopIS)*, volume 1901 of *LNCS*, pages 162–173, 2000. Springer.

[306] C. Perkins. Mobile IP. *IEEE Communications Magazine*, 35(5):84–99, May 1997.

[307] J. L. Peterson. Petri nets. *ACM Computing Surveys*, 9(3):223–252, September 1977.

[308] G. P. Picco, G. Cugola, and A. L. Murphy. Efficient content-based event dispatching in the presence of topological reconfiguration. In P. McKinley and S. Shatz, editors, *Proceedings of the 23rd International Conference on Distributed Computing Systems (ICDCS 03)*, pages 234–243, 2003. IEEE.

[309] P. R. Pietzuch. An Event Type Compiler for ODL. Computer Science Tripos Part II Project Dissertation, University of Cambridge Computer Laboratory, Cambridge, United Kingdom, June 2000.

[310] P. R. Pietzuch. *Hermes: A Scalable Event-Based Middleware.* PhD thesis, University of Cambridge, Cambridge, United Kingdom, February 2004.

[311] P. R. Pietzuch and J. Bacon. Hermes: A distributed event-based middleware architecture. In Bacon et al. [21], pages 611–618. ISBN 0-7695-1588-6. Published as part of the ICDCS '02 Workshop Proceedings.

[312] P. R. Pietzuch and J. Bacon. Peer-to-peer overlay broker networks in an event-based middleware. In Jacobsen [210].

[313] P. R. Pietzuch and S. Bhola. Congestion control in a reliable scalable message-oriented middleware. In M. Endler and D. Schmidt, editors, *Proceedings of the 4th International Conference on Middleware (Middleware'03)*, volume 2672 of *LNCS*, pages 202–221, June 2003. ACM/IFIP/USENIX, Springer Verlag.

[314] P. R. Pietzuch, B. Shand, and J. Bacon. Composite event detection as a generic middleware extension. *IEEE Network Magazine, Special Issue on Middleware Technologies for Future Communication Networks*, Jan/Feb 2004.

[315] D. Platt. The COM+ event service eases the pain of publishing and subscribing to data. *Microsofts Systems Journal*, September 1999.

[316] C. G. Plaxton, R. Rajaraman, and A. W. Richa. Accessing nearby copies of replicated objects in a distributed environment. In *Proc. of the 9th Annual ACM Symposium on Parallel Algorithms and Architectures (SPAA'97)*, pages 311–320, 1997. ACM.

[317] A. Pnueli. The temporal semantics of concurrent programs. *Theoretical Computer Science*, 13:45–60, 1981.

[318] A. Pope. *The CORBA Reference Guide.* Addison-Wesley, Reading, MA, USA, 1997.

[319] D. Powell. Group communication. *Communications of the ACM*, 39(4): 50–53, April 1996.

[320] R. Prakash and R. Baldon. Architecture for group communication in mobile systems. In *The 17th IEEE Symposium on Reliable Distributed Systems (SRDS '98)*, pages 235–242, October 1998.

[321] B. Quinn and K. Almeroth. RFC 3170: IP multicast applications: Challenges and solutions, September 2001. Status: Informational, http://www.ietf.org/rfc/rfc3170.txt.

[322] K. Ramamritham, P. Deolasee, A. Katkar, A. Panchbudhe, and P. Shenoy. Dissemination of dynamic data on the Internet. In S. Bhalla, editor, *Databases in Networked Information Systems (DNIS 2000)*, volume 1966 of *LNCS*, pages 173–187, 2000. Springer.

[323] S. Ratnasamy, P. Francis, M. Handley, R. Karp, and S. Schenker. A scalable content-addressable network. In *Proceedings of the 2001 Conference on Applications, Technologies, Architectures, and Protocols for Computer Communications (SIGCOMM)*, pages 161–172, 2001. ACM.

[324] Real-Time Innovations (RTI), Inc. Network data distribution services (NDDS), 2006. http://www.rti.com/products_ndds.html.

[325] S. P. Reiss. Connecting tools using message passing in the Field environment. *IEEE Software*, 7(4):57–66, July 1990.

[326] A. Ricci, A. Omicini, and E. Denti. Objective vs. subjective coordination in agent-based systems: A case study. In Arbab and Talcott [15], pages 291–299.

[327] D. Riehle. The event notification pattern—integrating implicit invocation with object-orientation. *Theory and Practice of Object Systems*, 2 (1):43–52, 1996.

[328] L. Rizzo. pgmcc: A TCP-friendly single-rate multicast congestion control scheme. In *Proceedings of ACM SIGCOMM'00*, pages 17–28, August 2000. ISBN 1-58113-223-9.

[329] L. Roberts. Rate-based algorithm for point to multipoint abr service. ATM Forum Contribution 94-0772R1, November 1994.

[330] M. T. Rose. *The Simple Book: An Introduction to Internet Management*. P T R Prentice-Hall, 2nd edition, 1994.

[331] A. Rowstron and P. Druschel. Pastry: scalable, decentraized object location and routing for large-scale peer-to-peer systems. In R. Guerraoui, editor, *Proceedings of the 18th IFIP/ACM International Conference on Distributed Systems Platforms (Middleware)*, volume 2218 of *LNCS*, pages 329–350, 2001. Springer.

[332] R. Sandhu, E. Coyne, H. L. Feinstein, and C. E. Youman. Role-Based Access Control Models. *IEEE Computer*, 29(2):38–47, 1996.

[333] S. Saroiu, K. P. Gummadi, R. J. Dunn, S. D. Gribble, and H. M. Levy. An analysis of internet content delivery systems. *ACM Operating Systems Review*, 36(SI):315–327, 2002.

[334] S. S. Sathaye. ATM Forum Traffic Management Specification 4.0. ATM Forum af-tm-0056.000, April 1996.

[335] B. Schilit, N. Adams, and R. Want. Context-aware computing applications. In *IEEE Workshop on Mobile Computing Systems and Applications*, pages 85–90, 1994.

[336] D. Schmidt and S. Vinoski. Time-independent invocation and interoperable routing. *C++ Report*, 11(4), April 1999.

[337] D. Schmidt, M. Stal, H. Rohnert, and F. Buschmann. *Pattern-Oriented Software Architecture: Patterns for Concurrent and Networked Objects*. Wiley, 2000.

[338] D. C. Schmidt and S. Vinoski. Programming asynchronous method invocations with corba messaging. *C++ Report*, 11(2), February 1999.

[339] S. Schwiderski. *Monitoring the Behaviour of Distributed Systems*. PhD thesis, 1996, Cambridge, United Kingdom, University of Cambridge.

[340] B. Segall, D. Arnold, J. Boot, M. Henderson, and T. Phelps. Content-based routing in Elvin4. In *Proceedings of AUUG2K*, June 2000.

[341] W. Segall and D. Arnold. Elvin has left the building: A publish/subscribe notification service with quenching. In *Proceedings of the 1997 Australian UNIX Users Group, Bris-*

bane, Australia, September 1997., pages 243–255, 1997. http://elvin.dstc.edu.au/doc/papers/auug97/AUUG97.html.

[342] S. Shah, S. Dharmarajan, and K. Ramamritham. An efficient and re-silient approach to filtering and disseminating streaming data. In *VLDB 2003, Proceedings of 29th International Conference on Very Large Data Bases*, pages 57–68. Morgan Kaufman, 2003.

[343] Z. Shen and S. Tirthapura. Self-stabilizing routing in publish-subscribe systems. In Carzaniga and Fenkam [66].

[344] S. Shi and M. Waldvogel. A rate-based end-to-end multicast congestion control protocol. In *Proceedings of 5th IEEE Symposium on Computer and Communication (ISCC'00)*, pages 678–686, July 2000.

[345] M. D. Skeen and M. Bowles. Apparatus and method for providing decou-pling of data exchange details for providing high performance communi-cation between software processes. United States Patent No. 5,557,798, September 1996.

[346] M. D. Spiteri. *An Architecture for the Notification, Storage and Re-trieval of Events*. PhD thesis, University of Cambridge Computer Lab-oratory, Cambridge, United Kingdom, January 2000.

[347] M. Stal. Web services: Beyond component-based computing. *Commu-nications of the ACM*, 45(10):71–76, 2002.

[348] W. Stallings. SNMP and SNMPv2: The infrastructure for network man-agement. *IEEE Communications Magazine*, 36(3):37–43, March 1998.

[349] J. Steffan, L. Fiege, M. Cilia, and A. Buchmann. Scoping in wireless sensor networks. In *2nd International Workshop on Middleware for Pervasive and Ad-Hoc Computing*, pages 167–171, 2004. ACM. ISBN 1-58113-951-9.

[350] J. Steffan, L. Fiege, M. Cilia, and A. Buchmann. Towards multi-purpose wireless sensor networks. In P. Dini et al., editors, *Proc. of SENET'05*, pages 336–341, August 2005. IEEE Computer Society.

[351] I. Stoica, R. Morris, D. Karger, M. F. Kaashoek, and H. Balakrish-nan. Chord: A scalable peer-to-peer lookup service for Internet appli-cations. In *Proceedings of the 2001 Conference on Applications, Tech-nologies, Architectures, and Protocols for Computer Communications (SIGCOMM)*, pages 149–160, August 2001. ACM.

[352] M. Stonebraker, E. N. Hanson, and S. Potamianos. The postgres rule manager. *IEEE Transactions on Software Engineering*, 14(7):897–907, 1988.

[353] R. Strom, G. Banavar, T. Chandra, M. Kaplan, K. Miller, B. Mukherjee, D. Sturman, and M. Ward. Gryphon: An information flow based ap-proach to message brokering. In *Int'l Symposium on Software Reliability Engineering*, 1998.

[354] D. Sturman, G. Banavar, and R. Strom. Reflection in the Gryphon message brokering system. In *In Reflection Workshop of the 13th ACM Conference on Object Oriented Programming Systems, Languages and Applications (OOPSLA'98)*, 1998.

374   References

[355] K. J. Sullivan and D. Notkin. Reconciling environment integration and component independence. In R. N. Taylor, editor, *Proceedings of the 4th ACM SIGSOFT Symposium on Software Development Environments*, pages 22–33, 1990. ACM.

[356] K. J. Sullivan and D. Notkin. Reconciling environment integration and software evolution. *ACM Transactions of Software Engineering and Methodology*, 1(3):229–269, July 1992.

[357] Q. Sun. Reliable multicast for publish/subscribe systems. Master's thesis, Massachusetts Institute of Technology, 2000.

[358] Sun Microsystems, Inc. Java AWT delegation event model, 1997.

[359] Sun Microsystems, Inc. JavaBeans API specification version 1.0.1, 1997. http://java.sun.com/products/javabeans/.

[360] Sun Microsystems, Inc. Jini Specification, Version 2.0. Specification, Sun Microsystems, June 2003. http://java.sun.com/products/jini/.

[361] Sun Microsystems, Inc. Distributed Event Specification, 1998.

[362] Sun Microsystems, Inc. Enterprise JavaBeans specification, version 2.0. Proposed Final Draft, 2000. http://java.sun.com/products/ejb.

[363] Sun Microsystems, Inc. Java management extensions. Instrumentation and Agent Specification, v1.2, October 2002.

[364] Sun Microsystems, Inc. Java Message Service (JMS) Specification 1.1, 2002.

[365] Sun Microsystems, Inc. Java 2 Platform Enterprise Edition Specification, v. 1.4, July 2003.

[366] Sun Microsystems, Inc. JavaSpaces Service Specification version 2.0, June 2003.

[367] Sun Microsystems, Inc. Java Remote Method Invocation (RMI) Specification 1.5. Sun Microsystems, 2004. http://java.sun.com/j2se/1.5.0/docs/guide/rmi/spec/rmiTOC.html.

[368] P. Sutton, R. Arkins, and B. Segall. Supporting disconnectedness — transparent information delivery for mobile and invisible computing. In *First International Symposium on Cluster Computing and the Grid*, pages 277–287, May 2001. IEEE/ACM.

[369] C. Szyperski. *Component Software: Beyond Object-Oriented Programming*. Addison-Wesley, 1997.

[370] A. S. Tanenbaum. *Computer Networks*. Prentice Hall, 3rd edition, 1996.

[371] A. S. Tanenbaum and M. van Steen. *Distributed Systems: Principles and Paradigms*. Prentice Hall, 2002. ISBN 0-13-066102-3.

[372] W. Tang. *Scalable Trigger Processing and Change Notification in the Continual Query System*. Oregon Graduate Institute, 1999.

[373] D. L. Tennenhouse, J. M. Smith, W. D. Sincoskie, D. J. Wetherall, and G. J. Minden. A survey of active network research. *IEEE Communications Magazine*, 35(1):80–86, January 1997.

[374] W. W. Terpstra, S. Behnel, L. Fiege, A. Zeidler, and A. P. Buchmann. A peer-to-peer approach to content-based publish/subscribe. In Jacobsen [210].

[375] Thales. Splice data distribution service, 2006.
http://www.prismtechnologies.com.

[376] P. Thapliyal, Sidhartha, J. Li, and S. Kalyanaraman. LE-SBCC: Loss-event oriented source-based multicast congestion control. Technical report, Rensselaer Polytechnic Institute ECSE, Troy, NY, 2001.

[377] The Eclipse Foundation. Eclipse. http://www.eclipse.org, 2005.

[378] D. Thomas. Message oriented programming. *Journal of Object Technology*, 3(5):7–12, 2004.

[379] D. Thomas and B. M. Barry. Model driven development: The case for domain oriented programming. In R. Crocker and G. L. Steele, Jr., editors, *Companion of the 18th Annual ACM SIGPLAN Conference on Object-Oriented Programming, Systems, Languages, and Applications*, pages 2–7, 2003. ACM.

[380] TIBCO, Inc. TIB/Rendezvous. White Paper, 1996.
http://www.rv.tibco.com.

[381] TIBCO, Inc. *TIBCO Rendezvous: Concepts*, software release 7.4 edition, 2004.

[382] P. Timberlake. The pitfalls of using multicast publish/subscribe for EAI. IBM MQseries Whitepaper, 2002. also published on messageQ.com.

[383] Trolltech. The Qt class library. http://www.trolltech.com/products/qt, 2005.

[384] J. van't Hag. "Data-centric to the max", the SPLICE architecture experience. In *23rd International Conference on Distributed Computing Systems Workshops*, pages 207–212. IEEE, May 2003. doi: 10.1109/ICD-CSW.2003.1203556.

[385] L. Vargas, J. Bacon, and K. Moody. Integrating databases with publish/subscribe. In Dingel and Strom [114], pages 392–397.

[386] S. Vinoski. More Web services notifications. *IEEE Internet Computing*, 8(3):90–93, May-June 2004.

[387] S. Vinoski. Web services notifications. *IEEE Internet Computing*, 8(2): 86–90, March-April 2004.

[388] J. Vitek, N. Horspool, and A. Krall. Efficient type inclusion tests. *ACM SIGPLAN Notices*, 32(10):142–157, October 1997. ISSN 0362-1340.

[389] W. Z. Vivek S. Pai, Peter Druschel. Flash: An efficient and portable web server. In A. Rubin, editor, *Proceedings of the 1999 USENIX Annual Technical Conference*, pages 199–212, 1999. USENIX.

[390] C. Wang, A. Carzaniga, D. Evans, and A. L. Wolf. Security issues and requirements for Internet-scale publish-subscribe systems. In *Proceedings of the Thirty-fifth Hawaii International Conference on System Sciences (HICSS-35)*, pages 3940–3947, January 2002.

[391] D. J. Watts and S. H. Strogatz. Collective dynamics of small-world networks. *Nature*, 393:440–442, June 1998.

[392] M. Welsh, D. Culler, and E. Brewer. SEDA: An architecture for well-conditioned, scalable Internet services. In K. Marzullo and M. Satya-

narayanan, editors, *Proceedings of the 18th ACM Symposium on Operating Systems Principles*, pages 230–243, October 2001. ACM.

[393] J. E. White. Mobile agents. In J. Bradshaw, editor, *Software Agents*. MIT Press, 1997.

[394] J. Widom and S. Ceri, editors. *Active Database Systems: Triggers and Rules For Advanced Database Processing*. Morgan Kaufmann, 1996.

[395] R. Wiener. Delegates and events in C#. *Journal of Object Technology*, 3(5):78–85, 2004.

[396] R. J. Wieringa. *Design Methods for Reactive Systems*. Morgan Kaufmann, 2002.

[397] A. Woo, S. Madden, and R. Govindan. Networking support for query processing in sensor networks. *Communications of the ACM*, 47(6): 47–52, 2004.

[398] World Wide Web Consortium (W3C). XML path language (XPath) version 1.0. Technical Report, November 1999. http://www.w3.org/TR/xpath.

[399] World Wide Web Consortium (W3C). Extensible markup language (XML) 1.0 (second edition), 2000.

[400] World Wide Web Consortium (W3C). Simple object access protocol (SOAP) 1.2. Recommendation, June 2003. http://www.w3.org/TR/SOAP/.

[401] J. Wu, editor. *23rd International Conference on Distributed Computing Systems Workshops (ICDCSW'03)*, 2003. IEEE.

[402] P. Wyckoff, S. W. McLaughry, T. J. Lehman, and D. A. Ford. T Spaces. *IBM Systems Journal*, 37(3):454–474, 1998. http://www.research.ibm.com/journal/sj/373/wyckoff.pdf.

[403] T. W. Yan and H. Garcia-Molina. Index structures for information filtering under the vector space model. In A. K. Elmagarmid and E. Neuhold, editors, *Proceedings of the 10th International Conference on Data Engineering*, pages 337–347, February 1994. IEEE Computer Society.

[404] T. W. Yan and H. Garcia-Molina. Index structures for selective dissemination of information under the Boolean model. *ACM Transactions on Database Systems*, 19(2):332–364, 1994.

[405] Y. Yan, Y. Huang, G. Fox, S. Pallickara, M. Pierce, A. Kaplan, and A. Topcu. Implementing a prototype of the security framework for distributed brokering systems. In *Proceedings of the International Conference on Security and Management (SAM'03)*, pages 212–218, June 2003.

[406] S. Yang and S. Chakravarthy. Formal semantics of composite events for distributed environments. In *Proceedings of the 15th International Conference on Data Engineering (ICDE '99)*, pages 400–407. IEEE Computer Society, 1999.

[407] Y. R. Yang and S. S. Lam. Internet multicast congestion control: A survey. In *Proceedings of the International Conference on Telecommunications (ICT'00)*, May 2000.

[408] A. Zeidler and L. Fiege. Mobility support with REBECA. In Wu [401], pages 354–361.

[409] X. Zhang, K. G. Shin, D. Saha, and D. D. Kandlur. Scalable flow control for multicast abr services in atm networks. *IEEE/ACM Transactions on Networking*, 10(1), February 2002.

[410] Y. Zhao, D. Sturman, and S. Bhola. Subscription propagation in highly-available publish/subscribe middleware. In *Proceedings of the 5th ACM/IFIP/USENIX International Conference on Middleware*, pages 274–293. Springer, 2004. ISBN 3-540-23428-4.

[411] D. Zhou and K. Schwan. Eager handlers — communication optimization in Java-based distributed applications with reconfigurable fine-grained code migration. In *3rd International Workshop on Java for Parallel and Distribute Computing (held in conjunction with the IPDPS 2001)*, page 110, 2001.

Index

A-ToPSS   64
abstract scope   161, 184
abstract window toolkit   310
access control   256
access control list   256
ACL   *see* access control lists
action   24
activation   *see* quenching
active bat   232
active object   329
active office   6, 232, 254
activity   135
additive increase, multiplicative
    decrease   286
administrator   151
advertisement   13, 23, 31, 107, 138
advertisement routing table   109
always   25
anonymous request/reply   15–16
AOP   *see* aspect-oriented program-
    ming
API   *see* application programming
    interface
application programming interface
    137
approximate matching   64
AQ   *see* Oracle Advanced Queuing
aspect-oriented programming   143
atomic step   69
attribute   37
attribute filter   36
AWT   *see* abstract window toolkit

BDD   *see* binary decision diagram

binary decision diagram   61–62
    ordered   61
    reduced ordered   62
blackout period   296
border broker   22, 332
broker   67
    border   *see* border broker
    inner   *see* inner broker
    local   *see* local broker
    virtual   *see* virtual broker
brute force   59

callback   16
cambridge event architecture   337–340
capability   257
causal ordering   31
CEA   *see* Cambridge Event Architec-
    ture
cell   324
CGM   *see* clustered group multicast
channel   19, 68
channel-based   309
client   67
clustered group multicast   122
COBEA   *see* CORBA-based event
    architecture
Common Object Request Broker
    Architecture   305
communication medium   197
component   3, 12, 14, 17, 21, 26, 167,
    187
    complex   *see* scope
composite event   234–235
    core language   238

detection   242
detection automata   236
requirements   234
composite event detection   236–242
composite event detection automata
  236
composite event type   235
congestion   277
congestion collapse   277
congestion control   276–287
  receiver-based   286
  sender-based   286
congestion metric   279
conjunctive filter   37
consolidation noise   287
construction phase   246
consumer   3, 12, 18
consumption policy   251
content-based filtering   20
content-based routing   80–107
context   145
control phase   246
CORBA   see Common Object Request
  Broker Architecture
CORBA-based event architecture   338
CORBA Event Service   19, 307–308
CORBA Notification Service   19, 36,
  137, 308–310
core-based tree   116, 329
core composite event language   236
counting algorithm   59
coupling point   184, 192
covering-based routing   23, 105, 106
covering algorithm   50
covering of conjunctive filters   44
curiosity   325

data-centric publish/subscribe   314
database management system   198,
  199, 205, 250
Data Distribution Service   146,
  313–317
data local reconstruction layer   316
data model   35, 37, 53, 332
DBMS   see database management
  system
DCPS   see data-centric publish/sub-
  scribe
DDS   see Data Distribution Service

decision tree   60–61
delivery policy   179
describable event set   237
destruction phase   246
DHT   see distributed hash table
directed diffusion   8, 121, 132, 270
distributed hash table   115, 123, 142
distribution policy   246
DLRL   see data local reconstruction
  layer
domain specific language   139, 147
DSL   see domain specific language
duplicate avoidance   120
duplicate notification   158, 166, 173
durable subscription service   326

ECA   see event–condition–action
Elvin   36, 296, 302, 340
engineering   183
  requirements   129
  with scopes   182–196
epidemic multicast   125
evaluation policy   251
event   3, 11
event–condition–action   7, 185, 186,
  193, 250
event-based   16–17
event-based style   3, 17, 140
event-based system   3, 11
event broker   21
event broker congestion   277
event channel   307
event composition   182
event dispatcher   329
event input sequence   237
event mediator   338
event notification service   3, 13
event pattern   250, 328
event sink   337
event source   337
event stream   325
event system   14
event type owner   259
eventually   25
eventual monotone remote validity   76
eventual superset validity   75

fairness   68
fairness property   68

fan-out   225
fast retransmit   286
fault masking   265
fault tolerance   264–276
feedback suppression   286
FIFO-producer ordering   30
filter   13, 37
  covering   39
  disjoint   37
  identical   37
  merging   47–49
  overlapping   37, 107
filtering   19–20
  channel   19
  content-based   20
  subject-based   19
  type-based   19
filter model   13, 35, 37, 53, 332
flooding   22, 105, 107

GEM   7, 251
generative state   237
generative time state   237
group leader   329
Gryphon   36, 65, 126, 302, 324–326
guaranteed delivery service   325

Herald   338
Hermes   117, 334–337
heterogeneity   2, 135, 173
hierarchical routing   112
hierarchical routing algorithm   112
high-level architecture   317
higher-level language   236
HLA   see high-level architecture
human processable   241
hybrid routing   113

IBM WebSphere MQ   318
ideal multicast   122
identity-based routing   23, 105, 106
identity of conjunctive filter   43
IFG   see information flow graph
imperfect merger   98
imperfect merging   49
implementation   17
implicit invocation   7, 142
InfoBus   34
Information Bus   320

information flow graph   324
initial routing configuration   72
initial state   237
inner broker   22, 332
input domain   237
input interface   164
interaction   17
interaction model   14–18
  anonymous request/reply   15
  callback   16
  event-based   16
  request/reply   15
interface   24, 26, 164, 181, 190
intermediate broker   325
interval timestamp   235
IP multicast   198, 199, 203, 225, 230

Java Event-Based Distributed Infra-
  structure   329
Java Message Service   33, 36, 137,
  311–313
Java Remote Method Invocation   310
Java RMI   see Java Remote Method
  Invocation
JavaSpaces   35, 311
JEDI   35, 126, 329–331
Jini   310–311
JMS   see Java Message Service

key class   261
knowledge   325
knowledge graph   325

Linda tuple space   18, 35
link bundle   325
liveness   29, 155, 164, 171
local broker   22, 332
local routing configuration   72
local subset validity   75
local validity   76
loose coupling   3, 132

machine processable   241
management   186, 196
mapping   see notification mapping
matching
  approximate   64
  by name   36
  by position   36

of XML documents   63
matching algorithm   50, 57–64
    binary decision diagram   61
    brute force   59
    counting algorithm   59
    decision trees   60
MDA   *see* model driven software
    development
MDSD   *see* model driven software
    development
merger   98
merging-based routing   23, 106, 107
merging algorithm   51
merging of conjunctive filters   47
message   11
message batching   68, 121
message selector   311
meta object protocol   181
middleware   2
mobile composite event detector   245
mobility   287–303
model driven software development
    185
monotone valid routing algorithms
    76–77
multicast   122, *see* IP multicast
    epidemic   *see* epidemic multicast
    IP   *see* IP Multicast

NACK window   285
name/value pair   37, 309, 312, 331, 338,
    340
name/value term   332
Narada Brokering   340
negative acknowledgment   325
neighbor matching multicast   123
network congestion   277
next   25
notification   3, 11, 37
notification classification scheme   13
notification delivery   72
notification forwarding   57, 72
notification mapping   169–176
notification matching   57
notification service   21
notification summary   146

OBDD   *see* ordered binary decision
    diagram

object   56
Object Definition Language   339
object serialization   310
Ode   250
ontology   184
open implementation   139
Oracle Advanced Queuing   229
Oracle Streams Advanced Queuing
    322
orchestration   183, 184
ordered binary decision diagram   61
ordering   118, 120
    causal   *see* causal ordering
    FIFO-producer   *see* FIFO-producer
        ordering
    total   *see* total ordering
ordinary state   237
output interface   164
overlapping of conjunctive filters   46
overlay network   21

peer-to-peer routing   112
perfect merger   98
perfect merging   47
permission   257
PHB   *see* publisher-hosting broker
PHB-driven congestion control
    algorithm   279
Point-to-point communication   311
port   *see* coupling point
principal   257
producer   3, 12, 18
pubend   325
publication endpoint   279
publish–register–notify   337
publish/subscribe communication   311
publish/subscribe
    concept-based   20, 184
    content-based   20
    self-stabilizing   265–266
    subject-based   19
    topic-based   33
    type-based   19
publisher   *see* producer
publisher-hosting broker   325
publisher endpoint   325
publishing policy   177
pull mode   307
push mode   307

QoS   *see* quality of service
quality of service   138, 181, 226
quenching   147, 148

READY   229, 340
REBECA   20, 22, 36, 296, 302, 331–334
record
    flat   36
    hierarchical   36
    semistructured   52–56
    structured   36–52
reduced ordered binary decision
        diagram   62
reflection   182
relational subscription service   326
remote procedure call   2
remote routing configuration   72
rendezvous-based routing   115
rendezvous node   115, 125
request/reply   15
    anonymous   *see* anonymous
        request/replay
restriction   259
RMI   *see* Java Remote Method
        Invocation
ROBDD   *see* reduced ordered binary
        decision diagram
role   184, 257
role-based access control   257
routing
    content-based   *see* content-based
        routing
    hierarchical   112–115
    in cyclic topologies   120–122
    rendezvous-based   115–117
    self-stabilizing   266
    using multicast   122–123
    with advertisements   107–112
    with joining and leaving clients
        119–120
    with topology changes   117–119
routing algorithm   74
    covering-based   91–98
    evaluation   126
    flooding   81–82
    framework   69–74
    hybrid   113
    identity-based   85–89
    merging-based   98–104

monotone valid   *see* monotone valid
        routing algorithm
    simple   82–85
    valid   *see* valid routing algorithm
routing configuration   72
    initial   72
    local   72
    remote   72
routing entry   72
routing framework
    extensions   107–125
    valid instantiation   77–80
routing optimization   43
routing table   72
    scoped   *see* scoped routing table
    update   73
RPC   *see* remote procedure call

safety   29, 155, 163, 171
SAMOS   250
scalable internet event notification
        architecture   326
scope   152
    abstract   *see* abstract scope
    context   183
    defining   183, 188
    deployment   185
    engineering   *see* engineering
scope architecture   197, 209
    addressing scope   205
    broker scope   210
    central hub   205
    client-side filtering   203
    collapsed filters   204
    integrated routing   213
    static deployment   202
scope attributes   161
scope distribution   200
scoped routing table   215
scope graph   153
    combining   182, 225
    composition   184
    descriptive   182
    engineering process   182
    instantiated   182
    management   *see* management
scope language   187
scope model   153
scope overlay   215

security   253–264
SHB   *see* subscriber-hosting broker
SHB-driven congestion control
    algorithm   279, 285
SIENA   36, 38, 39, 126, 264, 296, 302,
    326–328
simple event system   28
simple routing   23, 105, 106, 121
slow start   286
small-world network   124
Snoop   250
software engineering   *see* engineering
software pattern   141
source quenching mechanism   340
space redundancy   265
specification   25
    of an event system   23–33
       trace-based   28–30
       with advertisements   31–33
       with ordering requirements   30–31
    state   24
strawman approach   330
strong transition   238
structured event   308
subcriber   *see* consumer
subend   325
subject   19
subject-based filtering   19, 320
subscriber-hosting broker   325
subscriber endpoint   325
subscription   3, 13
    uncovered   92
subscription routing table   109
subtrace   25
summary   *see* notification summary
supertype subscription   339
system model   67–69

TCGM   *see* threshold clustered group
    multicast
template   35
temporal operator   25

always   25
eventually   25
next   25
Threshold Clustered Group Multicast
    122
TIBCO Rendezvous   320
time redundancy   265
timestamp message   120
topic   33, 311
topology change   117
topology maintenance   123
total ordering   31
trace   25, 72
transaction   135, 147, 182
transmission policy   176–182
traverse policy   180
TSpaces   311
tuple   35–36
type- and attribute-based routing   335
type-based filtering   19
type-based routing algorithm   334

UML   *see* Unified Modeling Language
Unified Modeling Language   143, 184,
    185

valid routing algorithm   74–76
virtual broker   324
virtual link   325
visibility   152, 199

weak transition   238
wireless sensor network   6, 132, 229
WS-Addressing   317
WS-Eventing   317
WSN   *see* wireless sensor network
WS Notification   317

XFilter   63
XML   20, 63
XPath   20, 52, 63

YFilter   63

Authors' Biographies

Gero Mühl is a postdoctoral researcher at the Berlin University of Technology. His research interests include middleware, event-based systems, self-organization, and mobile computing. He received a master degree in Computer Science (Dipl.-Inform.) and a master degree in Electrical Engineering (Dipl.-Ing.) from the FernUniversität in Hagen, and a Ph.D. degree (Dr.-Ing.) in Computer Science from the Darmstadt University of Technology.

Ludger Fiege works as senior engineer at Siemens Corporate Technology. His interests include middleware and software architecture, asynchronous messaging, event-based systems, and model driven development. He received a master degree from Universität Bonn and a Ph.D. degree in computer science from Technische Universität Darmstadt.

Peter Pietzuch is a postdoctoral fellow in the Systems Research Group at Harvard University. His research interests focus on large-scale distributed systems, including publish/subscribe infrastructures, global stream-processing applications, and peer-to-peer overlay networks. Prior to joining Harvard, he received his Ph.D. degree from the University of Cambridge in England, working on scalable event-based architectures. In 2000, he obtained a B.A. degree in Computer Science, also from the University of Cambridge.

Gustavo Alonso, ETH Zentrum,
Zürich, Switzerland;
Fabio Casati, Harumi Kuno, Vijay Machiraju
Hewlett-Packard, Palo Alto, CA, USA

Web Services

Concepts, Architectures and Applications

XX, 354 p. Hardcover
ISBN 3-540-44008-9

Like many other incipient technologies, Web services are still surrounded by a tremendous level of noise. This noise results from the always dangerous combination of wishful thinking on the part of research and industry and of a lack of clear understanding of how Web services came to be. On the one hand, multiple contradictory interpretations are created by the many attempts to realign existing technology and strategies with Web services. On the other hand, the emphasis on what could be done with Web services in the future often makes us lose track of what can be really done with Web services today and in the short term. These factors make it extremely difficult to get a coherent picture of what Web services are, what they contribute, and where they will be applied. Alonso and his co-authors deliberately take a step back. Based on their academic and industrial experience with middleware and enterprise application integration systems, they describe the fundamental concepts behind the notion of Web services and present them as the natural evolution of conventional middleware, necessary to meet the challenges of the Web and of B2B application integration. Rather than providing a reference guide or a "how to write your first Web service" kind of book, they discuss the main objectives of Web services, the challenges that must be faced to achieve them, and the opportunities that this novel technology provides. Established, as well as recently proposed, standards and techniques (e.g., WSDL, UDDI, SOAP, WS-Coordination, WS-Transactions, and BPEL), are then examined in the context of this discussion in order to emphasize their scope, benefits, and shortcomings. Thus, the book is ideally suited both for professionals considering the development of application integration solutions and for research and students interesting in understanding and contributing to the evolution of enterprise application technologies.

Contents: Part I: Conventional Middleware 1) Distributed Information Systems 2) Middleware 3) Enterprise Application Integration 4) Web Technologies Part II: Web Services 5) Web Services 6) Basic Web Services Technologies 7) Service Coordination Protocols 8) Service Composition 9) Outlook Bibliography; Index